RUTHERFORD

RUTHERFORD

Being the Life and Letters of the
Rt Hon. Lord Rutherford, O.M.

By

A. S. EVE, C.B.E., D.Sc., LL.D., F.R.S.

formerly Macdonald Professor of Physics
McGill University

With a foreword by

EARL BALDWIN OF BEWDLEY, K.G.

CAMBRIDGE
AT THE UNIVERSITY PRESS
1939

CAMBRIDGE
UNIVERSITY PRESS

University Printing House, Cambridge CB2 8BS, United Kingdom

Published in the United States of America by Cambridge University Press, New York

Cambridge University Press is part of the University of Cambridge.

It furthers the University's mission by disseminating knowledge in the pursuit of education, learning and research at the highest international levels of excellence.

www.cambridge.org
Information on this title: www.cambridge.org/9781107678811

First published 1939
First paperback edition 2013

A catalogue record for this publication is available from the British Library

ISBN 978-1-107-67881-1 Paperback

DEDICATED TO

LADY RUTHERFORD

WITHOUT WHOSE CARE AND FORESIGHT
THIS BOOK
COULD NOT HAVE BEEN WRITTEN

It is the great beauty of our science, that advancement in it, whether in a degree great or small, instead of exhausting the subject of research, opens the doors to further and more abundant knowledge, overflowing with beauty and utility. FARADAY

* * * *

In 1911 Rutherford introduced the greatest change in our idea of matter since the time of Democritus. EDDINGTON

* * * *

In watching the rapidity of this tide of physics I have become more and more impressed by the power of the scientific method of extending our knowledge of Nature. Experiment, directed by the disciplined imagination either of an individual or, still better, of a group of individuals of varied mental outlook, is able to achieve results which far transcend the imagination alone of the greatest philosopher.... From time to time there arises an illuminating conception, based on accumulated knowledge, which lights up a larger region and shows the connection between these individual efforts, so that a general advance follows. RUTHERFORD

PREFACE

In this book it has been my aim to hold up a mirror in which Rutherford may reveal himself, just as he was, in lectures, books, papers, speeches, portraits, letters, and casual talk.

All his books and papers, and many of his letters, were typed by his wife and it is due to Lady Rutherford's care that both the letters written to her husband and the innumerable reprints of papers sent to him have been preserved. The collection of reprints is now in the Cavendish Laboratory.

This careful collection and arrangement of material has greatly facilitated my work; I am also indebted to many friends of Rutherford for valuable help, in particular to Mr S. C. Roberts, to Professor J. Chadwick, to Professor M. L. E. Oliphant, and not least to my wife, who encouraged me at those times when the task seemed beyond my powers.

A.S.E.

May 1939

FOREWORD

BY

EARL BALDWIN OF BEWDLEY

The world knew Lord Rutherford as a man of science, knew—perhaps without fully appreciating its significance—of his work on nuclear physics: it was my privilege to know him as a man. Others, therefore, may write of the unique part he played in the rapid advancement of the frontiers of knowledge which has been so striking a feature of this century, and of the unfailing source of inspiration which his collaborators and pupils found in him. I write of him with the memory of his ever fresh and boyishly eager personality, of his patient kindliness and of his almost uncanny gift for singling out from a mass of confusing detail the one significant fact.

To me, as Chancellor of the University of Cambridge, there came many opportunities in the last few years to see how fruitful was his spirit. The Cavendish Laboratory under his guidance and through the boundless enthusiasm which radiated from him to inspire even the youngest of those about him, had reached even higher and more sustained levels than ever before. Indeed the Cavendish Laboratory *was* Rutherford, bone of his bone. He was no remote or inaccessible being: the students were his familiars, they were his friends. He encouraged them in their difficulties, he was quick to praise their success. Frank yet kindly, his criticism always had a force and vigour that suggested new and promising ideas to replace those it had demolished.

His was no weakling spirit demanding for its exercise seclusion from the world of every day, nor was he careless of the great and increasing change that science was working in every aspect of the life of man. With a vigour no whit diminished by his scientific labours he devoted much of his time to secure the application in industry of the methods and results of scientific research. He was convinced, none more so, that continued industrial prosperity in this country could only come from such an application and, as Chairman of the Advisory Council for Scientific and Industrial Research during the last seven years of his life, he never spared himself in working for its achievement. Perhaps they are but few who are aware just how much British industry owes to him, but it was my good fortune on many occasions to receive the benefit of his advice in this and other ways. His shrewd insight and practical mind rarely if ever failed him nor was his aid ever invoked in vain.

And yet, having said this much, how little has one really described him! His refreshing personality, his dauntless spirit, the merry twinkle of his eye, the exuberance of his ever-youthful, ever-joyful, enthusiasm: how can they be recaptured and confined within the limits of mere words! As well might one essay to distil the essence of the wind. One can only say he was a man, a peer among men: he was Rutherford.

CONTENTS

ILLUSTRATIONS

PLATES

TEXT-FIGURES

CHAPTER I

LIFE IN NEW ZEALAND

Ernest Rutherford was born on 30 August 1871, at Spring Grove, later called Brightwater, thirteen miles south of the town of Nelson, in the northern part of the South Island. The house has been pulled down, but there is a reproduction of it, in its derelict days, overleaf. His grandfather George Rutherford was a wheelwright at Perth, Scotland, and was engaged by Captain Thoms to migrate to New Zealand in order to erect a mill at Motueka. The whole family sailed in the *Phoebe Dunbar* from Dundee in 1842, and one of the sons named James, then aged three, was destined to be the father of Ernest.

About 1855 a widow named Caroline Thompson, with her little girl Martha, aged twelve, left Hornchurch, Essex, together with Caroline's parents, the Shuttleworths, and landed in Auckland after a long voyage in the sailing ship *Bank of England*. From Auckland they sailed in the brig *Ocean* to New Plymouth, on the south-west coast of the North Island, and remained there until trouble with the Maoris caused the women and children to be sent to Nelson in the South Island, and finally it happened that on 20 April 1866, Martha Thompson, at one time a school teacher, was married to James Rutherford.

In 1875 the Rutherfords moved ten miles to Foxhill in the same district of Nelson. A large family was born, seven sons and five daughters. One boy, Percy, died as an infant. Two boys, Herbert and Charles, aged nine and eight, were drowned in Pelorus Sound when sailing with an older boy of another family. The four sons who grew to manhood were George, Ernest, James and Arthur. Ernest was fourth child and second son. The names of his sisters were Nellie (Mrs M. P. Chapman), Alice (Mrs A. T. Elliot, deceased), Florence (Mrs H. G. Streiff), Ethel (Mrs H. G. Sergel) and Eva (Mrs L. T. Bell).

Rutherford's father, while at Foxhill, earned his living as a wheelwright, by small-scale farming and by bridge-building or contracting in connection with the railway under construction. He was a man of great energy, very straightforward, reliable, versatile and much liked by all classes. The mother valued education, could play the piano, was thrifty and hard-working.

At the age of five Rutherford was sent to the primary school at Foxhill where, with great good fortune, he was under a fine man and able teacher, Mr Henry Ladley, who taught Ernest as far as the fourth standard and should "get the credit due to him for starting the great man on his way". This remark is due to Mr R. Cameron of Seddon who recorded that "some years later at a social function Rutherford pointed out how his parents had sacrificed, how they had gone without many comforts, so that he might be educated, so that he might have a chance, adding 'I should never have been where I am today if it wasn't for my mother and father'".

Rutherford's mother preserved a small science text-book with his name in it, and a date which shows that he was then ten years old. The book was written by Balfour Stewart, Professor of Physics at Manchester, and the preface lays down the very principles which Rutherford followed in his career:

This book has been written, not so much to give information, as to endeavour to discipline the mind by bringing it into immediate contact with Nature herself, for which purpose a series of simple experiments are described leading up to the chief truths of each science, so that the powers of observation in the pupils may be awakened and strengthened.

We are fortunate in having many sources of information about the early days of Rutherford, collected by friends and relations and by those who were at school or college with him. We are specially indebted to Prof. Coleridge Farr, F.R.S., of Canterbury College, New Zealand and to Dr E. Marsden, now Secretary of the Department of Scientific and Industrial Research, formerly a distinguished research student at Manchester with Rutherford.

In 1882 the family moved by boat to Havelock at the head of

Rutherford's grandfather

Rutherford's birthplace

Rutherford's home at Foxhill

Pelorus Sound. James Rutherford set up a flax-mill where he prepared the native flax growing in neighbouring swamps. He also erected a saw-mill where he cut railway sleepers from the black and brown birch, and these were shipped to Lyttelton by sailing ships, so that sometimes a dozen of them together made a fine sight at the head of the Sound when held up by adverse winds. Ernest Rutherford now went to Havelock primary school and again came under the influence of an enthusiastic teacher, Mr Jacob H. Reynolds, who taught Latin to some of his pupils for an hour each morning before ordinary school work began; with the result that Rutherford, aged fifteen, won a fifty-guinea Marlborough Educational Board Scholarship, with the remarkable record of 580 marks out of a possible 600. This award enabled him to go to Nelson College where he was at once placed in the Fifth Form.

Nelson College was a good school, run on English public-school lines. The headmaster was a notable man, W. J. Ford of Repton and St John's College, Cambridge, who had been for many years a house-master at Marlborough. He was a genial character, a good classical scholar and a great cricketer. A man of outstanding height and weight, he was a great 'slogger', and few men, if any, have hit a cricket ball farther and higher. It was not possible to catch out a man who per-sistently hit the ball over the boundary.

Rutherford recalled a story of a day when Ford had a cricketing friend to visit him from England and the two of them took on the boys and some masters. Ford and his partner batted vigorously all the after-noon and utterly wore down the bowlers, but one of the masters, Littlejohn, persisted in using every wile in a vain endeavour to dislodge the batsmen, and Rutherford vowed that he was the real hero of this Roman holiday.

Dr W. S. Littlejohn, M.A., Aberdeen, was a famous teacher and disciplinarian, and an enthusiast in running the Cadet Corps, in which Rutherford obtained his certificate for proficiency as a sergeant. In due course Littlejohn became principal of Nelson College and later of Scotch College, Melbourne. He was really a classical master, but he also became a thorough and proficient teacher of mathematics and science, and much of Rutherford's later success depended upon this training. Only a Scot can do all that Littlejohn did.

Rutherford was in Forms V and VI and became head of the school. He had a liberal education which was not unduly specialised. Although mathematics was his strong subject, he had no difficulty in obtaining scholarships and prizes for Latin, French, English literature, history, physics and chemistry. The modern cry of 'too many subjects' does not seem to have perturbed him.

In June 1888, he was top in all subjects in the Sixth Form and in his report his work in classics is described as being "as good as ever", his English work as showing "capital style and great power of reproduction". In mathematics he was "easily first, very quick and a very promising mathematician".

A schoolfellow, Mr C. H. Broad, sent some interesting notes to *The Times* (22 Oct. 1937):

Rutherford and I were in the fifth and sixth forms together during the whole of his stay. He entered into the full life of the school in every way, games as well as study, and although he could never be described as an athlete, he was a forward in the first Fifteen in his last year. What always particularly struck me about him were his extraordinary powers of concentration, even in the midst of the greatest turmoil. Some of us used to take full advantage of his abstraction in various boyish ways, banging him on the head with a book, etc., and then bolting for our lives. Another thing I well remember was his habit of strolling about with Mr W. S. Littlejohn on the half-holiday, up and down little frequented streets near the College, Littlejohn drawing diagrams in the dust of Hampden Street and discussing them with Rutherford. In those days chemistry was optional with French, consequently Rutherford began his science late in his career at Nelson, and then he was practically taken alone in physics and chemistry by Dr Littlejohn.

At Nelson, in addition to prizes for mathematics, his strongest subject, Rutherford gained the Stafford scholarship for history, the senior English literature scholarship, the French scholarship, the Latin prize, Form VI, and the Simmons prize for English literature, Form VI.

Rutherford's parents

The strenuous, heroic life of the early colonists is well pictured by Dr Marsden:

In the meantime, the family had suffered two serious hardships. In the first place, his two younger brothers were drowned in an accident in the Sound. The father and brothers scoured the shores for three months in a vain effort to find the bodies. This accident so seriously affected his mother that she, for a long time, lost her sunny, cheery nature, and never again did she turn to her favourite music or play again her cherished Broadwood piano. A little later, James Rutherford had a serious accident on the small jetty from which he loaded his sleepers, unfortunately fracturing five of his ribs. Shortly after his recovery, the Atkinson Government cancelled orders for railway sleepers, and the family per-force looked round for fresh avenues of occupation. His father crossed to the North Island and proceeded North from Wellington on horseback, looking for suitable areas for flax. Eventually he arrived at Pungarehu, Taranaki, near the coast, 30 miles south of New Plymouth. Flax-milling had not started in this area, so he was able to obtain suitable land at £3 per acre, near other flax swamps which he was afterwards able to cut under royalty. Returning to Havelock he chartered the *Murray*, under Captain Vickerman, and loaded it with his whole family, three extra operatives, and all his household furniture, his horses, his flax-milling machinery and a quantity of timber. The charter cost him £100. It took three days to get to New Plymouth, where the whole outfit was unloaded and the effects transported by the rough track to Pungarehu. Leaving his wife and younger children at New Plymouth, he proceeded to carve out a home and establish his mill. Soon he was relatively prosperous, although his flax was sold in Melbourne at only £13 per ton and had to be transported to New Plymouth for shipment over a road which took nearly two days to travel, with a five-horse team drawing only three tons.

It is interesting to realise the energy and ability which James Rutherford put into his flax-milling operations. He harnessed water-power to drive his mill. He experimented and developed a method of soaking the fibre after stripping and subsequently a special scraper to remove the vegetable matter so as to minimise the labour and time of paddocking. He looked ahead and planted specially selected native varieties. Never-theless, he relied mostly on ready grown swamp-flax, and such was the success of his operations that the flax he produced was reckoned amongst the best in the Dominion, and he was later able to retire to New Plymouth, all his children having married and settled in different parts of the country.

There is a good photograph of the Rutherford house at Pungarehu, with a long line of white flax in the background and the top of the mill in the left rear with the houses for the mill-hands on the right.

Rutherford seems to have been in every way a normal boy and as a youngster to have taken his share in the chores around a farm, chopping wood, running errands, gardening and helping to milk the cows.

There was a tale which Rutherford used to tell with gusto, varying it a trifle to suit his audience. He told it to a group of young children who called him 'Grandpa' and later on promoted him to 'Lord Grandpa'.

"My mother sent me out to bring the cow home to the paddock and to collect some fire wood as well. So I drove the cow and pulled a big branch of a tree behind me. Then I thought, why shouldn't the cow help me? So I tied the branch with a rope to the end of her tail and she went quietly home till she came to a narrow gate. Here the branch jammed and the last bit of her tail broke off!"

"Whatever did you do?"

"I put a plaster on her tail and when I heard her in the byre later on mooing comfortably, I knew that she was all right."

"What did you do with the bit of tail?"

"I heard that cows grew from cuttings so I buried the tail in the ground."

Long after, the youngest child used to explain how cows came from cuttings.

There was another story of how Rutherford as a boy learnt to throw the *bolas*. He would hold one ball in his hand, whirl the other ball on the end of a string round his head and then let them go. He gradually learnt to throw them higher and farther, until one day they went right over the top of the house. From the other side came loud screams and he rushed round to find his young brother had been hit and was going

Havelock, showing the school house

The house and flax-mill, Pungarehu

off to tell his mother. "But after I'd sat on his head for half an hour he began to see reason."

As he grew older Rutherford began to enjoy other occupations; he photographed with a home-made camera, took clocks to pieces, made model water-wheels (like Newton), imitating his father, who was highly skilful in developing water-power for his mills. Rutherford was fond of music as a background, but it cannot be suggested that he was musical. So too he liked to sing, but with no great adherence to any tune. Reading was his great recreation, and he loved throughout his life all forms of literature. As a boy he was specially fond of Dickens and would enjoy reading him aloud to the younger members of the family, "joining most heartily with his huge laughter at the various humorous incidents of Pickwick". There were expeditions in the bush, catching brook trout or spearing eels, and occasionally a profitable hop-picking holiday. Ernest had a narrow escape while bathing with his brother Jim in the Wai-ti River. Ernest got out of his depth, but Jim just managed to reach him and get him out. This was before they had learned to swim. At Pungarehu, there were expeditions on horse-back in the bush to shoot wild pigeons at dawn when they assembled to feast on the berries of the lofty miro-trees.

Dr F. Milner, C.M.G., Rector of Waitaki Boys' High School, who was at Nelson School and at Canterbury College with Rutherford, has stated that "Rutherford always said that but for his gaining a scholarship to Nelson College from a country hamlet he would have been a farmer and never realised his special gifts". It is more probable that he would have come to the front and become a leader in any walk of life.

However that may be, the next great step in the ladder of his life was taken: he won a Junior University Scholarship (1889) and matricu-lated at the University of New Zealand.

When Rutherford went to Canterbury College, Christchurch, part of the University of New Zealand, it was a small institution with about seven professors and 150 regular students. We have a fair picture of the new undergraduate from a fellow-student, Mr R. M. Laing: "Rutherford was a boyish, frank, simple and very likeable youth, with no precocious genius, but when once he saw his goal he went straight to the central point." He joined in the life of the place with zest, played for the Football Fifteen as a forward, and took a share in the debates of the Dialectical Society. In 1891 he passed the first section of the B.A. degree in Latin, English, Mathematics and Mechanics, and a year later completed the requirements for that degree with French and Physical Science, and won a Senior University Scholarship in Mathematics. There is a good photograph of him when twenty-one. He seems to have come largely under the influence of two men—Prof. A. W. Bickerton, a man of original but heterodox views, who was rather a free lance in physical science; and Prof. G. H. H. Cook, who was a sound and orthodox mathematician and an able teacher. At the end of his fourth year he qualified for his M.A. degree with the rather unusual distinction of a double first—in Mathematics and in Physical Science. His mother with reasonable pride wrote the good news to his first schoolmaster at Foxhill.

Rutherford returned for a fifth year to Canterbury College, devoting himself largely to science subjects with a view to a B.Sc. degree. For his thesis he undertook a large section of research work on the "Magnetisation of Iron by High-frequency Discharges". This was read before the Philosophical Institute and published in the *Transactions* of the New Zealand Institute in 1894. Like many other young men, he had been immensely impressed with the discovery of electric waves by Hertz, which confirmed the electro-magnetic theory of Clerk Maxwell. Rutherford set up a Hertz oscillator in a "miserable, cold, draughty, concrete-floored cellar, which was usually known to students as the 'den' and in which they were accustomed to hang up their caps and gowns". He could thus generate a high-frequency, alternating electric current, which he could pass round insulated fine copper wire, wound into a coil in the middle of which he could insert steel sewing-needles, or short strips of iron or steel wire. He found, contrary to the generally accepted views of that time, that it was quite possible to magnetise, or demagnetise, iron and steel by rapidly alternating currents. It should be noted that he excited his oscillator with sparks from an induction coil or a Voss machine, and both were equally effective. As he had a series of damped oscillations, each half-wave of current in his wires was greater than the second half in the reverse direction, so that the total current for a single spark was greater in one direction than in the opposite. Hence the magnetisation, which was often only skin deep, as he showed by dissolving away the outer part of a needle with acid.

It is not possible to describe all the ingenious experiments carried out by Rutherford at this time, but it seems clear that he then began to develop a magnetic detector by which he could detect wireless waves passing through a distance of sixty feet down the length of the den and through opaque obstacles. He found that the most sensitive detector consisted of highly magnetised steel needles which lost much of their magnetism by a high frequency current in a coil round them. It is interesting to note that Henry, the famous American physicist, had independently detected lightning flashes many miles away by means of an aerial from roof to ground passing round a coil with steel needles in the centre. Rutherford afterwards told me that he had not then heard of these experiments but he did refer to Henry's magnetisation of steel needles by Leyden-jar discharges. His own published conclusions were "that iron is strongly magnetic in rapidly varying fields, even when the frequency is over a hundred million a second". We are here indebted again to Mr R. M. Laing, and to Mr S. Page, who was Demonstrator in Chemistry and Physics during this very period (*The Times*, 22 Oct. 1937). Indeed Mr Laing stated in *The Press*, New Zealand:

21 *Oct.* 1937. I saw him send a message by Hertzian waves from one end to the other of Professor Bickerton's laboratory, the old tin shed, as it was called, through various intermediate walls.

It is clear that the work he developed at Cambridge was initiated at Christchurch.

Rutherford's second paper on 'Magnetic Viscosity' was published in the *Transactions* of the New Zealand Institute (1896) and among other schemes he had a time-apparatus capable of measuring intervals as small as a hundred-thousandth of a second. Rutherford also enjoyed meetings of the Science Society, founded in 1891, at which he prophetically discussed the Evolution of the Elements. He also presented a paper on electrical waves and oscillations, about which the minute-book records:

The paper was very fully illustrated by experiments performed by Mr Rutherford...the most striking being a reproduction on a small scale of Tesla's experiments on rapidly alternating currents.

During his first term or so Rutherford shared rooms with Marris, in Montreal Street, Christchurch. W. S. Marris, now Sir William, was one of the greater lights of Canterbury College, and a contemporary of Rutherford. He passed into the Indian Civil Service—first, with a large margin. He went to Christ Church, Oxford, holding an I.C.S. Scholarship. He had a distinguished career in India, becoming Governor of the United Provinces. Rutherford and Marris are good evidence for the excellence of New Zealand education.

Sir William Marris writes:

Generally, I recall Rutherford as very modest, friendly but rather shy and rather vague—a man who had not yet found himself and was not then conscious of his extraordinary powers.

I suppose that he read Physics for his B.A. but his strong suit was Mathematics. He was easily the best mathematics student in College. He and I were in a separate Honours class for three years—in fact we were the whole class and we got some intensive cultivation from Professor Cook. Rutherford was immensely the better man, but I believe we shared the yearly exhibition of £20 twice if not three times.... The explanation is largely in the fact that Rutherford was nervy and excited in examinations, and also (I expect) that the exam. was based too largely on 'bookwork'. The professor was very insistent on thorough bookwork and did not comprise enough of the testing problems which must have disclosed Rutherford's superiority.

During his University days Rutherford used to coach students so that he might meet the expenses which exceeded his scholarship. Soon after going to College, he went to live with Mrs De Renzy Newton, a widow with four children. The eldest, Mary, and Ernest later became deeply attached to one another. She spent many holidays with the Rutherfords, and they were unofficially engaged before he left for Cambridge; but their engagement was not announced till 1896.

For a brief period Rutherford was a substitute master at the Boys' High School, Christchurch, and like many young men he had, at first, two great difficulties—his teaching was beyond the powers of the boys and he had trouble in keeping good order in his classes. No doubt he would have overcome, as others do, these obstacles, but he was reserved for a kindlier fate. The Prince Consort had organised the Great Exhibition of 1851; a large profit had been made, which was to be devoted to education. Scholarships were founded, with a view to bringing able men to British Universities and Rutherford was among the fortunate.

When his mother came to tell him of his good fortune he was digging potatoes. He flung away his spade with a laugh, exclaiming: "That's the last potato I'll dig."

Mr Evelyn Shaw, Secretary of the 1851 Royal Commission, has kindly extracted the facts of the award of a scholarship to Rutherford. There were two examiners, Prof. A. Gray, F.R.S., and Prof. J. E. Thorpe, the Government Chemist. Two candidates were presented by the University of New Zealand—Mr Maclaurin, who submitted a published paper on the 'Treatment of Gold', and Mr Ernest Rutherford whose "theses cover a very wide extent of experimental work on Magnetism and Electricity".

Gold won, as usual, and Maclaurin was placed first, but the examiners were so much impressed with Rutherford's work that they urged the award of a second scholarship to him. The Board of Commissioners thought that such action would set a bad precedent and suggested that his name might be put forward in the ensuing year. At that time the

University of New Zealand received a nomination every other year, but had missed a year in 1894. In the meantime Maclaurin, who was married, had been offered an appointment, and so Rutherford was elected.

J. C. Maclaurin became the New Zealand Dominion Analyst and did notable work in the development of the cyanide process for the separation of gold.

A. W. Bickerton, Professor of Chemistry and Physics, commended Rutherford's work in these words:

Rutherford has conducted long and important investigations into the time effects of electric and magnetic phenomena in rapidly alternating fields and by means of an ingenious apparatus of his own design was enabled to observe and measure phenomena occupying less than 1/100,000 second.

He continues:

Mr Rutherford has great fertility of resource, a very full acquaintance with both the analytical and graphic methods of mathematics and a full knowledge of the recent advances in electrical science and methods of absolute measurement. Personally Mr Rutherford is of so kindly a disposition and so willing to help other students over their difficulties that he has endeared himself to all who have been brought into contact with him. We all most heartily wish him as successful a career in England as he has had in New Zealand.

Ernest Rutherford
(Aged 21)

CHAPTER II

RESEARCH WORK AT CAMBRIDGE

When Rutherford was starting from New Zealand to go to Cambridge he had to borrow the money to pay for his passage. On his way to England he called at Adelaide University and saw Prof. W. H. Bragg. He called Bragg from a dark-room where he was trying to get a Hertz oscillator to work and enthusiastically showed him the magnetic detector that he was taking to England. So there was a remarkable fourfold conjunction, Bragg, Rutherford, oscillator and detector. On arrival in London he paid the not unusual penalty of those who leave the good air of the sea for the evil substitute in the city and suffered from sore throat and neuralgia.

Prof. J. J. Thomson wrote from 6 Scroope Terrace, Cambridge, to Rutherford in London:

24 *Sept.* 1895. I shall be very glad for you to work at the Cavendish Laboratory and will give you all the assistance I can. Though it is not absolutely necessary, I think you will find it advantageous to become a member of the University. We have now instituted a degree for research so that anyone who resides for two years and does an original investigation, which receives the approval of the examiners, receives a degree; if he resides for one year and does the investigation he is given a diploma. If you could spare the time to come to Cambridge for a few hours, I should be glad to talk matters over with you; so much depends upon the requirements and intentions of a student that a personal interview is generally more efficacious than even a long correspondence. In case you decide to visit Cambridge, if you will let me know when to expect you, I will arrange to be at the Laboratory; if equally convenient to you, Mondays, Wednesdays or Fridays are the days I should prefer, though just at present I could arrange for an interview on any day. I am much obliged to you for your paper, I hope to take an early opportunity of studying it.

Rutherford had sufficiently recovered from his indisposition to go to Cambridge, where he took a cab—the driver drove him to Cavendish College instead of the Cavendish Laboratory.

However, Rutherford was the first research student to arrive at the Cavendish and to be welcomed warmly by Prof. J. J. Thomson, best known to students by his first two initials; and the next arrival was J. S. E. Townsend, now Wykeham Professor of Physics at Oxford. Although the senior members of the University accepted the new arrivals with enthusiasm, there were one or two demonstrators with the ancient prejudice that no good things can come from the Colonies. Rutherford told the writer that these men used to pass his door at the Laboratory with a snigger. He politely asked them in, and told them he was in some difficulties with his experiments and would be grateful for their help. They quickly realised that they had not the faintest idea of what he was doing, for his famous detector was then new and unknown. "After that, they gave me no more trouble; they had got on my nerves a bit."

Rutherford's fame spread quickly among the younger men at Cambridge and Dr Andrew Balfour, later of the Egyptian Government Medical Staff, writing of him at the time, said: "We've got a rabbit here from the Antipodes and he's burrowing mighty deep."

It happens that we are fortunate enough to possess a most intimate knowledge of Rutherford in those days, not only from the letters that he wrote to his mother, but still more from those he wrote to Mary Newton, the girl to whom he was engaged, and whom he married a few years later. In reading these letters it must be remembered that he was more candid in them than in any other written record that he has left behind, that he was anxious to 'make good' as a young man coming to an old and famous University from the youngest of the Colonies, that he was anxious to make money so that he could marry the girl to whom he was writing, and above all, perhaps, that he was aflame with scientific curiosity and enthusiasm with an unusual vision of the possibilities before him in the subject which he loved.

RUTHERFORD TO MARY NEWTON

Cambridge: 3 *Oct.* 1895. I wrote my last letter about a week ago and have had some vicissitudes since that time. On Thursday I woke up at 3 a.m. with a bad attack of neuralgia in the left side of the face and head. Had a pretty bad time of it all day, but went to a chemist and got some stuff that didn't affect it. Next day I had an appointment to go and see Thomson at Cambridge. I had recovered a good deal so went down by the express taking about an hour and a quarter to get there. The country is pretty enough but rather monotonous. I went to the Lab. and saw Thomson and had a good long talk with him. He is very pleasant in conversation and is not fossilized at all. As regards appearance he is a medium sized man, dark and quite youthful still: Shaves, very badly, and wears his hair rather long. His face is rather long and thin; has a good head and has a couple of vertical furrows just above his nose. We discussed matters in general and research work, and he seemed pleased with what I was going to do. He asked me up to lunch to Scroope Terrace where I saw his wife, a tall, dark woman, rather sallow in complexion, but very talkative and affable. Stayed an hour or so after dinner, and then went back to Town again. I have forgotten to mention *the* great thing I saw—the only boy of the house—$3\frac{1}{2}$ years old— a sturdy youngster of Saxon appearance but the best little kid I have seen for looks and size [G. P. Thomson, now Professor of Physics at the Imperial College and a Nobel Laureate]. Prof. J. J. is very fond of him and played about with him during lunch while Mrs J. J. apologised for the informality. I like Mr and Mrs both very much. She tries to make me feel at home as much as possible, and he will talk about all sorts of subjects and not shop at all. On returning to London I paid for my trip by an extra powerful attack of neuralgia, and for the next three days I had a very bad time. It went to the left side of my head and I had a devil of a cold in left side of throat, so that I lay in bed a good part of the time on Friday, Saturday and Sunday—just going for a short walk in the afternoon. The servants were very kind and brought me anything I wanted to my rooms so that I had not much to complain of. On Monday, finding myself not much better, I went to a doctor and got a prescription and, whether due to Nature or the medicine, I was getting well again Tuesday, so that I went up to Cambridge with all my belongings.

Mrs Thomson had been very kind and looked me out some lodgings with a widow and gave me directions where to find her. They also asked me up to dinner in the evening which was expressly not a dress affair. Beside myself there was present a Miss Martin from Sydney University, who has been doing research work here last year but is going for a spell on the continent with her mother Lady Martin, and also present a Mr Townsend, a young fellow graduate, of Dublin University, who didn't get a fellowship he'd been working for and so came over to Cambridge to do some research work. As he has no friends here, he and I knock about together a good deal. I have now been in Cambridge for three days and have worked for two days in the Lab. I cannot pass an opinion about Cambridge, as I have not yet been about enough. The University term does not really begin till the 10th but there are a lot of freshmen up just in for their 'little go'. They are mostly very young and innocent looking, and so a good many are attended by their papas and mamas the first few days. They knock about in cap and gown and one meets them everywhere.

Now for a few remarks on my temporary lodgings for I have not yet settled my future movements. The place is in a street full of houses and is inhabited by an old lady of about 55. She has two rooms to let for £7. 10/- a term, the regulation price for such type of lodgings. Everything is so different as regards arrangements that I will go into particulars. Rent of bedroom and sittingroom 15/6 a week, coal and firewood 1/6, use of crockery 1/6. Besides that I pay for oil and all my own tucker, which she cooks. It is pretty expensive, the rent is so enormous, but I believe I am in an extremely moderate show. I will know at the end of the week about how much things mount up to. Thomson wants me to go into residence for the reason that by a new regulation research degrees are offered to students (resident). At the end of the first year, if the thesis is approved by the examiner, one gets a diploma, at the end of the second year a degree (of B.A. of course)—if the thesis is again approved. Research students, if they reside, will be in the same position as the graduates of Cambridge and will dine with them. The expenses will however be rather high, I am afraid, unless they can be cut down considerably. I will know for certain in a week or so at any rate. For the present I will stay in these lodgings till I see my way clear.

I have seen a good few of the colleges externally but I have not been through any of them yet, as I work all day in the Lab. I have now got

rid of the neuralgia, but not completely of its after effects. To judge by
the odour, my head the last few days has been turned into a chemical
factory for sulphuretted hydrogen (you know the smell). It is most
disagreeable; but I suppose is a product of neuralgia. My cold is quite
right and I am feeling in much better form now. London was about
80 F. all day for the ten days I was there and was very oppressive. It has
now suddenly cooled down and is quite wintry, so that I have got on
my thick things again; but the cool weather is much better for me.
I suppose the sudden change from the sea upset my equilibrium a little.
Dawson says he was troubled in much the same way the first month in
Edinburgh. It was lonely in London but I didn't mind much as long as
I could keep moving round. I saw a good deal, but don't feel inclined
to give a long dissertation....I have not yet seen Marris but have got
a letter from him. You will know by this time he is top of the I.C.S.
by over 1000 marks—a very big thing indeed. He is going to Oxford
for a year and I hope to see him before long....So much for my news
and don't imagine from my discursion on my health that I am not
alright. I am recovering my normal state rapidly and will soon be at
work pretty hard, as I don't intend to waste time. I am not going to do
much bookwork and so should keep in good form. It is of course a bit
strange at first with all one's friends across the sea....My success here
will probably depend entirely on the research work I do. If I manage
to do some good things, Thomson would probably be able to do
something for me. I am very glad I came to Cambridge. I admire
Thomson quite as much as I thought I would, which is saying a good
deal. They have both been very kind to me, as you may judge from
what I have written.

Trinity College: 20 Oct. 1895. ....I have at last crossed the Rubicon and
am now a regular undergraduate, or rather graduate of the University.
I have been waiting for the last fortnight to see what allowances they
would make for research students, as this is the first term the regulations
have come into force. J. J. Thomson looked after our interests and
obtained a substantial reduction in fees, and, as he strongly advised me
to join a college, I at last decided to do so. He wished me to join Trinity,
which is his own college and also the best as well as the dearest in the
university. It will not however make much difference to me which
college I join, and Thomson seemed to think I would probably be able

to command more influence from the people of Trinity than from any other college. Townsend and I went on Saturday and saw the Tutor Mr Ball—a very courteous and affable man—and he gave us all the details of what has to be done. Unless we joined on Monday we would lose this term, so this morning, which is the general matriculation day, as it is called, I went and got a cap and gown before 8.30. Research students wear a B.A. gown without strings. At 8.30 all the freshmen of Trinity, about 200 in number, assembled in the Hall, had our names read out and then trooped off to the Senate House, where we had to sign our name in a book kept for the purpose. By 10 o'clock I was formally initiated and then had to pay £5 matriculation fee, and £15 caution money to be returned at the conclusion of the University course. Today has been the general matric. day and more than 2000 freshers have been enrolled.... The expense of joining a college will not be much more than if I were a non-collegiate student for I shall stay in the lodgings where I am but can turn up to dinner when I like and only pay when I do so. The great advantage of joining a college is of course the number of men you come to know and the social life which as a non-coll. you would miss entirely. Thomson very strongly advised me to join and reckoned the extra expense would be not thrown away in the time to come. So far I have knocked round with Townsend a good deal. I told you I got a letter of introduction from Love of Melbourne to his brother at St John's, who is a lecturer there and a very noted man, an F.R.S., etc. He is a great authority on Elasticity and is a very prominent man in his line. I have heard indirectly that Mrs Love, mother of the above, has been enquiring whether I have arrived, so I will have to present myself the next day or two. I am not going to be backward in the ordinary social life. If I can do it cheaply, it will probably be more use to me than anything else. Have been working steadily in the Laboratory but have nothing very definite done yet.... The univ. here takes a tremendous amount of care of the undergrad....

Trinity College: 28 Nov. 1895. You have heard me talk of the Physical Society, J. J.'s pet society. Well, he has asked me to give an account of some of my work before it, and I am to occupy the whole meeting. I am to show some experiments for the interest of the vulgar. Usually it is only wellknown people, Profs. and such like, who shine before the society, so I appreciate the honour of being asked. It is my chance of getting a little lift up on the scientific ladder, and I intend to make as good use of it as possible. Among so many scientific bugs knocking about one has a little difficulty in rising to the front, and naturally I am very pleased at having the opportunity of asserting myself. You mustn't think I am egotistical in talking like this, but the sooner I rise up the sooner I will get a decent appointment and the sooner.... A great thing about it is that it is a sign J. J. thinks a fair amount of my work, and if one gets a man like J. J. to back one up, one is pretty safe to get any position for which he will put himself out to aid you. (What a change of pronouns!)

8 *Dec.* 1895. ...In my last letter I told you that J. J. had asked me to give an account of my work at the Physical Society. I am the first member of the Cavendish who has given an original paper before it, so I may consider the honour is greater than I can bear....Instead of taking only part of the time, J. J. stuck the following announcement on the notice board "A method of measuring waves along wires and determination of their period", with experiments by E. Rutherford, and left me to fill up the whole time. The term 'with experiments' rather knocked me sideways, but I went to work and rigged up a good few interesting experiments, which all came off very well. I had quite a distinguished audience including J. J. and Mrs J. J. and several other ladies—Sir G. Stokes—a good few lecturers and demonstrators besides the usual vulgar herd. I think the paper was quite a success and J. J. was pretty pleased, I think. No one but myself made any remarks, as it was rather beyond most of them. I really had to give a lecture and did not read anything at all. My friends all reckoned I did very well indeed, so I suppose I may consider it a success. Mrs J. J. in speaking to me afterwards complimented me rather neatly in a way (which of course I took at a proper valuation, *cum grano salis*) that struck me as decidedly good, "You kept us ladies very interested indeed, and I am sure it was sufficiently deep for the more scientific members of the society", or words to that effect...a very *à propos* way of conveying a neat compliment.

As I have got along very well so far, I hope to continue. I am at present investigating a new detector I have for electrical waves of high period and find it works very well....These detectors are even more sensitive than I thought, I believe they will be far and away the best metrical detector for electrical waves. The best method used before will only show the effect for a couple of yards but I can get quite a large effect at twenty yards. J. J. is very interested and comes round very often and gives what help he can. By the way I have not told you that I will be publishing some of my work before long. I spoke to J. J. about it and he said I had better send it to the Royal Society. As only the best papers, or at any rate the papers of eminent men, are chiefly found there, I have nothing to complain of. As a matter of fact very few papers are recommended by Thomson for the Royal Society. I must apologise for the amount of ego that fills this letter, but human nature will out, you know—I know you will be interested in the "vain outpourings of an idle mind". I have been working very hard lately and intend to keep on till within a few days of Xmas when hey presto—and I am in Edinburgh among friends I hope. Most of the undergrads have gone down and only the workers are at College so the town seems a little empty. Townsend, my particular friend at present, is going up to his cousins at Yorkshire for Xmas....I don't know whether I have described him to you. Imagine a middle-sized man, very fair hair rather scanty on top, very fair moustache and a true Irish complexion and a merry blue eye— rather good features and a very pleasant appearance altogether. He is a very fine mathematician and is a good deal of assistance to me in that way. I think it probable he and I will research together on abstruse subjects next term, for in some of the work I am doing, it is very difficult for me to do all without assistance.

As to his new detector, Rutherford described it himself, much later (1910), in chapter VI of the *History of the Cavendish Laboratory*.

By using very fine magnetised steel wires surrounded by a fine solenoid, the demagnetising property proved a very sensitive quantitative method for detecting electrical waves. Using large Hertzian vibrators, the electrical waves emitted were observed by means of the magnetic detector for a distance of about half a mile. These experiments were made before Marconi began his well-known investigations on signalling by electrical waves. This effect of electric oscillations of altering the magnetism of iron is the basis of the magnetic 'detectors' developed by Marconi and others, which have proved one of the most sensitive and reliable of receivers in radiotelegraphy.

The Cavendish Laboratory

The original Rutherford detector is preserved and revered in the Cavendish Laboratory.

There is one point in his letter of 8 December which does not accord well with some statements in the previous chapter, for in the letter he claims "I can get quite a large effect at twenty yards", as though this result was obtained for the first time in Cambridge, not New Zealand.

In order to form a clear picture of Rutherford's early success with wireless, you have to think of the transmitter, consisting of two large metal plates, side by side and in the same plane, with two short metal rods protruding from them and ending in polished brass knobs about half an inch or so apart. These were 'excited' so that sparks passed between the knobs, no matter whether by an induction coil or by an electrostatic machine, so long as there was a fairly steady shower of sparks, each of which caused electricity to swing to-and-fro between the plates, perhaps a few million times a second. It will be noted that this was the same scheme that Hertz used, and such an oscillator sends out electric oscillations into space, or into the ether—whichever term one prefers to use.

Half-a-mile away, with houses in between, was Rutherford with his receiver, made of two metal rods in line, each two feet long; joining their near ends was a fine wire going round a small coil wound about a minute bundle of fine magnetised steel wires. When a signal came, the needles lost their magnetism and this was easily shown by the deflection of a mirror with a little magnet behind it.

In this way Rutherford proved that the electromagnetic waves travelled from the transmitter to the receiver for a distance of half a mile or so, and through brick walls if they were in the way. This result greatly interested, but did not surprise him, though it certainly amazed many who saw his experiments.

He evidently guessed at future and further possibilities, but was ready to turn aside to other more interesting scientific problems. It was lucky that he did so, for otherwise he might have become an electrical engineer rather than the physicist who was destined to unravel some of the chief mysteries of Nature.

RUTHERFORD TO HIS MOTHER

1895. I have been for several walks six miles or more from Cambridge in various directions. One comes across some very old villages with mud or stone houses, thatched and very dilapidated, but picturesque. The great difference one observes is the amount of cultivation one sees around; turnips etc., seem to be growing everywhere for the cattle and sheep.

As I have been working pretty steadily at the laboratory I have got about little except on Sundays. My study is a pretty comfortable one and I have various photos. Christchurch views adorn various parts of the room.

Today has been wet and cold, but I have had quite a gala day. In the afternoon the Cavendish Society met, which is composed of a few of those interested in science. Two papers were read—one by a Mr Griffiths and another by Miss Marshall. The latter is doing a lot of experimental work with Professor Ramsay, the great chemistry man.

At a second meeting of the Cavendish Society Professor Ewing gave a little sketch of an instrument of his. You remember Ewing's magnetism which I used to set so much store by?—he is the author....

My own research work is progressing satisfactorily, but is pretty slow, of course. Thomson, I think, considers my work original and interesting, for when Lord Kelvin came to the laboratory here he told one of the demonstrators he wished he had been taken to see my work.

I work on the third floor of the laboratory and have a few fellows in adjoining rooms for company. The place is heated throughout with hot-water pipes and is quite warm all day long. The more I see of the laboratory the better pleased I am with it, for, although it is not as well fitted up as I expected, it has a fine collection of instruments.

RUTHERFORD TO MARY NEWTON

Cambridge: 15 Jan. 1896. ...I only saw the sun once in Edinburgh. My landlady had everything ready for me when I came back, so I soon felt at home again. I started to work at once in the Lab. on my electric waves and made my first experiment on long distance transmission of signals without wires. I set up the vibrator in the Prof's room at the Cavendish and my detector a 100 yards away in Prof. Ewing's Lab. and got quite a large effect through that distance, traversing three thick

walls by the way. The Prof. wishes me to continue at the work and
see how far I can detect the waves, so I have been working lately to try
and find the best conditions of sensitiveness of my detectors. I have
every reason to hope that I may be able to signal miles, without con-
nections, before I have finished. The reason I am so keen on the subject
is because of its practical importance. If I could get an appreciable effect
at 10 miles I would probably be able to make a considerable amount of
money out of it, for it would be of great service to connect lighthouses
and lightships to the shore, so that signals could be sent any time. It is
only in an embryonic state at present, but, if my next week's experiments
come out as well as I anticipate, I see a chance of making cash rapidly in
the future. I cannot say I am exactly optimistic over the matter, but
I have considerable hopes of being able to push it a good long distance....
I remarked before that J. J. is very interested in my work and makes all
sorts of suggestions to help me in getting a clear space for signalling.
My next experiment will be, I think, from the tower of the Cavendish
to St John's tower nearly half a mile away.... Townsend, my special
friend, has not been successful so far in research, but J. J. has appointed
him part-time demonstrator which is very good indeed. I could be
offered a similar position, only my schol. will not allow me to do so.
If I keep on as I am going, there will be no difficulty in getting enough
to keep me here even if my schol. is not continued. Townsend and I have
placed the Research Student on a pinnacle of honour. I by my paper
and work on waves, and T. by being appointed a demonstrator after
three months here. The three demonstrators are all extremely friendly
now they see we have made a strong position for ourselves and I grimly
rejoice for they did not take any notice of us the first two months
although they knew we were strangers and had no friends in Cambridge.
My paper before the Physical Society was a heavy blow to their assumed
superiority, and now they all offer to help us in any way they can and
tell me confidentially about their own little researches—so wags the
world.

25 Jan. 1896. This letter will deal with a dinner which I attended tonight,
which seems so unreal to me at present, that I had better put my
impressions down before they fade. I know you will be interested in
these somewhat trivial descriptions as you consider social intercourse a
thing that I sadly lack, but verily I have been plunging deeply lately.

But to lead up gradually to my discourse. My friend Townsend knows Sir Robert Ball.... Well! Sir R. developed a tremendous interest in my experiments on the detection of electric waves for long distances and must needs make an appointment to come down to the Lab. and see the effect and apparatus in general. He turned up one morning and I showed him how easily I could detect a wave through half a dozen walls and rooms, and he was very much interested. He is specially keen on it, as he thinks that experiments of the nature I am doing, will solve the difficulty of lighthouses in times of fog, when the light does not penetrate any distance. His idea is to fix up a vibrator in a lighthouse, and as a fog does not stop an electric wave a suitable detector on board a vessel should tell her when she is within say a couple of miles of a lighthouse. Of course the arrangement would be very useful for signalling at sea at night between vessels, and informing each other of their close proximity in times of fog. Sir Robert wanted to know about the whole matter, and ended by asking Townsend and myself to dinner at King's College tonight. He is an honorary Fellow of King's and of course you know what a name he has made in astronomy. We turned up in cap and gown at 7 p.m. and first of all met Sir Robert and all the Fellows in the Combination Room, as it is called. I was introduced to those whom he considered most interesting from my point of view, and he spoke of me in such flattering terms that I felt inclined to disappear out of sight. However I recovered my normal modesty and walked into Hall with Sir Robert marshalling me in front as an honoured guest. The 150 students of King's and the Fellows all dine in the large Hall together. The Fellows, i.e. all the professors and lecturers, etc., of the College dine at a separate table at the top of the room. All the students stood up as we entered and naturally they all wanted to know what in the deuce a youngster like me was doing among the Fellows. To add to the shock which I had already suffered I was placed in the position of honour at the table, on the right of the Provost as he is called. Grace was said, and dinner was served. Conversation was kept up pretty well, but the Provost [Augustus Austen Leigh] himself had an impediment in his speech and was not particularly brilliant. I was able, however, to keep up a conversation with some interesting men on the other side of the table. I really felt a great deal like an ass in a lion's skin after the way I was treated, for I did not see what I had done to be one of those whom "the King delighteth to honour" but Sir Robert Ball's remarks had

evidently fallen on good ground, for I was looked upon as a scientific
expert. Dinner lasted just an hour at the Fellows' table, and then we all
retired to the Combination Room, and, after a little desultory talk, all
sat down at a long table and chestnuts and cakes were placed on the
table as well as wine and cigars. Conversation was very interesting and
all the men were of course more or less famous in various branches,
and their conversation was worth listening to. Here again I was seated
alongside Sir George Humphry, a great medical man about seventy-five,
although he looks much younger, and we had an interesting chat on
various topics. I was sandwiched between Sir G. Humphry and Sir R.
Ball and tried to look as dignified as the circumstances permitted and
really felt at ease notwithstanding my usual shyness or rather self-
consciousness.

Seated opposite to me was a Mr Browning—a lecturer at King's—
and he made himself very agreeable. He seemed an extremely well
informed man—in appearance he was a good deal like the typical John
Bull one so often sees in Punch. He evidently knew such men as
Goschen and Chamberlain personally very well, so I thought he must
be rather a prominent character. In the course of conversation he asked
me to lunch tomorrow, where of course I am going. I believe he has the
most beautiful set of rooms in King's and is a very clever man as well
as a good conversationalist. Excuse the way I put the above. It reminds
me a good deal of Boyle—you remember? "He was the Father of
Chemistry and Brother to the Earl of Cork."

About 10 o'clock we rose to leave the Combination Room and a
Professor Oldham—Professor of Geography—asked us to his rooms
where we stayed half an hour telling stories. Sir Robert has a true Irish
sense of humour and tells a yarn well. I have not yet described him to
you and find it rather difficult. He is a man about forty-five, thickset,
with a red face of large dimensions, and a most humorous quizzical look.
He is a true Irishman and has the gift of the brogue when he likes. I got
back to my rooms after 11 and wondered whether all had been real, or
whether I had been a tin god for a time. I felt such an impostor
masquerading before the learning of King's, but it is a good thing my
modesty does not allow me to take all the above as my right.

It has really been a very eventful evening to me, both as regards the
people I met and the general impression I have received of the life of
these friars of learning, for really the system of dining in hall is a relic

of the time of the old monasteries which has been kept intact to the present day. Some of the men I met at dinner strike one as very capable, especially in their conversation, and it is a pity so many of them fossilize as it were in Cambridge, and are not that use in the world they might be.

...The Professor has been very busy lately over the new method of photography discovered by Professor Röntgen and gives a paper on it on Monday to which of course I will go. I have seen all the photographs that have been got so far. One of a frog is very good. It outlines the general figure and shows all the bones inside very distinctly. The Professor of course is trying to find out the real cause and nature of the waves, and the great object is to find the theory of the matter before anyone else, for nearly every Professor in Europe is now on the warpath....

Sunday. ....I went out to lunch with Mr Browning who is a History lecturer. There were three others present besides myself mostly of the very young type. One young fellow Goldsmid by name, and a Jew, has just come into a fortune of £30,000 a year. A Howard of Howard Castle was also present and seemed a very intelligent type. Conversation was very shoppy and I was very glad to get away. Browning is quite a character here. He is a bit snobby. From what I have heard he professes to know all the people worth knowing in Europe. It is a common yarn about him that he said "The German Emperor was about the pleasantest emperor he had met", which is I should judge quite characteristic of him. His rooms were very fine and he has a very good library of which he is very proud.

RUTHERFORD TO MRS NEWTON

6 *Feb.* 1896. I was working the other day at the top of the Lab. when Murray came up and told me an old man was waiting to see me down below. I was very surprised to find Sir G. Humphry awaiting me. I mentioned having met him at the dinner at King's the other Saturday. He came to ask Townsend and me to dinner on Sunday at 7.30 with his family circle. It seems he had written to me before, but had misdirected the letter, so he came to deliver the invitation in person. It is very decent of him to ask me to dinner at the very slight acquaintance with me.

RUTHERFORD TO MARY NEWTON

Trinity College: 21 *Feb.* 1896. ...I have had a little dissipation lately. You will remember my description of the dinner at King's and the great time I had there. In your mother's letter I have given a short account of the dinner at Sir George Humphry's: my dissipations in high life I keep to myself and don't tell my friends, as it wouldn't do at all. The cause of this sudden boom of the Research Student is due to the talk of Sir R. Ball on the subject of my wave-detector, which I believe he enthuses over whenever he gets the chance. What excessive virtue he sees in it I can't comprehend, but I can't stop a man from thinking absurdly on matters. He reckons the work I have done shows the wisdom of the University in providing for research students and the fame of Townsend and me has travelled to all the colleges, so that they are all opening their doors in their anxiety to welcome the research student of which we are the first examples. J. J. I believe openly declares that the new Research Students are a great success, and as I am at present the most prominent in that respect, I take a little of the praise unto myself. The more I boom of course the better it will be for me, for verily I must do something somehow, and verily the better it will be for thee since my fortunes are thy fortunes! The air here is full of the new photography, till I myself feel rather tired with it....The meeting in the Physical Society today was on this subject, but as I had on my Lab. clothes and was unshaven (for verily I have not the inducement to keep smooth I once had) I discreetly did not go down to the front for tea but paid my devoirs to Mrs J. J. afterwards. J. J.'s experiments did not go off very well in the lecture, so he got a bit mad, and made those sit up who asked silly questions. I was too wise to do so, but some fellows think they show their smartness by asking questions, whether he had tried this, and that, and so on, and J. J.'s dander rose and he turned the laugh against them most skilfully—whereat I rejoiced, for many of them are my enemies. I am of the opinion that the demonstrators regard the research students here with very little favour and try and put little obstacles in our path, but verily we rise superior to their machinations and they gnash their teeth with envy. There is one demonstrator on whose chest I should like to dance a Maori war-dance, and which I will do in the future if things don't mend. Adieu for the present to these warlike fancies, but the fact is my dander rose today and

I have not yet quite recovered. The man, from whom I was to get my cells for my midnight excursion, failed me at the last moment, and I, whose temper is on ordinary occasions most angelic, gave way to my feelings and did some tall talking for a few moments....I am going to try my experiments tomorrow night on the Common. I hope they will turn out successfully. Townsend and McClelland are going to assist me. I am to use Townsend's room for my detector and go across the Common about half a mile with my vibrator and accessories. I will leave this letter open till tomorrow so that I may add particulars....My experiment came off tonight and was rather unfortunate. I am writing this tonight about 1 o'clock just after my return from the expedition. I fixed up the receiver in T's room at 6.30 in the evening, and had arranged for two men to come at 7.30 to take the apparatus on to the Common with a small go-cart. I waited till 8, and they did not arrive, and then I went and hunted round for the firm who had got the men for me, hunted up a clerk, who found that the pumpkinhead, to whom I had given the order, had put time down at 7 instead of 7.30. The men had come, waited and gone, and I did not know of it. However we hunted up two men from the street, got a hand-cart and got started about 9.15, instead of 7.30. We got the things ready and set them up at 500 yards at which distance I anticipated a good large effect, but we could not get a sign. We tried different distances up to 100 yards and not a sign. I knew then the internal economy of my detector had gone wrong at the critical moment for I had tried 100 yards with the same detector before and had got a large effect. This detector I had arranged in a specially sensitive method, so when it had gone wrong somewhere I had nothing to replace it. However I had another rough detector handy and on trying that we got a large effect up to 350 yards. We did not try any further as it was getting very late, and I had to take the apparatus back to the Lab. before 11.30. So we hurried back, and got clear about 11.45. I was very tired, as I had been running about from 6 to 12 and feel very much like bed now.

It is a pity my detector went wrong at the critical time. I had tried it in the afternoon and it was all right but one of the wires had evidently got broken somewhere inside. I may try the experiment again under more favourable conditions and am certain in my own mind I could get an effect up to a mile easily. My friends were very disappointed at first, but they rejoiced when they saw it was evidently due to a break in the detector. They both worked like Britons, and I could not have done without them.

29 *Feb.* 1896. When I left off last time I had just been out on the Common trying to detect waves at long distance. The next day I tried and got an effect from the Lab. to Townsend's diggings a distance of over half a mile through solid stone houses all the way. The Professor is exceedingly interested in the results, and I am at present very useful when he is writing to various scientific pots as he can mention what his students are doing at the Lab. Some good startling effects with waves suit him down to the ground. The morning after my midnight excursion, Sir R. Ball came down to see how I got along. He called in the other day and told me I could make use of the Observatory which is about a mile out of town, if I wanted to try some more long distance experiments. It is a very good offer and I will probably go up there before long. I am at present trying to clear up some of the work I have in hand in order to get ready for publication. J. J. told me he had written to Lord Kelvin about my work, and he had been very much interested in it. I hope to see the old fellow before his days are done, as he is very old now, although he is still very active and works away as hard as ever it seems.... I have sent a couple of the new photographs home to Taranaki—one of a frog, one of a hand, both taken at the Lab. I am sorry I can't send you some, but these things are rather expensive and I have to be careful of the bawbees these days.

The new buildings at the Cavendish were opened with a conversazione and among many other exhibits there were:

> 29. Mr W. N. Shaw (Emmanuel).
>   Experiments on the Formation of Clouds.
>   Pneumatic Analogue of the Potentiometer, and
>   Experiment in Ventilation.
>
> 30. Mr E. Rutherford (Trinity).
>   Experiments with Hertzian Waves.
>
> 31. Mr Hutchinson (Pembroke).
>   Experiments with Röntgen's rays.

RUTHERFORD TO MARY NEWTON

Cambridge: 15 *March* 1896. ...The great event of the last few weeks has come and gone, and was a very great success—the Science Conversazione at the Cavendish Lab. I told you it was to be a very big affair and no pains or expense was spared to make the thing a success. From the entrance to the Free School Lane to the Lab., about 60 yards, was covered in with an awning and lighted with glow lamps, and carpet laid down. Inside the Lab. itself carpet was laid down all over the staircases and everything was prettily lighted up. The new room where the guests were received had the usual display of biological, geological and physiological apparatus besides the physical experiments, but it did not look anything like as well as the big Canterbury College Hall on a similar occasion. The demonstrators presented Mrs J. J. with a bouquet for the evening, which she held in her hand while receiving her guests. At 8.30, people began to arrive in great numbers, and soon the place was pretty well filled up. There were between seven and eight hundred present, all more or less the very select part of Cambridge Society. I was one of the stewards and wore a pretty favour of blue and pink. I had my new dress suit for the occasion, wherein I felt more at home, not to say at ease, than in my old ones. I think it was about time I went in for a new suit, on account of reaching the limits of elasticity. Mrs J. J. looked very well and was dressed very swagger and made a very fine hostess. J. J. himself wandered round looking very happy and grinning at everybody and everything in his own inimitable way. We first of all wandered round and had a look at the show generally, and I then proceeded to show my experiment to anyone who so desired. I had my vibrator in one room and my own special receiver in a room 40 yards away with five solid walls in between, and showed quite a large effect at that distance. Some people were very interested and reckoned it was *the* thing worth seeing in the Lab. I had quite a large number of distinguished people who came to see it and were very keen over it, including Sir George Stokes, Sir Robert Ball, Professor Vernon Boys and a good few other pots whom I did not know by name. Sir R. Ball was going round expounding the merits of the same in great style and some of the ladies who came round were tremendously interested or professed to be. Townsend assisted me and we both took turns at explaining and working the vibrator. My part went off very well, so

I have nothing to complain of. In the morning before the conversazione Mrs J. J. brought her mother Lady Paget to see the experiment, and I had to explain everything minutely. Lady Paget is a very active old lady who travels about a good deal. Mrs J. J. seems to have got an absurd idea about my experiments and is nearly as good an advertiser as Sir R. Ball. Mrs J. J. told anyone she got a chance, that they must go to see the wonderful experiment of Mr Rutherford and it was the only new thing being shown that night. It is lucky I am of a modest disposition, but these things don't affect me in the way you might expect. I really think Mrs J. J. regards me with considerable favour for several reasons. She always introduces me as Mr...who has come all the way from New Zealand. I think she appreciates the compliment to her husband by my coming straight from New Zealand to the Cavendish....I think I have mentioned Mr Wilberforce, one of the head demonstrators at the Cavendish. He is a very decent fellow, and a married man to boot, and his wife asked us round, so Townsend and I made a call. Conversation was rather slack....Mrs Routh came in just as we were going and enquired very affectionately about me. She is rather of the gushing type is Mrs Routh, but makes a very good one at conversation. Tonight I went to dine with Hutchinson, Ph.D., Fellow of Pembroke. I dined at the Fellows' table, there were only four besides myself as several had gone down. Hutchinson is a chemist chiefly, and is a smart fellow, in appearance tall and thin with rather sharp sallow features. I went to the Combination Room after dinner, then into the rooms of Prior—the Senior Tutor—and afterwards to Hutchinson's own rooms and left about 10.30. He is very interested over the Röntgen photography and is working a good deal at it at present.

30 *March* 1896. ...I think I told you I was going for my vacation with Maclaurin and some friends to Lowestoft...our party consists of five fellows, two Afrikanders, one Armenian, and two New Zealanders which is just a little mixed as regards nationality. The two Cape men are De Villiers and De Waal who are both studying Law here to go back to the Cape. De Villiers, as you may judge from his name, is descended from a mixture of a Huguenot and Dutch, but at present is more Dutch than French. He can speak Dutch as well as English. His pater is what corresponds to a parson in England and is a member of the Cape House of Representatives. He is a young fellow about 21 and has been very

successful so far in his Law exams. The other Afrikander is De Waal...a
broad shouldered good tempered fellow sunburnt and pretty dark. His
pater is also a member of the House of Representatives.... De Waal and
De Villiers are both very much in favour of the Dutch element in the
Cape and of course are very hard against Jameson in stirring up the
Dutch against the English. So much for my companions. We are very
comfortably settled in a house facing the sea, and live very comfortably
for about 30/- a week.... We all lead a very lazy and happy life, going
for walks and riding on bikes when the weather is fine.

Cambridge: 10 *April* 1896. I have just returned to Cambridge from
Lowestoft, and have spent the morning visiting the shops and dressing
as I went along, for I left most of my wardrobe at Lowestoft and it will
not arrive till later. The most interesting subject to me at present is the
great bicycle ride we made yesterday, from the effects of which I am
not quite recovered, as I am still a little stiff, though not nearly so tired
as I expected. The distance traversed was 90 miles. We had very fine
weather, and left Lowestoft at 6.30, leaving Maclaurin behind to come
today with the luggage of all of us. Unfortunately we had a head wind
all the way, which made the work much harder. We went straight on
for the first 40 miles to Bottesdale, and, as the wind was very strong for
the 10 miles before that, we were all completely knocked out of time
when we got there. De Waal, a big strong beggar in good training,
was more knocked up than any of us. The reason of our sudden collapse
was very obvious after consideration. We had a rather small breakfast,
and should have had something to eat after the first 20 miles, however,
I will be much wiser in the future. At Bottesdale we had a heavy dinner
(at 11.30 this was) and De Waal who was in a rather bad way, slept for
a couple of hours while De Villiers and I loafed about a bit till 1.30,
when we all set off again in pretty good trim, and arrived at Bury
St. Edmunds at 3.30, where we had a small tea. We passed the ruins of
the Abbey, but I am going there for a spin some Sunday and will have
a good look round. At about 4.30 we set off for Newmarket, and had
another tea there, and then on to Cambridge another 14 miles, arriving
there about 7.30 in pretty good form and not specially tired—thus ending
a big day's ride of 90 miles. My first consideration was a warm bath,
when I went down to tea in rather a light costume—my landlady
discreetly not observing the same, and read the newspaper and turned in

at 10. I felt a good deal like getting up at 6.30 but stuck in till 8, and loafed round this morning getting studs, braces, boots, etc. in order to appear fairly respectable as I left all my things at Lowestoft. Spring is just coming on here, the trees are beginning to bud, and I expect in another fortnight all nature will be smiling and gay. I was nearly three weeks at Lowestoft altogether, and had a pretty good time—quite lazy enough even for me. My usual routine was to rise at 7.30, when we slipped on pants and coat and went down to the sea and had a morning dip much to the wonder of everyone abroad at the time. We had some rather interesting times over the bathing. The first morning we went in, in front of our diggings, dressed *en règle*. A policeman came next morning saying he had been specially sent by the superintendent to tell us to go further down, as one of the landladies in the neighbourhood had objected. After two mornings a bit further down the beach, another policeman gave us the hint to move still further down, as another modest female had raised objections. So we finally had to walk about 1/3 mile to get a dip in the tide. It was pretty cold of course, but one felt in pretty good trim after it. The alarming modesty of the British female is most remarkable—especially the spinster, but I must record to the credit of those who were staying there, that a party of four girls used to regularly do the Esplanade at the same hour as we took our dips and generally managed to pass us returning, but we were of course in eminently respectable attire by that time. So much for that portion of the programme. At 8.30 we had breakfast, 1 p.m. lunch, 6 p.m. dinner. Oh, how we did eat. The day was usually spent loafing about the beach, bicycling, sailing out on the Broads, where we tried our best to see how near we could go without capsizing, for it was safe enough even if anything happened, for boats were all about. We went out for a drive twice, and dangled our legs out behind in the usual holiday fashion, and played the goat generally. I managed however to do a certain amount of work in the Maths. line and didn't read many novels during the last fortnight.... The characteristic of the country round Lowestoft is the Broads or inland lakes, which lie near the shore. Small boats ply thereon and look very pretty with green fields all round.

24 *April* 1896.   ...From Erskine's letter I should think he would probably not come to Cambridge. It is very expensive, of that there is no manner of doubt. My £150 won't keep me going by any means

and I do things on as economical a basis as possible. Everything is about as dear as it can be, as they pile it up on all sides for the students. I hope however that I will be able to make a little extra cash to keep me going for some time to come yet. In order to do that, I told J. J. I was out-running my allowance and asked if I could get any 'pups' as people who get coached are called. He volunteered to do what he could, so at present I have one pupil for three hours a week, and may get one or two more. You must not smile at my very modest number, compared with days of yore, for coaching in Cambridge pays, which is more than it did at Christchurch. For seven weeks coaching at three hours per week I get £9 which is about 9/- per hour—not so bad you know. That is the regular fee in Cambridge for all coaching, so if I could get half a dozen men for three terms I could nearly make a living for one, if not for two. I hope to make both ends meet somehow, but I must expect to dub out my first year....My scientific work at present is progressing slowly. I am working with the Professor this term on Röntgen Rays. I am a little full up of my old subject and am glad of a change. I expect it will be a good thing for me to work with the Professor for a time. I have done one research to show I can work by myself. The Professor has an assistant Everett, a very smart manipulator and supposed to be the best glass blower in England. He fixes everything up and helps generally but of course doesn't understand the theory very much. I generally turn up at 10, and the Professor comes in from 12 to 1.30. I then start at 2.30, and continue till 6.30, the Professor coming in at 4 or 5. It is rather interesting work so I don't mind long hours.

4 *May* 1896. ...The Professor the other day invited me to go for a match of golf with him on Wednesday afternoon at Royston, which is about 12 miles by train. Of course I did not know anything about the game, but the Professor reckoned he could teach me. We left by train about 2.15 and started about 3. The Royston Heath is rather a good one for golf—rather undulating and very fine air. We knocked about a bit and I learnt to knock the ball a considerable distance, if not very straight. We got back about 7.30 after a very pleasant time. It was very decent of the Professor to take me. I don't think, however, I am quite old enough for golf yet—at any rate to take it up with much enthusiasm.

28 *May* 1896. ...I have just had what I suppose is considered rather an honour conferred on me i.e. I have been elected a member of the

Cambridge University Natural Science Club which has a membership of about a dozen and is fairly select at that. They meet every Saturday night in one of the members' rooms and get someone to read a scientific paper and then an hour or two is given to social converse. I have been once or twice, and it is rather interesting if the paper is decent.

May 1896. Breakfast with McTaggart, Hegelian Philosopher and Fellow of Trinity, but he gave me a very poor breakfast worse luck. His philosophy doesn't count for much when brought face to face with two kidneys, a thing I abhor....I have been working pretty hard all this week with the Professor on the X rays, so I want an occasional run in the open. The Professor has been using my methods of wave-detection in his classes, so I feel quite a big man in consequence, as of course the methods are my very own and evolved out of my own inner consciousness, and it is a little advertisement for me, but I am extremely modest. The Professor the other morning showed me a circular in regard to a science appointment in India, value £450 and asked me if I would like to go in for it. He was asked privately to recommend somebody, so if I had felt inclined I dare say I would have had a good chance, but I don't want to leave England just yet. It is extremely good of the Professor to offer it to me, and I hope I will be able to get a chance of a good appointment in a year or two. The Professor is far and away the best man in England to have at one's back, so I hope to come off some day. The above is of course quite between you and me.

On 18 June 1896 Rutherford appeared for the first time before the Royal Society—a great event in a young man's life. He there explained the principles of his magnetic detector of electric waves.

In the autumn he was at the Liverpool Meeting of the British Association and Prof. J. J. Thomson spoke on his joint work with Rutherford on the laws of the conduction of gases when ionised by Röntgen rays. Rutherford at the Monday morning meeting showed his magnetic detector actually working and explained that he had received signals at a distance of half a mile. He said that it was very sensitive to long waves but the coherer was better for short waves. At the same Meeting of the British Association Signor Marconi said that he and Kempe had detected electric waves at a distance of a mile and a half on Salisbury Plain. This improvement in distance was due to the fact that he had used for the first time a vertical aerial on a wooden mast.

RUTHERFORD TO MARY NEWTON

6 *June*. I have got some news to tell you about....J. J. has asked me to give an experimental lecture at the British Association meeting in Liverpool on September 6. I expect you know the British Association meets every year and all the scientific men of the country turn up to it, so it is rather an honour to give a paper before it. I hope to manage all right but it is of course a long way to look ahead. My paper would deal with my work on the wave detectors and I think I could make it interesting to any who turned up. I suppose I am specially honoured in being asked, and it will probably be very useful to me in the future.

12 *June*. ...I met Professor Lodge the other day. He is examining here for the Science Tripos. J. J. introduced me and we had a long talk. Mrs J. J. has very kindly asked me to dinner next Wednesday as Lodge will be there, so I shall probably meet him again. The Rede Lecture came off on Wednesday and was a great success. J. J. was in very good form though a little nervous at first. My name came in the lecture and I hardly recognized my work so much did J. J. make of it.

18 *June*. ...My blushing honours are lying thick upon me. I have just sent my paper to the Royal Society, and got a proof of the abstract, or résumé of my paper, which I will forward. I will have to wait to see if they will publish the whole of it in the Philosophical Transactions of the Royal. It will be a great feather in my cap if they do, for only a few papers get into the Philosophical Transactions yearly, and they are supposed to be the best of the year. No sooner had my paper been sent to the 'Royal' than a letter comes to me through the Royal from the Royal Institution, asking me to exhibit my detector at a conversazione there tomorrow. I cannot accept, as it is rather much trouble and hard to fix up in a hurry, but it is quite an honour to be asked to exhibit there. The Royal Institution is rather a small place and has big scientific lectures every Friday....It will be just as well not to make any public experiments before the British Association meeting, where I am to give a lecture. My scientific work is progressing fairly well but it is rather a difficult subject I am on at present.

RUTHERFORD TO HIS MOTHER

July 1896. I have been working pretty steadily with Professor J. J. Thomson on the X-rays and find it pretty interesting. Everett, who is the professor's assistant, makes the bulbs which give out the X-rays. You know one can see the bones of the hand and arm, and coins inside with the naked eye.

The method is very simple. A little bulb is exhausted of air and an electrical discharge sent through. The bulb then lights up and looks of a greenish colour. The X-rays are given off and if a piece of cardboard, with a certain chemical on it, is held near it, metal objects placed behind can be seen through several inches of wood. The bones of the hand can be clearly seen and if one looks at a spectacle box, no trace of the wood is seen but only the metal rim and the glass. Aluminium allows the rays to go through easily.... I am not working at that side of the subject but at some of the actions of X-rays on substances, etc. I see by the papers the other day that a blind person or persons without any eyeball [lens] can see them when the rays fall on the retina.

We are all supposed to go into College if we come up for the long vacation. Only a small proportion of the Fellows and those who want to work come up. We are all packed into three Courts, as the others are under repair.

The rooms I now possess belonged to one Binnie, who went down for good a short time ago....He took most of his things with him. My bedmaker Mrs Walker, who looks after me generally, raked up enough things to keep me going. When one comes into College you are put into someone else's rooms and use his furniture, crockery and general belongings without ceremony. The crockery, knives and forks, coffee pot, etc., were bagged for the time from Lord Acton's rooms which are just below mine. I also have the privilege of reclining in his best armchair.

RUTHERFORD TO MARY NEWTON

July 1896. I went with Professor Haddon, the zoological pot, and some friends out to Barrington on an Ethnological expedition. The idea is to take as many as possible of the male inhabitants and measure up their heads, general height etc., in order to get the mean average of the East-Anglian type. Some old games of the children were photographed and photos taken of all those who were measured. You can't imagine how slow-moving, slow-thinking the English villager is. He is very different to anything one gets hold of in the colonies.

12 *Aug.* 1896. Things are going with me the same old way, working steadily and increasing my knowledge of physics generally. I have one piece of excellent news in regard to myself, viz., that I have received word that my paper goes in the Phil. Trans. of the Royal Society, which is an honour which befalls only the best of the papers that are sent in— so I congratulate myself on my success. I don't suppose I will get copies for some time, but they are generally done up very well, and you may be certain I will send you one to put in your archives. The work the professor and I have done this long vac. will probably be published in the Philosophical Magazine. We will also give an account of it at the British Association meeting at Liverpool. I am therefore steadily progressing to my desired aim, a lectureship or a fellowship. I believe the Cambridge degree will be of considerable assistance to me in the end, for it counts a great deal with the outside world.

Trinity College: 27 Aug. 1896. ...I have been very busy today running over my experiments for the British Association. Pye, the workshop man, has made some things very nicely for me and I will be very well equipped for my apparatus, and I only hope things will pass off successfully. Tomorrow will be my last day in the Lab. for work till October. J. J. is going for a short spell before he goes to the British Association and then to America. I have still got to write up my paper for the British Association but am putting it off till the last minute. I will start away on Wednesday for Lincoln, to stay a few days with Mrs Bell. I won't object to a holiday after my toil and labour of the year. Reviewing things generally, I think you ought to be pleased with my record for this year. I came up, a stranger in a strange land and finish up by taking a part in the British Association meeting. I have got some papers published and another on the way, not to mention a host of love letters and other correspondence that I have also completed. If my paper was a novel, I would dedicate it to you, but a dedication would look rather funny in a scientific paper. You must accept the will for the deed. I was speaking to J. J. today in regard to future prospects, and he remarked that if an 1851 Scholarship was continued at all, it would be continued in my case, so I am on the look out for another £150.

I sincerely hope that it is so, for if the Commissioners don't do so, I will be in a bit of a hole as regards cash. J. J. also remarked that I ought not to overlook the chance that after my B.A., I may probably stand a very good chance for a Fellowship at Tri..y. As the Fellowships are worth £300 a year for six years and the honour pertaining to them very great, it would be an extremely fine thing to have.

Trinity College: Aug. 1896. ...A good long time ago, I gave you a promise[1] I would not smoke and I have kept it like a Briton, but I am now seriously considering whether I ought not, for my own sake, to take to tobacco in a mild degree. You know what a restless individual I am, and I believe I am getting worse. When I come home from researching I can't keep quiet for a minute, and generally get in a rather nervous state from pure fidgetting. If I took to smoking occasionally, it would keep me anchored a bit, and generally make me keep quieter. I don't think you need be the least bit alarmed with regard to yourself. For I don't think I will ever become a confirmed smoker, but seriously I believe it would be a very good thing for me in many ways. Every scientific man ought to smoke, as he has to have the patience of a dozen Jobs in research work.

30 *Oct.* 1896. ...I am working very hard in the Lab. and have got on what seems to me a very promising line—very original needless to say. I have some very big ideas which I hope to try and these, if successful, would be the making of me. Don't be surprised if you see a cable some morning that yours truly has discovered half-a-dozen new elements, for such is the direction my work is taking. The possibility is considerable, but the probability rather remote. J. J. has not yet returned, but I am expecting him in a couple of days. We have now the Lab. quite full of people doing research work, in fact we haven't room for any more.

I generally work seven hours in the Lab. and then have my coaching and general reading in the evening. My research work takes up a good deal of energy, as I have to think out my programme for the next day every night....I am enclosing abstract of my Royal Society paper. The full paper is not out yet.

[1] Lady Rutherford explains the apparently arbitrary exaction of a promise not to smoke as follows: "Ernest suffered from a persistent irritation of the throat causing a slight cough which worried me in the light of the fact that his eldest sister was delicate and one always had the fear of tuberculosis."

6 *Nov.* 1896. ...I really am so busy of late I have hardly time to look round. My research is progressing at the rate of knots and everything going well. J. J. and his Mrs have returned. Horribly boisterous weather....I go to pay my respects on Saturday and, on the first occasion I get opportunity, I will make a confession—comprenez-vous?...On Tuesday J. J. has asked me to give a paper before the Cavendish Society, on our work, some more grind for me. We have also started, among Trinity men, another Physical Society, and I have to give the first paper in my rooms next Tuesday, so you can quite understand I am busy. The Cavendish is crammed with Research people and it is hard to find room for them. I am very glad I came last year, as I have got a good start.

11 *Nov.* 1896. ...Yesterday after my paper to the Physical Society, which passed off all right before quite a distinguished audience, I asked Mrs J. J. to be allowed to see her home, as it was quite dark, and J. J. was busy over some experiments. On the way home I broached the subject of our engagement, and she was very pleased to be told about it, and said some very nice things, and asked all sorts of questions about you, and says she is very anxious to see you when you come. I know she will do anything she can for you, so we will trust in the future that you may visit her when you arrive. She remarked that when I was giving my paper, not being learned enough to follow it completely, she had been wondering what was to be my fate in that respect. Truly the ways of women are not for men to understand....I think Mrs J. J. looks on me as a very nice boy and after I have taken her into my confidence, will take an affectionate interest in my welfare.

Last Saturday, Townsend, McClelland and I called on Mrs J. J.— the first time we had seen her since her return from America. As there was a large number of ladies present, we stayed over an hour assisting to entertain, for which Mrs J. J. was very grateful. I am now getting quite an expert at afternoon calls and don't feel at all shy usually. On Saturday night the Science Club met in Mr Shipley's rooms at Christ's College, which are reputed to be the finest at the University. Professor Marshall Ward gave a paper on "The effect of environment on plants", dealing chiefly with the recent work of a man named Klebs. I am sure you would have been interested, for even I was, and I am anything but a biologist. On Sunday, I went to lunch with Sir Robert Ball at 2 p.m.

and had a pleasant time till 4.30....Sir Robert is a great man for telling funny stories and jokes.

On Tuesday night, I gave an inaugural paper to a Physical Science Club, just formed among us researchers. It was held in my rooms and I supplied coffee, biscuits, baccy and cigarettes—the usual thing in these cases. The Laboratory is now full of 'researchers after truth' some of whom I know very well....I work in the professor's rooms—an unheard of thing up to the present, and it is very handy there to have Everett, the Professor's assistant, alongside to give me a hand occasionally. I am at present working at a most interesting and also most important subject, from the pure science point of view—the separation of the positive and negative ions of the hydrogen molecule, etc. It is a difficult matter to tackle but I have partially separated them already for ordinary air.

We have tea every afternoon at 4.30 in the Lab. in the professor's rooms, where he turns up, and discusses general topics.

2 *Dec.* 1896. In your last letter you ask whether J. J. is an old man. He is just 40 and looks quite young, small, rather straggling moustache, short, wears his hair (black) rather long, but has a very clever-looking face, and a very fine forehead and a most radiating smile, or grin as some call it when he is scoring off anyone. He has just been doing me another good turn. He generally examines the Sandhurst military candidates in Physics, but not being able to go on Saturday, he asked me to take his place. I am rather interested in seeing the chaps, etc. The fee will of course be two or three guineas for the day. J. J. has also dropped me a hint in regard to Fellowship exams. to get up some general philosophy as they are examined in that in the Fellowship exam. Don't let out anything about the Fellowship business, as it is all *sub rosa*....

9 *Dec.* 1896. I have been very busy this last week, as I think I told you in my last letter. J. J. asked me to take his place in the Sandhurst and Woolwich exams. On Saturday, Smith, my co-examiner and a Trinity man, and I went down to London, to the London University buildings, where the exam. was to be held. We had to arrange for the practical in Physics of twenty-four candidates in the three hours. We rigged up all the apparatus in the morning and examined them in the afternoon. We went out and had lunch in the middle of the day, and then came back and buckled to. The chaps to be examined were all between 17 and

19, and the ignorance of some of them was profound. The exam. ended about 6 and I then went and had tea and back to Cambridge again in the evening, arriving home about 9.30. J. J. also asked me to go down on Tuesday, when I examined fifty candidates, most of whom would go into Woolwich. It was interesting, but beastly hard work. We went down the Strand in a bus, then to the station, and another weary journey to Cambridge again. London, as usual at this time of year, was very foggy and miserable, but I like wandering round London at nighttime. I am leaving in ten days for Eastbourne. I shall probably visit Aunt D. before I go down. J. J. is very keen that I should work with him next term, so we will probably be in partnership again.

Eastbourne: Xmas Day, 1896. At the dinner I mentioned, some of the dresses were very décolleté. I must say I don't admire it at all. Mrs X., wife of a professor, wore a 'Creation' I daresay she would call it, which I thought very ugly, bare arms right up to the shoulders, and the rest to match. I wouldn't like any wife of mine to appear so, and I am sure you wouldn't like to either....Mrs J. J. is generally about half and half and looked very well indeed.

I think I told you in my last letter about my examining in London. J. J. paid me six guineas for that, which was not bad for me. J. J. seems very anxious to help me in every way he can. He had been sent some books from *Nature* to review, but asked me if I would do it. I of course agreed, and in consequence got a letter from *Nature* this morning asking me to review a couple of books. It is my first attempt at that sort of thing, and I hope I will manage to do it decently. The books are scientific ones dealing with X-ray business on which I reckon I am an authority.

As Sir J. J. Thomson said at a later date, referring to the ionisation of gases by X-rays:

Rutherford devised very ingenious methods for measuring various fundamental quantities connected with this subject, and obtained very valuable results which helped to make the subject 'metrical' whereas before it had only been descriptive.

It is clear from the above letter that Rutherford was fully conscious that Prof. J. J. Thomson and he, working together, had made a discovery of great importance. The fact is that between Easter and November 1896 they had laid the sure foundations of a new subject, an account of which is given in the *Philosophical Magazine* (Nov. 1896). They had explained by a simple theory, which has stood the test of time, how it is that Röntgen rays make air or any other gas able to conduct electricity. The radiation makes a large number of 'carriers' or 'ions', some positive, some negative, in equal numbers. Near a charged body, there is of course an electric field and this causes the positive ions to move in one way, and the negative ions in the opposite direction, carrying their charges with them, which by degrees neutralise the charged bodies in the paths. The ions also tend to recombine and neutralise one another to an extent dependent on the numbers present.

If ions are produced by Röntgen rays in a tube, they can be driven along by a current of air, gradually combining as they go. They will not pass through a little wad of cotton wool, because they give up their charges to it. If they are passed between two parallel plates, which have opposite electric charges upon them, the ions are swept out of the air, moving in opposite directions and giving their electric charges to the plates. Rutherford so clearly visualised the situation that later he said to me, "Ions are jolly little beggars, you can almost see them".

Prof. V. Bjerknes, an eminent Norwegian physicist, had met Rutherford at the Liverpool Meeting of the British Association, and later wrote to him:

University of Stockholm: 16 Nov. 1896. I was very glad to get your kind letter and to have the data of your researches which I find most interesting. When I hear of successful experimental investigations I always long to return to that kind of work, instead of 'paper work'. But do not misunderstand me: the more I penetrate any hydrodynamical subject, the more it interests me, and I would under no circumstances give it up.... I always look back upon the Liverpool days with great pleasure.

We have full winter here now with good skating. I am sure you would find the Scandinavian winter less 'enervating' than the Cambridge one.

I see you are still anxious to persuade me not to judge English Science altogether by the B.A. meeting. Your fear is quite superfluous. I judge it from the works of Faraday, Joule, Maxwell, Kelvin; and even if I did judge it by the B.A. meeting only it would not be as unfavourable as you seem to fear.

There is a story that when Rutherford was giving his address, with experiments, to the British Association at Liverpool, his apparatus refused to work. He looked up coolly and said, "Something has gone wrong! If you would all like to go for a stroll and a smoke for five minutes, it will be working on your return." This pleased all his audience; they went, came back and saw the experiment. A leading physicist said: "That young man will go a long way."

Writing in the last year of his life about his steadfast friend Sir Grafton Elliot Smith,[1] Rutherford put on record his views about the work of early research students at Cambridge:

Before coming to Cambridge, Elliot Smith had already shown his powers as an original investigator by his work on the anatomy of the brain, carried out under the supervision of Professor J. T. Wilson of the University of Sydney. I met Elliot Smith soon after his arrival in Cambridge, and we soon became close friends and saw much of each other outside laboratory hours. We had many interests in common, for we had both been brought up in the country, were very much of the same age, and both had to rely on ourselves to make our own careers. In addition, we were the only advanced students from our respective universities, and we were conscious that it devolved upon us to make good and to show ourselves worthy of the opportunities given to us....

Long afterwards Rutherford attended a lecture by Elliot Smith in which the famous Piltdown skull and its restoration was discussed. Rutherford chaffed him afterwards and said: "All that is necessary for a complete restoration of a 'Gigantosaurus' is one thigh bone and forty barrels of plaster of Paris."

[1] *Biographical Record*, ed. W. R. Dawson (Cape, 1938).

In the spring of 1897 Mary Newton came from New Zealand to England to visit relations and break a long separation. She spent a happy May Week with Prof. and Mrs J. J. Thomson at Cambridge, receiving the greatest kindness. There were visits to the races on the Cam, and to the Senate House to see Rutherford take the first B.A. Research degree. At a later date they went with two friends to Ireland and visited Killarney. Miss Newton returned to New Zealand in the autumn. There were obviously fewer letters written during the year so that the record is not so complete.

This seems a natural place to tell the story of the missing letters. From the time of his going to Cambridge until the year of the death of his mother, Rutherford used to write with great regularity every two weeks, and give her an account of the events of his life. After her death about a dozen of these letters were published in a New Zealand news-paper without Rutherford's consent. He did not approve of this publicity and wrote a request that all the letters written by him to his mother should be sent to him at Cambridge. They never came.

On his death the Trustees both wrote and cabled that the letters should be sent promptly to England, so that they could be incorporated in this authorised life. The letters were traced to a lawyer's office, but during a move they seem to have entirely disappeared. It need not be pointed out how grievous a loss this is to the world of letters and science. Indeed this brief account is here given in the hope that somewhere or somehow the missing letters may be found and published.

The following two letters were published in New Zealand.

RUTHERFORD TO HIS MOTHER

The Crystal Palace: May 1897. The chief point of interest to me was the horseless carriage, two of which were practising on the ground in front. One was capable of seating two persons and the other five. The engines were placed in the rear part of the vehicle and did not occupy much space, and the oil which is used for the motive power was kept in a cylinder at the back. They travelled at about 12 miles an hour, but made rather a noise and rattle. There was a collection of these horseless carriages in the Crystal Palace itself, some evidently very old and correspondingly clumsy and heavy. I was not very much impressed with the machines as vehicles, but I expect they will come into very general use shortly. They say the expense of running about 12 miles an hour is about one penny per hour, which is rather cheaper than a horse.

Cambridge: 1896. On Tuesday morning the Mathematical Tripos lists were read out at 9 o'clock in the Senate House. The Girton and Newnham girls and the visitors ranged themselves in the gallery of the second floor, the Newnhamites at one end and the Girtonites at the other. The fellows stand just inside the hall and face the examiners, who have a position in the gallery above.

Punctually as the clock strikes nine, the examiner began: Senior Wrangler—pause—silence—Fraser, of Queens' and there was great applause. A Queens' man has not been Senior since 1858 and so the rejoicing of Queens' was great. I believe a fund has been collecting since 1859 for the Queens' man who should become Senior Wrangler. A good many have joined purposely, but have failed, and Fraser will probably now scoop the money which amounts to about £400. All the Wranglers are then read out, then the Senior Optimes, and then the Junior Optimes, which correspond to second and third class respectively.

The next examiner then appeared on the scene and commenced: 'Women' when the undergrads interrupted him with 'Ladies' and 'Cap', thereby intimating that he had to lift his cap to the ladies. As he would not they kicked up a hullabaloo till he was forced to raise his cap and they then made him proceed with his reading. The first Girtonite rejoiced in the name of Longbottom and was between the 11th and 12th Wrangler.

As the names were read out the girls all clapped for their own college and made rather a noise much to the delight of a dog who got inside the Senate House and started barking for company and thus afforded considerable amusement for a few minutes till he was removed by a porter.

The second part of the Mathematical Tripos was then read out. The results are here arranged alphabetically in a division and Maclaurin (New Zealand) was in the highest class.

After the results are read out the ladies in the gallery roll up the papers on which the lists are printed and throw them to the undergraduates below, who scramble for them in a seething mass for about five minutes. Woe unto the straw hat that enters into the fray, for it generally emerges a battered wreck. It is considered a great honour to secure one of these crumpled papers. The custom is an old one and is perpetuated.

I sent in my paper to the Royal Society with an abstract and my paper was read before the Royal Society last Thursday. The abstract is published before the reading of the papers for the Fellows of the Society. I am in hopes my paper will be published in full in the Philosophical Transactions. This only happens to the best of papers every year and I hope mine will be one of them.

Not only did the Commissioners of the 1851 Exhibition renew Rutherford's scholarship for a third year but Trinity College awarded him the Coutts Trotter Studentship, value £250 for two years. In May 1898, Rutherford was 'taken to task' by the Assistant Secretary of the Commission for not reporting this award at the time. However Rutherford replied by stating that he did not realise that an additional scholarship was a 'position of emolument', since many 1851 Scholars held scholarships from their own Universities.

RUTHERFORD TO MARY NEWTON

12 Dec. 1897. ...I have a great piece of news to tell you, which you probably have heard about already. I went in for the Coutts Trotter Scholarship, and it has been awarded to me for two years at £250 a year, think of it—nearly enough to get married on. I collected together my published papers and sent a short MS. of my unpublished work and calmly awaited the result. On Friday J. J. casually remarked to me that the Council met that day and would probably fix up about the Coutts Trotter. So I trotted round, but said nothing about it....About 6, a Trinity College clerk came round & presented me with two letters, one from my Tutor congratulating me on getting the Scholarship and the other asking me to Trinity Commemoration dinner, which took place that evening. I handed my tutor's note to J. J., and after a minute he came round, and after congratulating me, calmly remarked that he had known about it all day, but couldn't tell me. There was great excitement among the researchers as I had kept quiet about going in. The best part

of the matter is that I can still get the rest of my 1851 Scholarship money, so I will be quite rich for the time. My chances for a Fellowship are of course greatly improved, if I work along steadily as I have been doing, I should stand a good chance after the end of my Scholarship. I shall probably go in this year, but if I should get one, I should drop £250 as I would only have one year of my Scholarship. It is a very fine thing for me, and I know that you will be even more pleased than I am myself. Dr Smith wrote to the Agent General, informing him of my success, but I don't know if it was cabled out. It would go as correspondent's news in any case, so expect an outpouring from me on a subject of which they know nothing whatever. I have, as I said, greatly improved my chance of a Fellowship and it looks as if J. J. is inclined to support me.

During 1897, Rutherford made rapid progress in the experimental investigation of the behaviour of ions, that is, of the little carriers produced in any gas by the action of the Röntgen or X-rays.

He proved the correctness of a conjecture, due to Prof. J. J. Thomson, that the rate at which the positive and negative ions joined up together, by their mutual attraction, was proportional to the *square* of the total number of ions present.

He also found that when X-rays ionised the air between two parallel plates, the current between them would at first increase at an equal pace with the increase of voltage, following Ohm's law. This would not long continue, for soon the current would tend to a constant value as the voltage was further increased, because *all* the positive and negative ions are swept out in opposite directions, as fast as they formed, provided the voltage was large enough. The current was then said to be 'saturated'. He proved that the speed of the ions was proportional to the strength of the electric field. He measured their mobility, that is, he determined for a field of a volt per centimetre the sum of the velocities of the positive and negative ions. He found for air 3·2 centimetres a second, and for hydrogen 10·4.

By an ingenious contrivance, screening off one-half of the region between the parallel plates, he showed that the positive and negative ions had about the same speed, but actually it was proved by Zeleny that the negative ions move about 20 per cent faster than the positive. A fine year's work, particularly when it is remembered that other very able men were also now beginning to work in the same field—besides J. J. himself, there were men like Langevin, Townsend, John Zeleny, C. T. R. Wilson and H. A. Wilson.

The next step, during 1898, in Rutherford's research work, was an attack on the ions produced when ultra-violet light falls on a sheet of clean metal, such as zinc. He proved that the ions produced in the neighbouring air were all negative ions, but had the very same properties as those produced by X-rays passing through air. He did one exquisite experiment. He rapidly varied the charge on the lower illuminated plate, positive and negative alternating, so as to make the ions dance up-and-down, and increased the alternating voltage till the ions *just* reached the upper plate and gave their charges to it. In this way he found their velocities.

At this period most remarkable discoveries were made in France. In 1896 Becquerel found that uranium gave out some new rays which would go through black paper and would affect a photographic plate. These rays could be bent by a magnet. Schmidt discovered a similar property in thorium. Mme Curie examined a number of ores of these two elements and noted that an ore of uranium was three or four times as 'active' as the uranium it contained. What a paradox! She decided that there must be something new present. Pierre and Marie Curie began their heroic experiments of extracting the unknown, and discovered minute quantities first of polonium, next of radium, which proved to be, as it approached purity, more than a million times as active as uranium.

Rutherford's curiosity was aroused and he set to work at Cambridge, in the Cavendish Laboratory, to find out whether the ions produced by uranium behaved in the same manner as those produced by X-rays and by ultra-violet light. He found it to be so. He quickly noticed that there were two types of rays present which he called, from the first letters of the Greek alphabet, alpha and beta rays. The alpha rays were cut off, or absorbed, by a stout sheet of paper, but the beta rays were a hundred times as penetrating. He published this work when he got to McGill. It was the beginning of his life's main work.

Prof. Ayrton in his final report to the Commissioners on Rutherford's work said

he would not be expressing an exaggerated opinion in suggesting that the Commissioners may feel justly proud that it was through their scheme of Science Scholarships and through their selection of the candidate, Mr Rutherford, that such a first-rate Physicist was able to come from New Zealand to the Cavendish Laboratory of Great Britain and prepare himself, so to say, for his present post—the Professorship of Physics in the McGill University of Canada.

RUTHERFORD TO MARY NEWTON

Trinity College, Cambridge: 22 April 1898. In my last letter I told you Callendar of McGill University, Montreal, had been appointed professor in London. His appointment will probably be open shortly, but I think it doubtful whether J. J. will want me to go in. There would probably be big competition for it, all over England, as the average man does not mind going to Canada, though he would bar Australia. It is a little uncertain what kind of a man will be wanted. The salary is not large, only £500, but still I would not mind taking it, but I think it is extremely doubtful that I will compete for it. I will not go in unless J. J. advises me to. I expect Townsend will go in anyhow as he is anxious to get something to do. Personally, next to New Zealand I would rather like Canada, as I believe things are very jolly over there.

10 *May*. Today the question of allowing a Roman Catholic Hostel was voted on, and was lost by two hundred votes. I don't think I shall go in for this Montreal Chair. J. J. does not appear to wish me to, and as my chance would depend entirely on his backing, it wouldn't be much use my trying....It is possible J. J. may want me to stay on here. It will be advisable to see who are going in, rather than to go in anyhow on one's individual merits, which in these days does not count much, unless one can obtain strong backing. I would like to stay here another year if my scholarship would keep going.

18 *May*. ...I have made up my mind to go in for the McGill Chair, chiefly as a business matter, because it will probably do me much good even if I don't get it. I settled to go in last Saturday, and have been very busy about it since.... Everybody says that Montreal is a very nice place to live in, and as there are quite a large number of professors, there would be any amount of University society. The salary is only £500, but still enough for you and me to start on, and we could manage to have a very good time on that, I think. I am not very keen on getting the appointment, as I have my scholarship going and the possibility of a Fellowship, but it is as well to keep to the fore in these things. There are at any rate two of us going up from the Cavendish.... If they want research a good deal I may be in it, but if experience in teaching, I must take a back seat. You mustn't put any faith in my chances, they are paradoxical, and a big man may crop up at any time and leave us all in the rear. I have been seeing about testimonials this week and seeing people who have influence in the Colonies. I have been to see Sir R. Ball and he is giving me a testimonial and going to work a friend at Montreal. I have also seen Dr Macalister who knows all the pots at Montreal and may say a good word for me. I will send you a copy of my testimonials, they have to be in Montreal by the 20th of next month, so there is not much time to lose. There is one good thing about my going in, it may make J. J. act in regard to getting me something to do in Cambridge. I will probably go in for a Fellowship this year, if I don't get the appointment, but my chances are very problematical. It would be a lovely idea, if I got the Canadian appointment, to go out next October and work for six months when they have nearly six months vacation, which would suit me very well to go out for you and get spliced, or if there was not cash enough, I could stay another year and come out for you about two years from now—if you could wait that long—but this is all in the air.

2 *June* 1898. ...I have been very busy over this McGill appointment, but have come to the conclusion that my chances are 'up a spout'.... I am just getting my testimonials printed. I got good ones from J. J. and Glazebrook, and I will send you one when they are finished, but it appears to me it will be money wasted. However I must not complain as I have my scholarship to fall back on, although it is not enough to marry on—unfortunately for us.... I find by a regulation

that we researchers cannot go in for the Trinity Fellowships this year, we have to be twelve terms in residence, and I have only nine. As far as I can see, my chances for a Fellowship are very slight. All the dons practically and naturally dislike very much the idea of one of us getting a Fellowship, and no matter how good a man is, he will be chucked out. There is a good deal of friction over this research business, which was intensified by my getting the Coutts Trotter. I know perfectly well that if I had gone through the regular Cambridge course, and done a third of the work I have done, I would have got a Fellowship bang off. My chances for this appointment and for a Fellowship are, I reckon, darned small—one has to face the situation squarely and not look always on the rosy side.

Cambridge: 14 *July* 1898. ...During this week I have been interviewed with regard to the Montreal Chair by Dr Peterson, Principal of McGill, and Cox, Professor in Montreal, who will be my colleague if I went out there. I was in Hall on Sunday night and saw J. J. pointing me out to some man and was told it was the Montreal delegate. Peterson and Cox came to the laboratory on Monday and I saw Peterson twice for about an hour altogether. I don't suppose I'll be appointed as I believe they want the man they appoint to be a lecturer with experience, as there will be a good deal of that type of work to do....I don't think J. J. has much to do with the appointment at all, but it will mostly depend on Peterson's report. I rather liked both Cox and Peterson, and they were both very pleasant. I expect to hear in about three weeks or a month about the result, but am not bothering my head about it. My chances are almost entirely due to research, and that is only a partial consideration with them. Besides I think they consider I am rather youthful. The weather has been very variable this week, hot and cold, and I have been rather toothachy and headachy, a rather unusual thing for me. I was annoyed today, when I came home to my rooms at 7, I found a letter from Griffiths, asking me to tea to meet Callendar the late McGill professor; I hope, however, to get another chance....By the way I got a £1 prize from the Trinity Rifle Club for shooting the other day.

30 *July.* I missed writing last mail as I did not feel like writing, until the McGill appointment was settled, but as I am still in a state of uncertainty, I cannot postpone for another week. I did not at first think I stood much chance, but from later reports I believe it is very probable that I will be

appointed. The first inkling I had of it was when I was with Griffiths at dinner in Sidney College. He remarked that he supposed I knew I was to be appointed to the McGill Chair, as all its arbitrators had sent in a recommendation in my favour. I was considerably surprised, but these results were confirmed later by Professor Ames of Johns Hopkins University who said that he had seen Peterson, President of McGill, who had said that my name had been sent over to Montreal in order to receive the official sanction from the Board of Governors, so if I don't get it after that, something has gone wrong with the works. On Monday I received a confidential letter from Professor Cox asking me to see him. Of course he couldn't say much but seemed to think I would be appointed, and so we talked over matters in general in regard to Montreal and my duties there. Mrs Cox came in later, so we talked of things in general. In case I am appointed Cox will be my colleague, and will assist me in the lecturing. He is a mathematician in the first place, and a Fellow of Trinity, and has done a good deal of extension-lecturing at which I believe he is very good. He doesn't know much about experimental work, so I will have to run the experimental part, and the laboratory generally. He is a man about 45, I should think, just getting grey. His wife is tall and somewhere about 40, I should think. She is pleasant, and so is he, and I think we should get on well together. I wish to goodness this matter was settled finally, as it is very upsetting to work in the interim. If I get it, I shall leave for Canada on September 8th by the *Yorkshire* and get to work by October. It is rather a sudden change, and will mean a great deal of work, as I have to start lecturing at once to classes of sixty or seventy, and generally have to run the show. I daresay you have in your imagination dubbed me a professor, but I must confess I did not think my chance would come so soon. I hope you can see your way to going over there in the near future. It seems rather absurd discussing these things before matters are settled, but I know you won't mind if I don't go after all. The University has two terms and finishes up early in April, so there is nearly six months vacation, which will suit me admirably to get over to New Zealand when the time arrives to fetch you. If I get appointed, I will have to start work at once, so the question is what am I going to do when my vacation arrives. Am I to go to New Zealand to fetch you to look after me and become Mrs Professor, or am I to wait another year to get enough cash to do it in style? I would very much want of course to go to New Zealand, but I suppose, as it requires a fund to start housekeeping, I ought to wait a little....McGill is a very important place to be at, for Callendar was a F.R.S., and a Fellow of Trinity, and quite a pot in the scientific world, so I will be expected to do great things. I am expected to do a lot of original work and to form a research school in order to knock the shine out of the Yankees!...From what I hear, I believe Montreal is a very fine place, and although there is plenty of cold in the winter, it is really very enjoyable weather, bright and frosty. The summers are hot and a good few of the residents leave in the hottest parts of the year. Living, I should imagine, is much the same as in New Zealand, but we should of course have to keep up a certain amount of style, and we ought to do it very comfortably on £400 and put by the rest. I believe I have told you, in previous letters, that the Physical Laboratory is one of the best buildings of its kind in the world and has a magnificent supply of apparatus which alone cost £25,000, so it is a place one can be proud of. Macdonald, a millionaire, has supplied all the coin and continues to do so, so I have got to keep in with him, if I have any large expenses. I believe it is possible that salaries may go up in future, and at any rate I will probably get a little extra by professional advice, so we oughtn't to do so badly. I will write as soon as it is settled, and I will see if I can't get a wire sent out by the Press Association, so if I have got it, you will probably have heard before this letter arrives. I suppose if I get it, I may consider myself a very lucky person, as professorships are not as a rule being thrown about for men of my age....I suppose I have to thank J. J. a good deal in the matter, but his position has been a very difficult one. Schuster and Lodge must also, I think, have supported me, for they were supposed to have a say in the matter. Peterson is however the man who runs the whole show.

Among those who wrote testimonials for Rutherford were his Tutor, Mr W. W. Rouse Ball, Sir Robert Ball the astronomer, E. H. Griffiths and Richard Glazebrook.

J. J. Thomson recited Rutherford's achievements, specially naming the 'Detector' with which he received signals by means of electric waves (and without wires) across about three-quarters of a mile of the most densely populated part of Cambridge—work done two-and-a-half years before and in advance of the recent attempts at wireless telegraphy.

He continued later:

I have never had a student with more enthusiasm or ability for original research than Mr Rutherford and I am sure that if elected he would establish a distinguished school of Physics at Montreal....I should consider any Institution fortunate that secured the services of Mr Rutherford as a Professor of Physics.

Principal Peterson and Prof. John Cox, who had visited Cambridge to find a worthy successor to Hugh L. Callendar, returned to Montreal well satisfied with the appointment.

RUTHERFORD TO MARY NEWTON

3 *Aug.* 1898. Rejoice with me, my dear girl, for matrimony is looming in the distance. I got word on Monday, from Dr Peterson, to say I was appointed to Montreal. All my friends are of course very pleased, and I have to submit to being called professor without having a boot to throw at their heads. I am very pleased that it is settled and that I am not left in the cold, but I am sorry for many reasons to leave Cambridge. ...I think it will be probably much better for me to leave Cambridge as, on account of prejudice in this place, it would be difficult to get anything to do, and if I did, it would be a rather unpleasant position, and at any rate it would be many years before I could raise £500 a year. My chances of something better afterwards will be much greater if I have already held a chair. Callendar's own career is a good example of possibilities of obtaining a better appointment. I will be practically boss man in the Laboratory, as Cox knows very little about Physics, though he does a good deal of lecturing. It sounds rather comic to myself to have to supervise the research of other men, but I hope I will get along all right. There are about four men doing research in the Lab. some of whom are as old as myself, so I will have to carry it off somehow. I am leaving England on Sept. 8th....Taking a first class passage cost £12—not so much....I will not have extra much cash when everything is settled, as I wish to buy a good few books and they are very expensive.

11 *Aug.* I have received a great influx of letters the last week from all over the place about my appointment. I also received a letter from you discussing the problem of how to get married on nothing...the cash is of course a consideration, but as I am paid monthly we should

have enough to get along on—but time will tell. I believe the professors who are living on their income are not expected to entertain much; afternoon teas are the extent of their expenses, but those who have rich wives do a good deal. We will be of the poorer variety, and will have to live to ourselves. I was interested in your account of the professorship at Wellington....I think if I were offered the New Zealand Chair at £700 and McGill at £500 I would take the latter as my chances of advancement are much better in McGill, than if I got out to New Zealand. There is also a certain amount of satisfaction in having a swell lab. under one's control, and probably in New Zealand my chances of research work would be very small. I am going out to Canada with the Coxes and MacBride the zoological professor, lately appointed from Cambridge. I met him at breakfast the other morning....I am very busy at present getting things in order and writing up papers. I find it takes a lot of energy to tackle to. I think my appointment is a very much discussed matter in Cambridge as Callendar was considered a very great man, so your acute mind will at once gauge my importance if they place me in his shoes when the beard of manhood still is faint upon my cheeks. I am only a kid for such a position and I will have rather a hard time the first six months getting things going.

27 *Aug.* ...I have got a very nice letter from Professor Bovey who is an engineering professor, and Queens' College man. He is what is known as Dean of the Applied Science Faculty, i.e., he is head of the show. He wants to know when I am coming and it appears from the letter he is intending to ask me to stay with him on my arrival provided he has got back from his holidays.

2 *Sept.* I am leaving England next week so this is the last letter I will write here. Elliot Smith and I are staying at my old digs. and working like Trojans. I have nearly finished my paper and feel very relieved. I am going to pack up my books and things generally tomorrow to get them off my hands....I don't know if you heard of the terrible accident to the Hopkinsons, whereby he and his three children lost their lives in the Alps...you met one of them at the dinner party at Mrs J. J. ...The accident has made a great stir among Trinity people.... We are expecting to hear of the battle of Omdurman tomorrow. The Colonel Henry suicide in Paris is making a great sensation, and is quite overshadowing the Czar's disarmament proposals. We live in interesting times.

RUTHERFORD TO HIS MOTHER

I know that you will be pleased to hear that I have got the Montreal post and so start up in life as a professor on £500 a year and a prospect of adding more in the days to come as far as regards coin, and an unlimited prospect of work. I got a letter from Dr Peterson saying I had been appointed by the Board of Governors at Montreal. All the other candidates had considerable experience in lecturing, and teaching, but I was appointed greatly for research and partly because they reckoned I could lecture when I tried. It is as good an opening for a start as I could wish.

The McGill University has a good name and they have a good lot of men as professors, all at £500. The salaries are small compared with the endowment of the laboratories and the enormous money spent on them, but that is chiefly due to the fact that the money has been advanced by Macdonald, a millionaire, who made his money in tobacco and he lives on £250 a year, so he reckons a professor should live on £500. However £500 is not so bad and as the physical laboratory is the best of its kind in the world, I cannot complain.

There will be two physics professors on an equality nominally—Cox, a fellow of Trinity, a mathematician who built the laboratory but is not an experimental man in any way. He will do a good part of the elementary lecturing and I will have to run the practical classes in the laboratory, the research work and do lecturing as well....

You must bear in mind that the competition in scientific circles is very keen in these days and men quite expect to have to lecture and demonstrate with not much over £200 a year for ten or fifteen years before they get an appointment such as McGill. As a matter of fact I am extraordinarily lucky to start so well, for not one man in 10,000 ever gets the opportunity, for I have had practically no experience in teaching for which you are supposed to pay. It may interest you to know that Callendar, who was a Fellow of Trinity and a F.R.S., as well, took the appointment at Montreal when it was not so good as it is at present....

The Master of Trinity, H. Montagu Butler, sent a gracious farewell letter with a wise forecast of the future:

Killin, N.B.: 1 Sept. 1898. We shall all be sorry to lose you at Trinity but I cannot help feeling that you have a noble position before you, both from the scientific and from the patriotic point of view. It is not everyone of our countrymen who has, so to speak, in his veins the blood of Scotland, New Zealand, Trinity and Canada.

Perhaps someday the same wave that has restored Professor Callendar to us may bring you also back over the Atlantic. Meanwhile you will be doing a grand work in Canada, as the representative of other good things besides your own subject.

CHAPTER III

McGILL

Professor E. W. MacBride has recalled the arrival of Rutherford in these words:

Rutherford's bachelor life in McGill was very short. He came out in the same cabin with me in September, 1898 and he left Montreal to be married in May or June 1900. I think I could claim that Walker, the Professor of Chemistry, and I were his two closest friends. I boarded in a house in McGill College Avenue kept by an Englishwoman— Miss Birch—and Rutherford also took a room in the same house. Miss Birch however closed her house, and Rutherford, Walker and I took rooms with breakfast in a house at the bottom of Union Avenue.

F. P. Walton, then Dean of the Faculty of Law at McGill University, writes of the early days in Montreal:

I remember the circumstances of Rutherford's appointment at McGill pretty clearly. Peterson wrote to J. J. to invite suggestions and J. J. wrote back that if they were willing to take a quite young man with no teaching experience he had a great belief in young Rutherford.

Peterson and Cox went over in the summer and interviewed Rutherford and it was done.

Cox, I always thought, behaved very generously in regard to Rutherford and Rutherford was, I think, fully conscious of it.

Cox was an accomplished and versatile man and a good teacher but, as I was always told, had no flair for research. When he realised what a young eagle they had got, he took over a lot of the drudgery which embitters the lives of professors and gave Rutherford much more leisure than a junior professor usually gets.

By the way, Cox made a remark to me before Rutherford came which has stuck in my memory and might well be mentioned. He said one day that he was feeling rather dispirited because there seemed nothing new going on in Physics. The main things, he said, had all been found out and the work which remained was to carry on a great number of experiments and researches into relatively minor matters.

When Rutherford got going, Cox was ready and glad to sing another tune. When Rutherford brought his wife to Montreal we invited them to stay with us until they found a home, and it was in this way we got to know them so well in the early days.

A. E. Taylor and MacBride were about the same period and we saw a lot of them all.

They were a very lively and intelligent crowd as you will remember.

After a stormy passage across the Atlantic Rutherford was greeted in Montreal with that genuine kindness and hospitality for which that city is famous, and he was soon plunged into his work. He was not a good lecturer at first, for he began lecturing over the heads of his students, assuming a knowledge of mathematics beyond the stage of their instruction. Callendar had at first made the same mistake before him, until the engineering students signed a round-robin requesting that the Calculus should not be used in Physics before it had been taught to them in their course of Mathematics. This protest seemed reasonable and was upheld by the Faculty. Rutherford, however, quickly understood the limitations and capabilities of the students, and a dogged enthusiasm pulled him through his early difficulties.

Rutherford soon had some electroscopes and electrometers in working order and he continued to investigate the singular radio-active properties of uranium and thorium. On 24 October he wrote a letter to Hayles of the Cavendish urging him to send some preparations of those elements as soon as possible, together with the account which would be settled on their arrival.

At Cambridge he had established a difference between the two types of radiation from uranium which he had named alpha and beta rays. It had already been proved by Becquerel that the beta rays could be deflected in a magnetic field and in such a direction that it was clear that they carried the type of electricity called negative—the kind produced on sealing wax when rubbed with fur. The beta rays (or

particles) were rightly identified with the *electrons* which Sir William
Crookes had discovered in a well-evacuated electrical discharge tube.
These electrons, investigated by Sir J. J. Thomson, have little weight
(or mass) and indeed it takes 1850 of them to weigh as much as a single
one of the lightest of the atoms, namely hydrogen. The electrical charge
of an electron is its notable feature and is enormous as compared with
its mass.

These nimble and faithful Ariels do our bidding in a thousand ways—
convey our electrical energy, help us to telephone, broadcast, and to
render television.

No one, however, knew the nature of the alpha particles and Rutherford determined to discover this, particularly because he found that the
energy of the alpha particles was far greater than that of the beta particles.
His investigations of the alphas, as we may call them, continued with
astonishing results over a quarter of a century and led, stage by stage,
to the most remarkable discoveries in the history of science, culminat-
ing in the proof of the atomic nucleus and in the transmutation of
elements.

A young electrical engineer, R. B. Owens, recently appointed at
McGill, had been awarded a research exhibition from Columbia
University. He asked Rutherford to suggest a subject, and set to work
to investigate the radiations from thorium, just as Rutherford had done
for uranium. Unexpected difficulties arose. Owens found that his
source behaved in a most erratic manner and traced the trouble to air
currents. Something that was neither thorium, nor alphas, nor betas
could be blown about! In the dilemma of a mystery it was called
thorium *emanation*, a non-committal term. Rutherford proceeded single-
handed to investigate this new discovery. He found that it was not
'ions', for the emanation would go through cotton-wool and past
charged bodies. He drew the thorium emanation along tubes and
found that it carried its radio-activity with it, but got weaker as it
travelled. He measured the rate at which the activity grew less, and
proved that half its radio-activity disappeared every minute, no matter

what the initial strength. This result was sufficiently remarkable, but it was only part of the story. This thorium emanation could make other bodies radio-active when they were in contact with it! This new effect was called at first 'induced or excited activity', but under Rutherford's influence the name was later changed to *active deposit*, for he found by degrees that the emanation was a gas, and the after-effects due to a material deposit. This active deposit of thorium was found to decay away to half value in $11\frac{1}{2}$ hours.

It is well to remember the startling character of the evidence that confronted Rutherford at McGill. It seemed necessary to believe that one atom gave birth to a new atom which wandered away from its original home and in turn gave birth to yet another new kind of atom, and so down a chain of changes. Such a prolific atom with its whole progeny of atoms stood in most striking contrast to the old idea that atoms were eternal, unchanged and unchangeable.

Parallel work was taking place in France where Dorn discovered radium emanation, and the Curies the ensuing excited activity or active deposit. If the reader has for the first time come across the ideas just set forth, he will, unless possessed of unusual ability and insight, be somewhat befogged, for he has to think about thorium and uranium, which are both elements; about emanations which are now known to be gases, but at first were complete mysteries; about the resulting active deposits, now known to be material (but not then), decaying away at different rates; about alpha and beta rays. The reader may rest assured that his fog is nothing to the darkness in which the early workers groped, and out of which Rutherford found his way. The successive steps in his scientific progress are so intimately bound up with his personal life that it would be difficult to divorce them and it seems best to unfold both together in the way they actually occurred.

RUTHERFORD TO MARY NEWTON

Montreal: 25 Sept. 1898. I am now fairly settled in Montreal as I have been here since Tuesday morning. Professor Bovey with whom I am now staying telegraphed for me to stay with him when I reached Quebec and I have been here since and am going to my lodgings tomorrow.... We left Liverpool on Thursday in the *Yorkshire*, a boat of about 5000 tons. Professor and Mrs Cox and their son Jack were on board...by a peculiar coincidence MacBride and I shared the same cabin...he is a good talker but very dogmatic and assertive....It started to blow hard as soon as we got off the coast of Ireland and we had heavy gales all the way and made a very slow trip. One day there was a heavy gale and great seas so the head of the vessel was kept to the wind for five hours without making any headway...misfortune dogged our footsteps up the St Lawrence for we were troubled with fog and smoke and took a day longer up the river than we should have. Got into Quebec at nightfall....I stepped on the soil of Canada there for the first time. Left at 4 next morning and went all day up the St Lawrence. It is a magnificent river very wide and in many parts the banks are very pretty. Professor Bovey, who is Dean of the Applied Science Faculty, met me and we went together to his house....I could not have been treated better....I have met Mr Macdonald several times, he is the millionaire who has given all the money for the Science labs. He is a grey-headed Scotsman, simple and unassuming, and I like him very well....I have got lodgings in McGill College Avenue near the University. I get board and one room for 30 dollars a month. It is an extremely difficult matter in Montreal to get a good sittingroom and bedroom except for exorbitant prices. I have a big bedroom and will make partial use of this as a sittingroom, until I can find something more suitable. I have of course a fine large private room at College, and will be there most of the day. I am very pleased with everybody and everything so far. I start my first lecture next week to a class of about a hundred—on Magnetism.

Now I have to consider the weighty question of when I am to come out for you. I have been collecting data since I have been here and have come to the conclusion that I will have to stay here a year before I go out to fetch you. I start about zero as regards cash and it would hardly be right for me to fix up till I have settled my bill at home. They say living is a third dearer in Canada than in England. Rents are very high in the town and a small pillbox of a house costs £100 a year. Supposing I stay here now for eighteen months and come out in April or May, I will probably have saved £400 and will then have enough to start housekeeping after paying off my educational expenses. If I went out there for you this year (as of course I should very much like to do, needless to say) I would hardly have enough cash after six months to go out and bring you back and we would start with short funds at once, so I really think, my dear, we will really have to postpone our partner- ship for eighteen months from now. It seems a very long time when the opportunity is so to speak at our hand, but I think it will be much better in the end. I will have had time to get settled and things in order, and so be able to devote more time to you when you do arrive. Moreover I am afraid I could not get away early this year, as they don't like the young professors running off at once. Next year my case would be different, and I could probably get away a month earlier than this, which will prevent my matrimonial duties being too rushed. I am sure you will like Montreal. A professor here occupies a very prominent position in Society far more so than in New Zealand, and one is not so buried as in New Zealand.

4 Oct. 1898. I am in such a rush of work that you will have to be content with a short letter from me today. I am now in full swing of work and have not a moment to spare as I am getting classes in order, lecturing, etc. I do not fortunately do much lecturing, one class of about fifty twice a week and then another class for two hours, but I have in addition the supervision of all the practical classes which come five times a week so I will have my time very fully occupied. My first lecture came off yesterday. Professor Cox introduced me and I then bucked in. Today I lectured to my advance class for two hours as a preparatory to practical work—a class of fourteen. I am also now very busy doing some work for Professor Bovey on vibrations on house properties on which a lawsuit is pending. It has to be done in a hurry, so I have been busy fitting it up. It may mean a considerable amount of time to complete it, as it will mean a good deal of sensitive apparatus to be made. I am very pleased with the Physics Building which is very large and fine, six storeys or rather seven, and filled with apparatus. Everything is very bright and polished, in fact, almost too much so for a building where work is to be done. I hope however to buck into work before long. I believe we are to go out a good deal this term.... I have been out to several dinners, also to see the Royal Electric works in town, and to see their works on the Richelieu river where they intend to supply 20,000 horse power to Montreal. The works are not yet completed, but they

are on a very large scale. The manager of the works paid our train fares, gave us dinner, and generally treated us royally.

7 *Nov.* 1898. I forget whether I told you that I am at present engaged on some outside work with Professor Bovey. Some people are suing the company, which runs the car service, for the vibration caused by the power house when the dynamos are running. I have been making some instruments to record these vibrations and to take a pen and ink record of the vibrations in horizontal and vertical directions. The vibrations are in general from 1/100 to 1/1000 of an inch, and the instruments have to be far more sensitive than earthquake recorders. It entails going down to the power house every day and taking the record off and putting another on and fitting up the instruments generally. I hope to get a fee of at least £50 before it is over, but that may be some months yet. We are on the side of the Electric Company and are going to show that the vibrations are too small to affect any houses in the neighbourhood. As a matter of fact the plaintiffs have no ground whatever for their complaint. I practically do all the work, and Bovey and Cox turn up as witnesses at intervals and sign the records. Callendar here was considered a universal genius and he also did something of this kind last year, and I also gain a sort of reflected glory for doing a thing which the great Callendar apparently alone was able to do. This extraordinary belief in Callendar is a little unfortunate for me, as however well I do I will never rise to the acme of greatness attained by him. These little things will all help to shorten our time of probation and to have a little more to spend when it does come off.

29 *Jan.* 1899. I had a rather amusing experience the other night. I think I have told you that about 5000 Russian emigrants called Doukhobors, a persecuted Russian community, have come to Canada. Mrs Cox and several others have taken a great interest in these people, and get up subscriptions for them, as they are coming out in the middle of winter to Winnipeg, where it is about minus 20 degrees pretty continuously. We had to go out by train a good distance, and then found we had to wait about three hours till the first train with the emigrants arrived. However, at 11 p.m., the first contingent drew up at the junction and we inspected them, or rather their feet, for all of them were asleep in their bunks with bare feet. They are rather a fine looking people and very clean I believe. The ladies were specially interested in a little baby,

born on the steamer, which was called Canada. We got back to Montreal about 1 a.m. and I had supper at the Cox's....I was very glad to see my paper on Uranium radiation appeared in the Phil. Mag. this month. It is the longest paper I have published so far, and I will send you a copy as soon as I get one. I am better pleased with it than any paper I have done, and it represents over a year's work. I was surprised to hear from W. S. Marris in Meerut last week, congratulating me on my appointment. He was good enough to state that he always knew I was a better man than he, and seems to have a very exalted and quite absurd idea of what I have done. It was very decent of him however to write.

RUTHERFORD TO HIS MOTHER

19 *Feb.* 1899. Three of us walked over the frozen St Lawrence last Thursday, and a long walk it is. The river is two miles wide at Montreal and it is frozen solid the whole way across. It is rather interesting to walk and to see sleds and sleighs going where three months before great ships were lying at the wharves. The river does not freeze flat like one would naturally expect, but like one's back lumber yard. Great blocks and hills of ice are scattered over it where the pressure of the river has piled up the ice blocks on top of one another.

There is a great industry here collecting ice for summer use and for export to the W. Indies. The blocks are cut out of the river, stored in warehouses and doled out to customers during the summer. Over 100,000 tons are collected every year for summer use. It is the custom in Montreal to have the iceman visit you the same as the baker. He, however, leaves the blocks of ice in front of your door and you can take it in, or let it melt there, as you like.

They say the river is a great sight in spring, just before the ice breaks up, when the great ice-shoves take place. They had to build a great guard embankment in the river to prevent the ice pushed by the river from invading the streets near where the wharves are in summer, for the piles are always removed before the river freezes.

RUTHERFORD TO MARY NEWTON

Montreal: May 1899. Since my trip to the States I have had a varied and busy time. I wrote up a paper for the Royal Society of Canada and did a good deal of vibration work for the Street Railway. I was surprised on Wednesday morning to get a long distance telephone message from

Cox at Ottawa asking me to deliver the Royal Society Popular Lecture at Ottawa. It appears the man, whom they originally asked, a prominent American Entomologist, was taken ill suddenly, so they were much put about to get a man at a moment's notice. Cox proposed me, and they jumped at it, as they thought that the subject of wireless telegraphy was a good one for the occasion. It was a distinct honour to ask me, so you may be as proud as you like as regard to it. I had only one day to get ready my experiments and all. I got there Thursday morning by the 9.45 train, and just had time to rig up the experiments for the evening. Everything went off very well and the applause was great. It was a very fair audience in a large hall, and was the biggest show I had attended so far....I think it is quite probable I may be elected a member of the Royal Society of Canada next year—an F.R.S. on a small scale, so you will wear, on your arrival, your blushing honours thick upon you.

23 *June* 1899. I am just beginning to reflect on the amount of just cussing I must have caused you all, by missing the 'Frisco mails with unfailing regularity. I got at the bottom of the matter this morning, when I found that the list of mails closing via 'Frisco only referred to 'Frisco, although it was issued by the P.O. Department here. The consequence is that all my letters have been posted about four days late for the mail. I have been caused to reflect on the matter by receiving a cable from home asking me if I was well. The cable I wouldn't have minded much, but having to stump up eleven dollars for the return wire was too much for me....I am working and playing tennis and bicycling at odd times. I forwarded 'David Harum' to you, which I hope you will like.

2 *July* 1899. Yesterday was Dominion Day—the Anniversary of the Federation of the different provinces of Canada, and as such was kept a high holiday. It was a windy day with a temperature of about 84 in the shade, so it was quite hot enough for yours truly without taking much violent exercise. I worked a little in the Lab. in the morning, had lunch at the Cox's and went for a small picnic on the other side of the mountain....

I have been working fairly steadily lately and have been fortunate enough to hit on some rather interesting things which occupy most of my time. I occasionally find time for a short bicycle ride or a game of tennis. The Lab. is very fine and cool in the hot weather and very pleasant to work in. Today, Vaughan, Clay and I went for a run over to the island (St Helen's) by boat, to a sort of Harry and Harriet's ground, which is much frequented on a Sabbath day by the French population. It is rather a pretty place, well planted, but spoilt by the host of papers thrown about by careless picnickers.

On 23 July 1899, J. J. Thomson at Cambridge wrote a long letter to Rutherford about the latter's paper on the 'emanation' from thorium, and on the active deposit obtained from the emanation. He was particularly surprised at the emanation's retaining its radioactive properties after being bubbled through strong sulphuric acid. He was no less astonished at the fact that the active deposit tended to collect on a *negatively* charged rod and not on a positively charged rod. He suggested the use of a C. T. R. Wilson cloud-chamber—then in its early stages of development.

J. J. added that, in his own work, he had found that the ratio of charge to mass, for the negative ions due to ultra-violet light, was the same as for the cathode rays, proving, as we should state now, that both types were electrons. "I am inclined to think that at low pressure negative electrification may be produced—while the positive charge remains the big atom."

RUTHERFORD TO MARY NEWTON

Sept. 1899. Lectures have already begun and the professors are collecting together for the fray. Walker, Owens, Carter, etc., are back.... Owens has just got back from Cambridge, where he had a good time, and met all my friends and brings back reports of their doings. From what he says, I evidently left quite a considerable reputation, but modesty forbids to tell the stories he tells. I believe Bovey when he was in England had everybody telling him what a good man he had got until he got tired of it at times. The trouble is that Callendar left such a reputation behind him, that I of necessity have to keep in the background at present, but I believe J. J. rather knocked the feet off Bovey, who started with J. J. by saying "Of course we were all very sorry to lose Callendar." The reply was "I don't see why you should be, you got a better man anyway." I am getting rather tired of people telling me what a great man Callendar was, but I always have the sense to agree. As a matter of fact, I don't quite class myself in the same order as Callendar, who was more an engineering type than a physicist, and who took more pride in making a piece of apparatus than in discovering a new scientific truth—but this between ourselves. By the time you arrive they may possibly begin to know the real state of affairs, as I have several friends on the staff who have strong views on the subject.

I am very busy with research work just now, I have three men going at research and one lady, I have to look after them at first pretty closely. I am giving a course of postgraduate lectures this year on Electrical Waves and Oscillations which will give me a good deal of trouble to arrange. This is the first thing of this kind ever done here and rather surprised them when I suggested it.

18 *Nov.* 1899. My work is progressing very satisfactorily. I am in the throes of composition of another paper....Professor Poynting of Birmingham was round here last week and I had a good time taking him round....My chief solace at present is to keep working. I hope to finish several papers before I get out to New Zealand, in order to rest my oars a little when I get back with you. I have an F.R.S. in my mind's eye and hope I won't have to wait too long for it.

2 *Dec.* In your last letter you mentioned that I hadn't mentioned if I was doing any research work. My dear girl, I keep going steadily turning up to the Lab. five nights out of seven, till 11 or 12 o'clock, and generally make things buzz along. I sent off on Thursday another long paper for the press which is a very good one, even though I say so, and comprises 1000 new facts which have been undreamt of...suffices it to say that it is a matter of considerable scientific moment. I have been working with the rare materials uranium and thorium, which give out a kind of X rays continuously for no apparent reason. I have found out that the thorium in addition gives out a lot of radiation. This makes the third paper I have sent for publication since my arrival, so I haven't done so badly. I have a great deal of work on hand some of which I hope to finish up before I leave for New Zealand, so that I can take a holiday with a clear conscience.

31 *Dec.* 1899. I am writing this on the last day of the old year before going out to lunch with the Rev. Barnes, Unitarian Minister, whose son is demonstrator in the Lab. He is one of the best preachers, and a

well educated man. His son, about 27, is the best man in physics here. We are giving him his D.Sc. degree shortly. I have at last buckled to and looked into the matter of the time-tables, fares, etc., to New Zealand and have settled my plan of action. Lectures end on March 28, and I have to be at San Francisco on April 18th to arrive in New Zealand on May 9th, so I shall not have much time to get my exams. through. We are not supposed to leave McGill till the end of April, when Convocation meets, but I will be allowed leave, so I am told, for such a unique occasion. I am not quite sure but I think I shall go to San Francisco, via Chicago and the States, as it will be shorter in time. I shall probably leave here April 11th. I have decided we shall come back via C.P.R., which I think meets your requirements; so I am taking a return fare for twelve months as it saves a good deal. My travelling fare (return) will be about 530 dollars, but this does not include Pullman berths, meals and so forth, which will amount to about another 100 dollars in all. By the time I get back to Montreal, and paid your fare back, we shall probably have blown about 1250 dollars altogether, as I intend to travel comfortably on our honeymoon. Anyhow I shall probably have about 1000 dollars in hand when we arrive for furnishing etc., so we shall be able to make a fair start. I expect you are anxious to start housekeeping your first winter, but we will see what we can do anyhow to prepare for a run to England pretty soon. I have been working very hard this vac., and hope to get an important paper done before I leave for New Zealand, as I don't expect to do anything more for six months at least, till I have got you comfortably settled somewhere. ... Work begins again in four days. It was about 10 below zero today, and yesterday, with a wind, was about the coldest day I have felt. In a seven minute walk to College I got my ears frozen. A cabman told me in time, so I rubbed them with snow and they didn't swell up much afterwards. I am wearing a fur cap today to keep my ears safe.

Everyone is much interested to have a look at the strange creature I am going to bring from New Zealand. I am sure you will find a number of friends you will like among the College people, who will do anything they can for you. Professor Owens, the electrical engineering professor, is a great friend of mine. He is taking a flat next year and his chief object in doing so is to prepare for suppers for you and me, etc. In fact he said he designed his quarters especially with that object in view.

THOMSON TO RUTHERFORD

21 Dec. 1899. Your second paper arrived safely and I have sent it to the Phil. Mag., your first paper ought to have been in the December number as I passed the proofs for it at the beginning of November. I think they must have intended to have a double number and then changed their minds. Your results are exceedingly interesting, the idea that I got on reading the experiments was that the radioactivity was due to thorium vapour or emanation which was carried by the ions— I mean that the emanation in the field tended rather to condense round the positive ions than the negative ones as we might expect an electro-positive substance to do. I see Giesel makes out that the radiation from polonium is affected by a strong magnetic field; if this is so, it might be worth trying whether your emanation from thorium were so affected....

Remember me to Owens and with best Christmas wishes from my wife and myself.

In 1900 Henri Becquerel and Rutherford had a dispute as to the stoppage of alpha rays by thin aluminium sheets. Rutherford contended that the alpha particles lost speed and were then more easily deflected by a magnetic field. Becquerel was led by experiment to a contrary opinion, but on further investigation he found that Rutherford was right and made a 'handsome acknowledgment'.

RUTHERFORD TO MARY NEWTON

Montreal: 4 *Feb.* 1900. J. Macmillan Brown wrote to me a day or so ago and tells me he is off again to England. He seems to think I might want to go for the chair at Christchurch but I don't feel very keen about it. There is no lab. and it would be very difficult to get much of a lab. and I would have to stay there all my days. I am afraid I would be very discontented there without a good lab. and I still have a certain amount of ambition....I don't think you will mind Canada for a few years for, between you and me, I don't regard myself as finally settled here, but hope to get over to England some day.

3 *March* 1900. Relief of Ladysmith. I am getting my work in order for my departure. The Faculty today gave me an Honorary M.Sc. Degree. I was not keen on it, but they evidently wanted to do it. I shall probably

be elected a member of the Royal Society of Canada soon, a very select body. I shall also soon be elected a member of the American Physical Society also a very select body.

TOWNSEND TO RUTHERFORD

27 March 1900. I am glad to hear from you and to have news of your doings. You seem to be pretty busy, I wish we had opportunities here of doing work outside the Cavendish as I must confess that eternally working at ions is beginning to be tiresome.

I have just come back from Aldershot having joined the volunteers, the war has induced a lot of men to join. We had a fine rough time of it, but it was so complete a change that I thoroughly enjoyed it. This is the first opportunity that I have of writing since I got your letter and have not had a minute to examine your problem and I suppose it is too late to try it as you wanted the solution before 1st of April.

There was a great rag on here the night of the relief of Ladysmith. There was a large fire in the market square and a band playing. The police arrested several undergrads (Westlake among them). They were had up next day and fined heavily ($£10$ each some of them) by the Mayor and other pro-Boer magistrates. The people kicked up a great row about it and a petition was sent to Parliament to have the under-grads given a free pardon and the Home Secretary has just written to say that he has recommended the Queen to pardon the 'felons'. It is a great score off the Mayor. There are all sorts of attacks made on his house and he has to be guarded by the police as the undergrads want his blood.

Everybody has been so excited about the war that there has not been much work done this term. I suppose you saw in the papers that Preston is dead. I am very sorry for him as he was an extremely decent chap.

H. A. Wilson has been away in Germany for two terms, he has got the Allen scholarship $£250$ for one year which is a very nice little requisition.

Here follows a letter by the illustrious G. F. Fitzgerald, who suggested that a shortening of the length of a body in its line of motion, to an extent dependent upon its velocity, would account for the failures or 'null' results of the Michelson-Morley experiment made to determine the motion of the earth through space (or ether) by experiments wholly conducted on the earth's surface. The Fitzgerald shortening became merged in the great principle of special relativity enunciated by Einstein.

FITZGERALD TO RUTHERFORD

Trinity College, Dublin: 5 May 1900. Let me call your attention to a note in "The Electrician" (May 4th p. 39) on a German Post Office experiment on Poulsen's telephone in which he records telephone messages by manufacturing consequent poles on a long steel tape in just the way I advised you to record Hertzian radiations when you were working at their magnetic effects at Cambridge. I would suggest your returning to those observations of yours and seeing whether it would not be possible to do as I suggested because it would be a very rapid way of receiving wireless telegraph messages, not involving coherers and decoherers etc., and besides, of recording the messages conveniently.

We were very much interested in your thorium experiments. There seems no doubt that there is some emanation from the thorium. I see in the same "Electrician" (p. 41) that Debierne says he has isolated a substance he calls actinium which he thinks is the active material in the thorium experiments. This actinium gives out something that is magnetically deflected, but I am not sure that this is not almost always present to a small extent in all these cases and that it is merely the very powerful ones in which it has been observed. This magnetically deflected emanation must be of the nature of Lenard and Cathode rays—probably disembodied electrons: J. J. prefers to call them corpuscles. I have got some thorium, but we are all too lazy here to do experiments and indeed between National Education Boards, Veterinary College Boards, Technical School Boards etc., etc., one gets sick of doing anything.

Poulsen's method of recording signals on moving steel tape or wire is the same scheme as that adopted by Marconi for recording signals in his magnetic receiver. It was later used for recording telephone messages and, in various forms and under various names, is in frequent use today for delayed speech, putting it into 'cold storage' until it is required again.

During the years 1899 and 1900 a number of papers appeared in the

Philosophical Magazine written by Rutherford, sometimes with his research students, such as Grier and Miss Brooks. These contributions indicate enormous industry. He showed that the rays from uranium were not reflected or polarised, as some had suggested. He tested the ionisation due to alphas in half-a-dozen different gases and found the total effect varied but little in them. In January 1900, he clearly stated that thorium emanation was *material*, consisting of very small particles, themselves radioactive. He had named the three types of radiation, alpha, beta and gamma. Villard had found that the last named radiation resembled X-rays in that they were not deviable by magnetic forces. As the betas were about one hundred times as penetrating as the alphas, so the gammas were about a hundred times as penetrating as the betas. Rutherford was sharpening his weapons for the attacks which were soon to follow—attacks on the main issues. His discovery of the emanation and active deposit of thorium placed him quickly in the first rank of investigators. His statements were not guesses; they were all deduced by repeated experimental tests. He had certainly earned his summer holiday, which was made long enough to enable him to visit New Zealand, see his parents, marry Mary Newton and return with her to Montreal, where they settled down in the autumn in a comfortable little house on an economical scale. They had both learnt the value of money in a hard school; it was a lesson that he did not forget, and perhaps it quickened his generosity.

In the summer of 1900, there came from Oxford to the Department of Chemistry at McGill, F. W. Soddy, a young man, versatile, full of energy and ambition. Rutherford naturally needed help on the chemical side of radioactivity, and he turned to the Macdonald Research Professor of Chemistry, who was not attracted by the work (he was an organic chemist), thereby missing one of the great chances of his life. But when the opportunity came to Soddy, he seized it with alacrity, so that a partnership was established in which Rutherford was naturally leader, but the combination was powerful and led to astounding results.

It must be remembered that after research work has been completed in the laboratory, time and care are necessary in writing, preparing tables and diagrams, and getting the material ready for the press. Rutherford always insisted on great attention to the preparation of papers, both in style and matter; there must be clarity and sequence. After the paper was prepared, it took yet more time for the post to carry it to England; then there was further delay in printing and finally some kind friend in England would read the proofs. Altogether, there is quite a lag in time between a discovery and its publication, particularly when an ocean-crossing is involved. At a time when the leading savants in Europe were competing in the acquisition of new results in radio-activity, it was difficult for Rutherford to hold the lead, but he did so by ingenuity and industry.

In 1900, Debierne discovered another radioactive element which he called actinium and this was rediscovered two years later by Giesel who called it the 'emanating substance'. It soon became clear that there were three emanations belonging to three different groups, or families—radium, thorium and actinium. These could be readily distinguished because they decayed to half-value in three widely different periods of time, namely, nearly four days, a minute, and three seconds respectively. These radioactive gases are today called radon, thoron and actinon, and they belong to the same class as the stable gases, such as helium, argon, neon and the like. This statement is premature as regards the historical order of discovery, but helps in clarity. Rutherford had already proved (*Phil. Mag.* Feb. 1900, p. 171) that the induced or excited radioactivity (active deposit) was due to the emanation and was not due to the thorium itself or to any sort of radiation from it.

There is little written record of Rutherford's life during the year 1900, and for the best of reasons. He went to New Zealand to marry Mary Georgina Newton, only daughter of Arthur Charles and Mary de Renzy Newton of Christchurch, New Zealand. He saw his parents at his old home, so that few letters passed. With his wife, he returned to Montreal in the autumn and lived in a house on St Famille Street.

THOMSON TO RUTHERFORD

Cavendish Laboratory, Cambridge: 15 Feb. 1901. I had a letter from Dorn lately, complaining that you had not sufficiently acknowledged his work in your paper, and wanting me to communicate with the Royal Society a paper asserting his claims. I wrote to him that all that had appeared was an abstract, and that I was sure he would find when he read the complete paper that you had suitably acknowledged his work. With regard to the Royal Society, Whetham is up again and will I hope get in. I think it would hardly be politic to run another candidate prominently connected with the Cavendish Laboratory as there would be very little chance of their both getting in, while one alone would be almost certain, so I should think the best policy would be to run one at a time. We are very busy and have a good many men working here, among them two Canadians—Patterson and Baker—who are both doing well. I am at present working at some experiments to decide whether the corpuscles have masses other than electrical. I am just finishing a paper developing the idea I published last February that the ionization in the discharge tube is due to the motion through the gas of other ions. I find that it links the various phenomena very well together and is I feel convinced the right explanation. Townsend, as you know, has left for Oxford, he could not find accommodation at the Clarendon Laboratory and so is working at the Observatory.

RUTHERFORD TO HIS MOTHER

Montreal: 2 April 1901. You have probably been aware for several days that you have now the honour of being a grandmother. I hope Father feels correspondingly dignified after reaching the stage of grandfather. The baby much to Mary's delight, is a she, and is apparently provided with the usual number of limbs. There is much excitement in college and on the night of her arrival I was toasted at a whist party. It is suggested I should call her 'Ione', after my respect for ions in gases. She has good lungs, but I believe uses them comparatively sparingly compared with most babies. The baby is of course a marvel of intelligence and we think there never was such a fine baby before. I hope Pater is well—enjoying life—and meditating on the probable number of his grandchildren.

The daughter, born 30 March 1901, was named Eileen Mary, and she was an only child.

Rutherford received several offers of positions in the leading Universities of the United States of America, and at one time Columbia University endeavoured to build up a great school of Physics there by sending tempting offers to secure J. J. Thomson and Rutherford together. However their offers were not accepted. On 20 March 1901, Rutherford wrote to J. J. Thomson asking his advice about P. G. Tait's chair at Edinburgh which had become vacant.

RUTHERFORD TO THOMSON

26 *March* 1901. I think you know fairly well my position here. The laboratory is everything that can be desired, and I am not overburdened with lecturing. I am very comfortably situated, and have practically a free hand in the laboratory. On the other hand the salary is not very great and does not go very far in Montreal, where I think the cost of keeping a household is considerably greater than in England.

After the years in the Cavendish I feel myself rather out of things scientific, and greatly miss the opportunities of meeting men interested in Physics. Outside the small circle of the laboratory, it is seldom I meet anyone to hear what is being done elsewhere. I think that this feeling of isolation is the great drawback to colonial appointments, for unless one is prepared to stagnate, one feels badly the want of scientific intercourse. I appreciate the fact that it is a very important Chair for a young man to apply for, but I thought that such an opportunity might not occur for years.... I have now had a good deal of experience in arranging and managing laboratory classes and also in lecturing to large classes. My class in Electricity and Magnetism this year was over a hundred....

Your corpuscular theory seems to take the field in Physics at present. I have been reading the Paris report on the subject. We are having a great discussion on the subject tomorrow in our local 'Physical Society' when we hope to demolish the Chemists. I was sorry to hear of your Mother's illness. With kind regards to Mrs Thomson.

THOMSON TO RUTHERFORD

12 *April* 1901. As I cabled you, I think you had better stand for the Edinburgh Chair, if you wish to return to England. I do not think

the chances of your getting the post very promising, as Knott has been doing the work for some time, and has a good deal of local influence— the election too is made by a body of local men who do not know anything of Physics, so that local influence will be especially powerful in this case: at the same time, I think the candidature will do you good, as it will let people know that you are willing to leave Montreal, and your name will naturally come up if other vacancies occurred. I think that Townsend gained considerably by going in for Glasgow though he was not successful. I think from what I have heard that the field for Edinburgh is likely to be considerably larger than that at Glasgow which was unaccountably small. I quite appreciate the isolation of scientific workers in the Colonies, it was that made Threlfall give up Sydney. I hope that by now your wife and yourself are quite strong again. Is there any chance of your coming over this year? We should be delighted to see you. We have a great many people working at the Cavendish and some interesting experiments going on on the electrical properties of thin films. I think on the corpuscular theory of metallic conduction a great deal of information is to be got from the study of these and I hope to be able to get in this way the mean free path of the corpuscles and hence the velocity of these under electric force in the metal. I suppose you have seen Debierne's work on actinium, a substance which is closely associated with thorium, and which has extraordinary powers of producing induced radioactivity; do you think there could have been any of this in your experiments on the thorium radiation? I am preparing for a new determination of using the ionization due to radium instead of Röntgen rays. I think the greater constancy of the radiation will greatly increase the accuracy of the determination.

You will I am sure be sorry to hear that my mother died at the end of last month, it has been a great sorrow.

25 *April* 1901. I congratulate you and Mrs Rutherford most heartily on your promotion to the responsible position of being parents. I know how much a child adds to one's happiness. My wife will be writing to Mrs Rutherford to congratulate her.

There is the prospect of an immense field for Edinburgh; among others, is McGregor of Nova Scotia and an old pupil of Tait's; he is coming over I believe to interview the electors. If things went well with your candidature I think it might be worth considering whether you could not run over so as to be in a position to see the electors—a practice which is universal in Scotland.

I was much interested in your experiments on the effect of temperature on the emanation which I was reading last night....

After due consideration no application was made for this Chair, and Rutherford, instead of being isolated, became himself a source and centre of attraction.

SIR WILLIAM CROOKES TO RUTHERFORD

18 *Dec.* 1901. I received your letter with enclosure for Dr Knofler. This I sent at once. His address is—Dr Knofler, Plotzensee, bei Berlin, Germany. I am sure he will send you some pure thorium nitrate, as he is large-minded and a thorough man of science. If, however, there is any delay or difficulty let me know and I will send you some of my own stock.

I have followed your experiments and theories with great interest and shall be very pleased to receive the copy of the paper you say is soon coming to our Chemical Society. I am preparing a paper which may contain some matters of interest to you, and will send you a copy as soon as it is in type.

M. Becquerel told me a curious circumstance a short time ago, and asked if I could verify it. He prepared some time ago an inactive uranium nitrate. Now, on repeating his experiments with that identical sample he found it had reassumed its radioactivity. I am at work on old compounds of my own to see if I can get similar results.

The last paragraph of this letter is full of interest. Crookes found that he could separate from uranium nitrate a small quantity of highly radioactive material (as judged by its action on a photographic plate) which gradually lost its efficiency, while the original uranium regained what it had at first lost. The period to half-value was ultimately shown to be about a month, while the radiation consisted of beta rays.

This remarkable result of Crookes preceded Rutherford and Soddy's fundamental experiment on the separation of thorium X from thorium, and their careful examination, by the *electroscope*, of their decay and rise curves, gave the clue to the *inevitable* theory of radioactivity which they so courageously formulated.

In December there was a letter from a friend suggesting that Rutherford should move from Montreal nearer to the 'hub of things' by becoming a candidate for the chair of Physics at University College, London, in succession to Callendar, who had been appointed to the Imperial College of Science and Technology. The application does not appear to have been made.

RUTHERFORD TO HIS MOTHER

Montreal: 5 Jan. 1902. I left for New York after Christmas in order to attend the meeting of the American Physical Society. The meeting was held in Columbia University, where I gave two papers which were pretty well received, as I am the only worker in the field of excited radioactivity in the English-speaking world. We had lunch in the body of the building, and there I met Sir Robert Ball, who is touring the States giving a series of popular lectures in astronomy. In the evening the mathematicians and the physicists joined together in a dinner at the Marlborough Hotel.

There I met two Cambridge men I knew, who are teaching mathematics in this country, one at Johns Hopkins University and the other at a leading women's college. Although it is one of the best of women's colleges in America, most of the professors are men, and generally young men too. The place is governed by a lady dean who is an autocrat of the toughest type. She dismisses professors right and left, and it generally happens that those she sends off immediately get better appointments elsewhere—such is the esteem in which the other universities hold her judgments. I believe it is quite sufficient for her to dislike the cut of a man's trousers or his hair for him to be sent about his business. In addition a man's position is untenable if he does not smile enough or smiles too much on the girls—so apparently things are difficult to adjust exactly to please the head.

I am now busy writing up papers for publication and doing fresh work. I have to keep going, as there are always people on my track. I have to publish my present work as rapidly as possible in order to keep in the race. The best sprinters in this road of investigation are Becquerel and the Curies in Paris, who have done a great deal of very important work in the subject of radioactive bodies during the last few years.

About a year ago I received an offer to go as Professor of Physics to the University of Columbia at a salary of 3500 dollars a year. I have not considered the proposition seriously as it would probably not be so good a position as I have at Montreal. I want a much higher salary than that to tempt me away.

Some of this information about the Women's College may have been gathered from Prof. James Harkness of Trinity College, Cambridge, who was for a time professor of mathematics at Bryn Mawr, and later was a colleague of Rutherford at McGill. Harkness is well known from his book *Theory of Functions*, the first to make this subject known to many English readers. He was a man of wide and sound learning, none too common at any time, and he had both sympathy and excellent judgment, so that Rutherford, like the rest of us, would often call on him and get his friendly and sagacious opinion whenever difficulties arose or openings occurred.

HARKNESS TO RUTHERFORD

Bryn Mawr: 17 April 1902. I should very much like to go to Lord Kelvin's reception on Monday night, but I am not quite sure whether I can get away. There is a journal club meeting at 2 on Monday afternoon followed by a Theory of Functions class. The first I must go to and as the class consists of graduate students it will not do to cut it. This will make it difficult to get away, and moreover there is the difficulty of getting back again. If the rush should not prove too appalling when the time comes, I shall turn up at the reception.

McKenzie and I were very pleased at Miss Brooks getting the Fellowship. I must say I envy her the year under J. J.

Look out for McKenzie at the New York meeting; he will count on seeing you.

Ions are in the air here: is it action at a distance from Montreal, or the more direct influence of Miss Brooks?

My chemical colleague Kohler is extremely keen about the matter (is it fair to call it 'Matter'?).

THOMSON TO RUTHERFORD

13 *May* 1902. I shall be very glad to give Miss Brooks permission to work in the Laboratory and to attend lectures and I am sure my wife and I will do all we can to make her stay in Cambridge pleasant and profitable. If she would like to live in Newnham College rather than in rooms in the town I will sound the authorities and see if it can be managed.

I got the papers about radioactivity yesterday, it seems to me that your explanation clears up a great deal of obscurity. I am glad it came before the chapters on Radioactivity in my book were printed off. I shall be able to introduce it and make the account much more connected. I am working at the increased conductivity produced in the air by bubbling it through water: it is extraordinary the analogy between this and the behaviour of an emanation. If you put a negatively electrified body in gas of this kind it becomes strongly radioactive—the activity dying away to about half value in half an hour, while, if the body is not negatively electrified, the induced radioactivity is very much smaller. When you have made a piece of metal radioactive in this way, it will stand washing in water, heating to a red heat without losing its property: the only difference between it and the ordinary induced radioactivity is that it is very much more easily absorbed, it will not go through paper or aluminium. These results make me doubt whether Elster and Geitel's induced radioactivity is really due to some rare substance; it seems to me it is probably made from wind and water! C. T. R. Wilson has discovered that freshly fallen rain is radioactive.

Rutherford wrote to *The Electrician*, 25 July 1902, quoting his famous paper to the Royal Society in 1897 entitled "A Magnetic Detector of Electrical Waves, and some of its Applications". He ends with a few points of particular interest, not generally known:

These results were obtained before Marconi began his well-known experiments in England. From experimental data on this subject, I had come to the conclusion that the 'magnetic detector' was inferior in delicacy to the coherer, and in the press of other scientific work, I had not paid much attention to the subject. I have however used for over a year a device very similar to that employed by Marconi in his latest form of receiver—namely, an endless band of steel wire passing through the solenoid in which the electric oscillations are set up....Marconi has apparently applied to the method the principle used by Poulsen for recording telephone messages with very successful results. It is to be hoped that further success will attend his efforts to use the magnetic receiver, which has many advantages over the erratic coherer.

Rutherford and Barnes also obtained wireless signals on a train running at full speed past a station where a transmitter was set up. A coherer was used to complete the circuit and when each signal arrived the current sounded an electric bell. This was done during a summer holiday, part of which was spent with his family in Canada at Kamouraska, on the south side of the St Lawrence river, not far from Rivière du Loup.

RUTHERFORD TO HIS MOTHER

Montreal: 1 Aug. 1902. I am at present engaged in writing a book on "Radioactivity" which I hope to get through in a year's time. You see I have not taken into account Solomon's injunction, "Oh, that mine enemy would write a book." I have been very busy of late with work, but have also managed to take a little relaxation in the way of theatres, etc. Last week we went to hear Mrs Patrick Campbell in "The Second Mrs Tanqueray". It was very well done and was the best thing we have had in Montreal for some time.

At my lecture on "Wireless Telegraphy" I had the largest audience they had ever raised at McGill. They were stored everywhere, including some who were looking through a ventilator in the top of the roof. The experiments went off very well. I have promised to give a lecture before a church society on Wednesday evening on "Atmospheric Electricity" which, however, will be quite a small thing. I dislike the trouble of outside lectures, and try to avoid them as far as I can.

I see New Zealand is again to the fore in the matter of contingents and is quoted here as a model colony.

This was a period of great scientific advance, for the joint work of Rutherford and Soddy, clearly stating the fundamental principles of the theory of radioactivity, appeared in the *Philosophical Magazine* in

Sept. and Nov. 1902. They had discovered that the emanation did not
come directly from the element thorium but from an intermediary
which was called by them thorium X. Rutherford and Soddy studied
their material electrically by its ionisation effects with an electroscope.
They found that thorium X, the immediate parent of the emanation,
decayed away or disappeared at such a rate that half the amount was
gone after four days. But as the thorium X lost its radioactivity the
thorium regained its strength to an equal degree, so that the sum total
of the two remained constant. This result is best illustrated by the rise
and decay curves shown in the diagram. The shape of the curves indicates

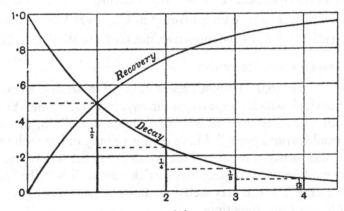

Recovery and decay curves

that the rate of decay is always proportional to the amount present and
this means that the law is one of *chance*, or of unknown internal causes
which cannot be affected by most violent physical and chemical changes.
Mme Curie had already indicated that radioactivity was an atomic
property, since chemical changes had no influence on its radiations.
The observers soon began to talk of the atoms themselves rather than
of the radioactivity due to the rays to which they gave origin. Prof.
H. E. Armstrong was nearer the truth than he thought when he
inquired, perhaps with a touch of scorn, why these atoms were seized
with an 'incurable suicidal mania'.

Rutherford and Soddy accounted for their results by stating, firstly that there was a constant production of thorium X by the thorium; and secondly that the radioactivity of the new matter decreased steadily by an exponential law from the moment of its formation. They extended this rule and showed it to hold for other consecutive radioactive substances. The statement actually made by them in italics reads: "The normal or constant radioactivity possessed by thorium is an equilibrium value, where the rate of increase of radioactivity due to the production of fresh active material is balanced by the rate of decay of radioactivity of that already formed."

Rutherford was a wizard in the measurement of these quantities because of his long apprenticeship with ionisation at Cambridge. He was equally expert with an electrometer or with that simple but tricky instrument, the gold-leaf electroscope. The chemical difficulties were not small, because some preparations of thorium give off their emanation freely, while others have the habit of occluding the emanation. The rise and decay curves were subsequently shown to hold as between thorium X and the emanation; uranium and uranium X; and equally between radium and the emanation from it.

The rate of decay of radium emanation was published by P. Curie in 1902, while Rutherford and Soddy published their result in April 1903. So too Becquerel had published in 1901 the rise and decay curves of uranium and uranium X, but the theory underlying these changes indisputably rests with Rutherford and Soddy. Every subsequent discovery in radioactivity has confirmed the laws they stated in 1902. Those who find more interest in the human rather than the scientific point of view may like to turn to page 342, where are the Rutherford arms, after he was made baron. See how the shield is divided into four by a St Andrew's cross of bent lines. Those are the decay and recovery curves discovered in 1902.

Rutherford and Soddy now directed their attention to the character of the emanations. Sir William Macdonald had presented a liquid air machine to the Physics Laboratory. It was duly installed and, as Rutherford stated, "Canàdian air was for the first time liquefied".

The emanation of thorium, or of radium, was allowed to pass freely through a coil of metal tubing. When this coil was immersed in a Dewar flask containing liquid air, the emanation was condensed in the metal tube! Clearly then the emanations were gases—radioactive gases, something quite new. The fact that they were not altered by chemical treatment, such as bubbling through various weak or strong acids, indicated that they belonged to the class of inert gases, such as argon.

It is of interest that we have a first-hand account of these thrilling days of discovery from Prof. Soddy (*Old McGill*, vol. 36, 1933, p. 19).

Rutherford and his radioactive emanations and active deposits got me before many weeks had elapsed and I abandoned all to follow him. For more than two years scientific life became hectic to a degree rare in the lifetime of an individual, rare perhaps in the lifetime of an institution. The discovery that the emanations were argon gases, followed by that of thorium X as an intermediate product between thorium and the emanation it produces, led rapidly to the complete interpretation of radioactivity as a natural process of spontaneous atomic disintegration and transmutation.

It is difficult amid the practically continuous vistas that were opened by these discoveries to single out the steps. The condensation of the emanations by liquid air, was one; the separation of thorium X was another. I remember how, in grim determination not to be outdone by a mere chemist, and to show him that physical methods were quite as good, Rutherford shook up some thoria with gallons of water till he was tired, then boiled all the water down triumphantly and produced from the residue a minute quantity of the new body. The first magnetic deviation of the alpha-rays by Rutherford, and later their successful electrostatic deviation, were also notable events. I recall seeing him dancing like a dervish and emitting extraordinary imprecations, most probably in the Maori tongue, having inadvertently taken hold of the little deviation chamber before disconnecting the high voltage battery and, under the influence of a power beyond his own, dashing it violently to the ground, so that its beautifully and cunningly made canalization system was strewn all in ruins over the floor.

One Saturday, at lunch time, I left him hard at it in the laboratory over a refractory active deposit that was not behaving itself. When

I re-entered the Campus after lunch he was pacing about in front of the Physics Building with a smile that carried farther than words. As I got within hail I received the inspiring message—"Soddy, the darned thing's going up." It appeared that instead of decaying quietly, as all good active deposits should, this one had elected to grow. It was the first observed instance of delayed growth of activity due to the intervention of a rayless product in the series, which is now so familiar.

It must be remembered that these early experiments at McGill were carried out with exceedingly feeble preparations of radium, only about 1000 times radioactively as strong as uranium, whereas the Curies had prepared radium salts a million times more powerful than uranium. For that reason Rutherford's first attempts at the magnetic deviation of the alpha particles were not successful. Success followed on his obtaining, through the kindness of Madame Curie, a preparation of radium 19,000 times as strong as uranium, and even this material was only one part radium to ninety-nine parts of barium!

On 3 October 1902, J. C. (afterwards Sir John) McLennan wrote to Rutherford from the University of Toronto a graphic and terse account of his experiments at Niagara Falls.

I found that a well insulated wire, extending well up into the Falls, at least 100 yards long, kept the charge up day and night negatively at from 8000 to 10,000 volts. This seemed to indicate that the water, in falling, acquired a negative charge and that the spray gave it up, in passing, to the suspended wire.

McLennan always found a great inspiration in Rutherford and determined to secure for Toronto a Physics Laboratory as good as that at McGill. He succeeded in getting a better, and he also attracted an able group of research students around him. He never secured for them the same large freedom of thought and action which was a chief distinction of Rutherford wherever he went. The two men continued on most friendly terms in peace and war work, for Rutherford appreciated the zeal and determination of McLennan in the face of opposition and difficulties of a formidable character, some of which were perhaps avoidable. When McLennan retired and lived in England, Rutherford and he worked together with enthusiasm in the promotion of a scheme for the treatment of cancer with the gamma radiation from radium.

These early days at McGill were perhaps the most remarkable of Rutherford's life. Here was a young man working at first all alone and almost the only help he got was to be told that he was wrong! Soddy was a staunch ally, and his chief, John Cox, the Director of Physics, was his loyal supporter.

Prof. Norman Shaw has told the story well (*McGill News*, Dec. 1937):

When Rutherford was working on the detection and isolation of the numerous members of the radium family and developing the theory of the disintegration of matter, there were several occasions when colleagues in other departments gravely expressed the fear that the radical ideas about the spontaneous transmutation of matter might bring discredit on McGill University! At one long-remembered open meeting of the McGill Physical Society he was criticised in this way and advised to delay publication and proceed more cautiously. This was said seriously to the man who has probably allowed fewer errors to creep into his writings and found it less necessary to modify what was once announced than any other contemporary writer. At the time, he was distinctly annoyed and his warm reply was not entirely adequate, for in his younger days he sometimes lost his powerful command of ready argument, when faced with unreasonable or uninformed criticism.

Immediately John Cox quietly rose to his support, and gave a clear review of the new ideas. Cox not only revealed incidentally his own grasp of the validity and significance of Rutherford's discoveries, but ended rhetorically with a stirring prediction that the development of radioactivity would bring a renown to McGill University by which in the future it would be widely known abroad. He ventured also to predict that some day Rutherford's experimental work would be rated as the greatest since that of Faraday—an opinion now held by the majority of those competent to judge.

RUTHERFORD TO HIS MOTHER

Montreal: 27 Oct. 1902. I think when I last wrote to you I was going to Toronto on the Grand Trunk to do some experiments in wireless

telegraphy. They came off very well with the train travelling at 60 miles
an hour. We got signals over eight miles from the station. I have been
very busy lately with experiments of all kinds. We have just installed
our liquid air plant. It is going first-class and will produce about a quart
of liquid air per hour. We were very successful in finding that the
radioactive emanations or gases from radium and thorium were condensed by the intense cold by passage through a tube immersed in liquid
air. This was at a temperature of about 200 degrees below freezing point.
The point is valuable as showing that the view I have held all along is a
true one. These gases for a time possess the power of giving out X-rays.
I think I have got several rather good things on hand. I think I have
proved the existence of a type of radiation from the earth which will
go through 100 feet of iron. The result has important bearings on the
problem of the production of electricity in the atmosphere.

CHAPTER IV

EARLY TRIUMPHS

Rutherford was always interested in the alpha rays because he quickly understood that their energy was far greater than that of the beta and gamma rays combined. The magnetic and electrical deflection of these rays was one of his great results. This was published in the *Philosophical Magazine* for Feb. 1903, and the actual experiments were carried out with remarkable ingenuity—even for Rutherford.

He arranged for the alpha rays to pass between a number of parallel plates, not far apart, into an electroscope which detected them. He then applied a magnetic field which caused the alphas to swerve sideways into the plates, and found, of course, that they did not then reach the electroscope. Yes!—but which way did they swerve? To find this out he covered one half of the openings between the tops of the plates with little lids. When the magnetic field was in one direction the alphas were caught by these half-lids, but on reversal of the field they escaped through the openings. Thus Rutherford made quite sure that the alpha particles bent the opposite way to the beta particles; the alphas were therefore positively charged; for the betas, which are swift electrons, are negatively charged.

He then made alternate plates positive and negative electrically and proved that the alpha particles were repelled by the positive plates and attracted by the negative, as was expected. He was able to measure fairly well the amount that the alpha particles were deflected both by his magnetic fields and by his electric fields, which were also measured.

Hence he found, by easy calculations, the velocity of the alpha particles to be about $2 \cdot 5 \times 10^9$ cm. per second, or about 15,000 miles a second; he also found that their electric charge per unit mass was 6000, which figure indicated that the alpha particles were either hydrogen or helium. This courageous experiment was carried out with great

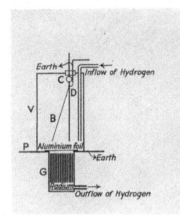

Alpha ray electroscope

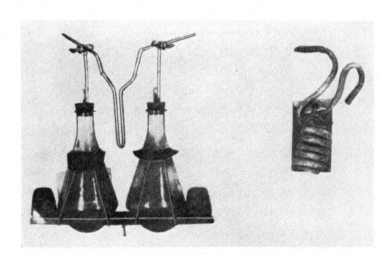

Apparatus for measuring
heating effect of radium

Lead tube for
condensing radium
emanation

difficulty, and for a first attempt was a magnificent achievement. Rutherford also found by experiment a fair estimate of the number of alpha particles which leave a gramme of radium every second—about a tenth of a full billion. This work, done at McGill, is sometimes overlooked, because later and better experiments made at Manchester by Rutherford and others have given far more accurate results.

My own first meeting with Rutherford was at McGill in January 1903. Rutherford was already famous, radium was the rage and the radioactive theory of the break-up of atoms was a topic of supreme interest. Journalists besieged the Physics Laboratory and wrote fabulous and fantastic stories until they were forbidden the sacred precincts. Authoritative accounts of papers and meetings were therefore issued; but, though the statements were sound, they were not generally understood.

At this first interview I realised that Rutherford's eyes had a curious fascination. As you looked at them, you began to understand the saying: "If thine eye be single, thy whole body shall be full of light." I asked whether there were any chance of joining him in October; he said there might be, and it was so. This was rather an impulsive step, but I have never regretted it.

The generosity of Sir William Macdonald was a large factor in Rutherford's success. He had built and equipped the laboratory, endowed his chair and presented special apparatus when it was required. Macdonald was a remarkable man, who gave many millions for various purposes to McGill University. A bachelor, he lived very simply; a merchant prince, he prepared and sold tobacco wholesale, cash before delivery, so that there were no bad debts and the little book-keeping required only a small staff. Smoking he detested as a 'filthy habit'.

He always felt regret that so many of the students left Canada for the States. When he attended a lecture by John Cox with displays of electricity passing through gases, he exclaimed, "How beautiful and how useless!" but later he said that all his expenditure was fully justified by Rutherford's results alone.

As a newcomer, I was working with gamma rays in the basement when Rutherford rushed in breathless. "Open the windows, put away your pipes, hide your tobacco." "All right, but what is the trouble?" "Hurry up! Macdonald is coming round the laboratory." We were all, men and laboratory, dependent on the sale of tobacco, but to use it was a crime!

Macdonald's gift of the liquid air machine led to the condensation of the emanation[1] and the actual tube in which this was effected is much treasured at McGill.

There was a further fruitful gift of five hundred dollars for the purchase of radium and Rutherford bought from Prof. Giesel of Brunswick 100 milligrammes of well purified radium bromide which contained about 75 per cent of the radium element. Rutherford and Soddy had already suggested (Phil. Mag., 1902, p. 582) that helium might be a product of the disintegration elements, because this gas was usually present locked up within radioactive rocks.

Many years afterwards, when he was awarded the Franklin medal in 1924, Rutherford gave an account of this period:

In 1903, Soddy left Montreal to work with Sir William Ramsay on chemical problems. In order to avoid unnecessary overlapping, I had outlined a scheme of future work before Soddy left Montreal. He was to tackle the important problem whether he could detect the growth of radium in purified uranium salts, while, as soon as sufficient radium was available, I was to examine whether helium was produced from radium. In 1903, I spent the summer in England and heard that practically pure radium bromide was being sold commercially by Professor Giesel at the very low price of about one pound per milligram. Ramsay and I both purchased some of this material. I remember well a visit I made to Soddy at University College on the day when Ramsay and he were to collect the emanation from about 20 milligrams of radium to test whether they could detect its spectrum. Soddy told me he would take this opportunity of noting whether any helium was released from the radium. That afternoon, the presence of helium was detected by its

[1] Rutherford and Soddy, Phil. Mag., May 1903.

spectrum. I loaned them my radium to confirm the important discovery. During the summer, I visited Professor and Madame Curie in Paris and found the latter was taking her degree of D.Sc. on the day of my arrival. In the evening, my old friend, Professor Langevin, invited my wife and myself and the Curies and Perrin to dinner. After a very lively evening, we retired about 11 o'clock in the garden, where Professor Curie brought out a tube coated in part with zinc sulphide and containing a large quantity of radium in solution. The luminosity was brilliant in the darkness and it was a splendid finale to an unforgettable day. At that time we could not help observing that the hands of Professor Curie were in a very inflamed and painful state due to exposure to radium rays. This was the first and last occasion I saw Curie. His premature death in a street accident in 1906 was a great loss to science and particularly to the rapidly developing science of radioactivity.

The value of the electric charge per unit mass of the alpha particles, determined by Rutherford, had left it an open question whether the alpha particle was a hydrogen or a helium atom. It had not then even been surmised that lead was the final product of radium, but all was in a state of uncertainty, so that the experiments of Ramsay and Soddy were awaited with interest.

In the first experiment they dissolved in water 30 mg. of radium bromide, prepared three months before, and found helium among the collected gases, which showed the characteristic line D_3 of helium.

In the second experiment, they used the radium bromide that Rutherford had lent them for this very purpose and after removal of oxygen and hydrogen, and the condensation of the emanation and carbon dioxide with liquid air, the remaining gas gave a practically complete spectrum of helium. This indicated that helium was a disintegrative product of radium, and increased the probability that alpha particles were helium, but it did not decide the question. Further evidence was however forthcoming that the helium was derived from radium emanation itself, which, newly separated, showed at first no helium but after four days gave the spectra of emanation and of helium.

On 13 July 1903, Soddy wrote to Rutherford: "You will have received our wire and heard the result of the second experiment with your radium." The growth of helium from radium emanation was observed by Ramsay and by Soddy, who added that "it was a veritable triumph and we are very grateful to you for making it possible".

This year the Cambridge University Press published *The Conduction of Electricity through Gases*, by J. J. Thomson. This book collected together all the wonderful work that had been carried out at the Cavendish Laboratory and elsewhere on ionisation, electrons, spark and arc discharges, and kindred subjects. It was a book constantly by our sides and contained many references to Rutherford's discoveries at Cambridge and Montreal.

The Secretary of the Royal Society wrote (3 April) that he was glad to learn that Rutherford was coming to England in May. "You may be the lion of the season for the newspapers have become radioactive. I see that you are again monopolising most of the Phil. Mag.!"

THOMSON TO RUTHERFORD

14 *April* 1903. I sail for New York on May 1st, so that I am afraid I shall not be in England when you arrive. I expect however to be back in England by the third week in June. I hope to find you still on this side and that you and Mrs Rutherford will be able to come and make a long stay with us. My wife had a little daughter about a month ago. I am glad to say both got on exceedingly well. Miss Brooks has been getting some very interesting results with thorium, she finds that the emanation is able to produce induced radioactivity long after it has lost its power of ionising the gas around it; this, as you showed, becomes inappreciable after a few minutes, the power of producing induced radioactivity lasts for about two days. One of the men in the Laboratory recently tried whether a mass of thorium was hotter than the surrounding air, he found that it was so by about $1/300°$ C. I have found a radioactive gas in Cambridge water, or rather in that part of the water which comes from deep wells. Dewar liquefied for me the gas extracted from the water, then we allowed the liquid to evaporate, collected the gas that first evaporated, and then separated the gas from the last portion of liquid—the latter was extraordinarily radioactive, while the former was

much less so than before liquefaction. We have got photographs of the spectra but have not yet measured them up. Hoping to see you.

In 1903, Harper Bros. of New York made proposals for an article on radioactivity to be written by Rutherford for their well-known magazine. Their letter contained a warning, to which Rutherford usually paid heed: "It is hardly necessary to suggest to you that our readers do not generally possess special knowledge of the subject you are treating, for which reason it is desirable to use as few technical terms as possible." Recent rapid advances in science have made this warning of greater importance, and at the same time rendered its observance more difficult. Rutherford wrote a good clear statement as requested, although he was exceedingly busy with his first book *Radio-activity*. In 1903, then 32 years old, he received the honour of election as a Fellow of the Royal Society of London.

H. L. Cooke was recommended by Rutherford to detect the gamma rays from the radium everywhere present in minute quantities in the earth. Cooke made a sensitive electroscope and found but little difference when he screened it with some lead. "Try more lead", said Rutherford. So Cooke piled up lead on all sides and found that the penetrating radiation was indeed present—only to create a new difficulty, for the radiation came from above, as well as from below. The exact words of Cooke in the *Phil. Mag.*, No. 6, 1903, p. 403 are:

Experiment showed that the amount of penetrating radiation was approximately the same in the laboratories, on the roof, and on the ground outside the building; in fact this radiation seems to be present everywhere with practically uniform intensity.

This report indicated that there must be radioactive matter in the atmosphere, which had indeed been detected by Elster and Geitel (1901).

It must not however be thought that these experiments dealt with 'cosmic' rays to any great extent, for Prof. Blackett states that the cosmic rays are but ten per cent of the gamma rays on the earth's surface, as measured by the ionisation they produce.

In the summer of 1903, Mrs Newton came to England and was joined by Prof. and Mrs Rutherford and their little daughter Eileen. They spent a month in Switzerland and some time in North Wales, where the rain was so constant that they moved to Chester, and here they were joined by Prof. Laing of Canterbury College, New Zealand, and by Mrs Rutherford's brother, Charles Newton. Rutherford spent a busy summer correcting the proofs of his book.

Walter Vaughan, Bursar and Secretary of McGill University, like the Principal, Sir William Peterson, was a warm supporter of Rutherford and his work. He wrote from Lake Placid, New York, where he had gone for a year of quiet and rest, suffering from tuberculosis:

VAUGHAN TO RUTHERFORD

11 *Aug.* 1903. I have heard lots of your trip from Dr Peterson and Cox, and give you all sorts of cordial congratulations, especially on the acceptance of your theories—which is a triumph. More power to your elbow! I am sure you must have got a great deal of pleasure and delight from the visits you have been making to your co-workers. But I almost wish there were less cause for congratulations, for I fear that some of these 'foreign devils' will be taking you up to the top of a high mountain, and showing you all the kingdoms of the earth—and you will succumb to the temptation. And thus we shall lose you!

You may have heard that I am hors de combat—off duty for a year. I see that Soddy is suggesting a trial of the inhalation of radium gas for tuberculosis. How would you like to experiment on me? I am willing to be a martyr to science if you can give the gas a tobacco flavour.

Physiologists wrote letters to Rutherford on the interesting effects produced when radium is brought near the eye. In a darkened room a rested eye, even with the lid closed, receives a sensation of light, no doubt due to the radiation going through the eyeball and affecting the retina.

It was stated that while ordinary daylight swiftly bleaches the pigment known as visual purple, yet that twenty hours' exposure to 50 mg. of radium bromide at 2 mm. distance, acting through thin mica, did not bleach the pigment at all.

J. J. Thomson wrote with reference to an article in *Nature* (10 Dec. 1903) by Rutherford on the Cavendish Laboratory:

It seems excellent although the part of it where you speak of my work is too appreciative. We are very sorry to hear that the weather was so shocking in Wales, it can be bad there as I know to my cost....I hope you will have a good time at the British Association; Lodge is to be there. Langevin came over to Cambridge for the presentation of the portrait. I was delighted to see him and to find how little he had changed.

At the Southport Meeting of the British Association a discussion was opened by Rutherford, before a crowded sectional meeting, on the emanations from radioactive substances. In his speech he summarised all the evidence available, and he also referred to Sir William Crookes's instrument, the spinthariscope, where individual alpha particles from polonium are seen to strike a fluorescent screen—"the first time probably that we have observed any single atom effect." He thought that the energy of radium came from the *inside* of an atom. He concluded by stating that "our theory explains everything that we know at present concerning the phenomena". Sir Oliver Lodge then congratulated Rutherford and supported his views, after which he communicated a written statement from Lord Kelvin.

Kelvin thought that radium received its energy "by absorption of ethereal waves", that while beta rays were electrons, the gamma rays were simply vapour of radium, and the alpha rays were atoms of radium or radium bromide! It is almost ludicrous today to put such views into print. However Prof. H. E. Armstrong expressed himself "astonished at the feats of imagination to which he had listened", yet he leaned towards Lord Kelvin's view of an "external source of energy" and added that "chemists certainly had no evidence of atomic disintegration on the earth". Rutherford in reply adhered to the internal, atomic energy source, and pointed out that the heat generation was proportional not to the mass of salt present, but to the amount of emanation.

By this time (1903) Rutherford had settled the "genealogical tree" of the first five members of the radium family. Radium was the father of radium emanation, who begat radium A, who begat radium B, who begat radium C. Then he met with a poser, which took him a year to solve. In the figure it may be seen that after the alpha particle leaves the radium atom that which remains is the emanation. Three successive products shoot out alphas, and as for radium C it is a question of fire-works, for there are all three radiations (alpha, beta and gamma). Let us concentrate our attention on radium B, which Rutherford found to be 'rayless', although beta rays have since been found. Here is a most remarkable case of a man who confidently demanded the existence of a substance—of an element—for which no physical or chemical evidence was forthcoming, except that the laws of radioac-tivity, as laid down in 1902, proved that radium A did not turn directly into radium C but that there must be an intermediate body. In the same way, many physicists demand the existence of an ether to con-vey electromagnetic radiation and waves, even when they find them-selves unable to prove its existence, in spite of the rapid motion of the

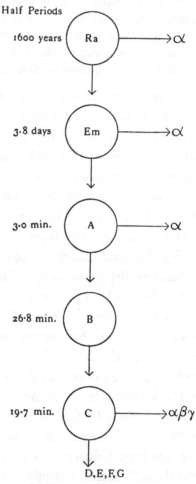

Radium transformations

earth through the medium. As Lord Salisbury said to the British Associa-tion at Oxford: "The aether is the nominative case of the verb to undulate."

Long afterwards, I recalled to Rutherford the boldness of his creation of radium B and he was himself impressed: "That's quite a point", he remarked.

When Soddy reached England he was full of enthusiasm and gave a series of public lectures on radioactivity which were fully recorded in *The Electrician* and it was feared that these might damage the sale of Rutherford's book. Soddy also undertook to write a book on 'Radioactivity' and, although he proposed to emphasise the chemical side of the subject, there was much material which he had obtained from his colleague at McGill at lectures and otherwise. There is no doubt that Rutherford felt strongly on this question and he wrote a firm letter expressing his point of view. It was finally agreed that Soddy's book should not appear until one or two months after the publication of Rutherford's.

Soddy freely and frankly acknowledged his debt to Rutherford in this statement in the preface to his book: "This task could hardly have been possible but for the training I have received in the subject from Professor Rutherford."

It says a good deal for both men that their friendship survived this strain and that Rutherford supported Soddy in his subsequent applications for appointments, in his election as a Fellow of the Royal Society and for his Nobel Laureate (1921), as Soddy gratefully admitted.

On 30 Dec. 1903, the 'feature of the day' at the American Association for the Advancement of Science at St Louis, Missouri, was a lecture by Rutherford on radium. Flasks of liquid air were sent by special mail from Washington so that the condensation of the radium emanation could be shown to a large audience. He successfully showed this experiment and explained that a pound of the emanation, if it could be procured, would initially pour out energy at a rate equivalent to 10,000 horse-power. He spoke in a guarded manner of the future possibilities of radium as a curative agent in special cases of cancer, and he touched on the prolongation of life on the earth owing to the heating effect of the radium content.

Early in the year a letter arrived from Prof. J. Larmor, the Secretary of the Royal Society, conveying an invitation to Rutherford to deliver the Bakerian Lecture. The same day there arrived an interesting letter from J. J. Thomson, who wrote:

It is exceedingly kind of you to dedicate your book to me, nothing could give me greater pleasure. . . . I understand that Soddy's lectures in *The Electrician* were the lectures he gave in London and Cambridge. . . . I don't think you have any reason to be afraid that people will be misled. . .it may however interfere with the sale of the book a little, though I should not expect to any serious extent. . . ."

The letter goes on to discuss the experiments relating to the detection of helium in radium emanation, stating "I always when speaking on the subject point out how you had foreseen and indeed planned the experiment and that if it had not been for you it would never have been made. . . ."

Another letter came to Rutherford at McGill which gave him the greatest pleasure. It was written by W. H. (now Sir William) Bragg at Adelaide explaining how, working with Kleeman, he had made a testing chamber gradually approach a thin layer of radium and had thereby discovered that the alpha particles from radium and its subsequent products had different but quite definite ranges in air. Thus radium C shot forth alpha particles with about twice the range of those from radium itself, indicating a much higher explosive energy for the former. It thus became possible to identify a radioactive element by the length of the range of its alpha particles. It also became apparent that products with a long life had short ranges, and that quick changes meant long ranges.

Rutherford found that the length of these ranges could be measured equally well by ionisation, fluorescence or photography, and that alpha particles from radium C lost the power of producing these effects when the energy fell below 60 per cent of its initial value.

Larmor, writing (29 March 1904) to Rutherford from St Helen's Bay, 'Belfast, raised the interesting and difficult question why, after the expulsion of the positive alpha particle, the residual atom should not be negatively charged. The opposite is actually the case. For example, when a radium emanation atom (Rn) ejects an alpha particle, the residual atom (radium A) has a positive charge and moves towards a negatively charged wire. The alpha particle not only loses its own satellite electrons but, directly or through recoil, tears away one or more of the outer electrons of the residual atom. The fact is indisputable, however halting the explanation.

About 1904 I was asked by Rutherford to make a sensitive, small-capacity electroscope the gold leaf of which would remain charged for two or three days. This I failed to do. So Rutherford said: "Lester Cooke used to make them; why can't you? Get Jost the mechanic to make you one!" So I went to Jost and repeated this. He said: "If I could not make a better electroscope than Cooke, I'd shoot myself!" So he made a beauty to look at, but a bad one to go. Its leaf collapsed in twelve hours. This puzzled me. One night I could not sleep and got up in my diggings and made an electroscope of a tobacco tin, an amber mouthpiece of a tobacco pipe, and some Dutch metal foil; charged it with sealing wax and went to sleep. The leaf of this home-made freak electroscope remained open and charged for three days, and solved the problem. An electroscope made of material outside the laboratory would remain charged for 48 hours inside the laboratory, but all the material inside the building was contaminated and coated with active deposit including the slow period transformations of radium. Rutherford said "Good boy!" though I was eight or nine years his senior in age. Precautions were then taken to prevent the escape of radon which had been at the root of the mischief, hitherto unsuspected.

Rutherford's book on radioactivity was pressed forward with all possible speed in order to forestall other authors who were writing on the subject both in England and abroad. W. C. D. Whetham revised the proofs and McClung made the index. The following extracts are from Whetham's letters from Upwater Lodge, Cambridge. The first deals with an experiment of Ramsay and Soddy who passed sparks through radium emanation in a sealed tube and observed its spectrum.

As this gas diminishes always to half-value in little less than four days, at the end of twenty-eight days there would remain but one part in seven twos multiplied together, or one part in 128. They did in fact find it disappear from observation, while helium increased.

WHETHAM TO RUTHERFORD

26 *July* 1903. Many thanks for your papers which will be most useful. I am sending you two little ones of mine which I do not think I have sent you before.

I am writing an article on radioactivity for the autumn number of *The Quarterly Review*, and shall be glad to have your papers before me.

In my first rough draft I mention your playful suggestion that, could a proper detonator be found, it was just conceivable that a wave of atomic disintegration might be started through matter, which would indeed make this old world vanish in smoke.

I don't know if you care to be thus reported in print, as I do not think you have published your views about the end of the world, but I make it quite clear that it was only a speculation, but, unless you give me permission to let it stay, I shall cut it out in my revised copy....

13 *Feb.* 1904. That is a pretty work of Ramsay and Soddy on the vanishing of the emanation, which you describe in your new manuscript. I have not heard of it before.

We were very much interested to hear you had been trying to detect N rays. Burke tried in the Cavendish Laboratory, also with negative results. Is it a physiological effect, or did Blondlot dream it?

In the first letter Whetham refers to a joke of Rutherford's that "some fool in a laboratory might blow up the universe unawares". However, most transmutations require the expenditure of more energy than they liberate. Not so the uranium-radium and thorium families, which are still liberating an energy which was built up possibly in the sun, long ago before the earth was broken away from it by the near approach of another star.

As for the N rays, they were discovered by Blondlot in France. Many men in many laboratories tried to repeat his experiments without success. The rays were not a hoax, but a mare's nest—unique in science.

Prof. R. W. Wood of Johns Hopkins University was persuaded to visit Blondlot and he ingeniously proved that the experiments and deductions had no relationship. He published a letter of his experience in *Nature*. Since that day there have been no N rays!

Rutherford's first book, *Radio-activity*, was issued by the Cambridge University Press early in 1904 and was dedicated to J. J. Thomson. Colleagues and reviewers alike received it with enthusiasm and perhaps the best brief criticism was that "Rutherford had not only extended the boundaries of knowledge on this subject, but had annexed a whole new province".

The interesting question arose whether radiation could evoke radiation. The answer was No. Radium emanation continued to transform at the same rate with lordly indifference as to whether it was condensed by liquid air on to a very small area, or whether the same amount of emanation was spread throughout a large container. Extreme heat, low temperature and enormous pressure had no detectable influence on the life history of the emanations; "in this respect the changes in radioactive matter are sharply distinguished from ordinary chemical changes." (*Radio-activity*, p. 191.)

Sometimes experiments are made which are never published; thus Rutherford and Douglas McIntosh—a clever chemist then at McGill—extracted some helium from the mineral thorianite, mixed it with radium emanation in a small discharge tube and sparked the mixture vigorously, using an induction coil, on the off-chance that the radium emanation and the helium would combine *atomically* to produce radium. This was not expected to happen and it did not take place, nor was it thought worth while to publish an attempt at a result so contrary to expectation.

Godlewski—a clever and charming Pole—had come from Lemberg to work with Rutherford at McGill. In the early days of radioactivity we tried, without much hope, to deflect the gamma rays of radium in a very powerful magnetic field. As expected, it was a null experiment, but Godlewski thought there might be a better chance with the softer

rays of actinium. One morning he showed me his photographic plate
with two distinct lines half-an-inch long, branching like the two horns
of an antelope. He had used a magnetic field and reversed it. He was
dancing with joy and greeted Rutherford: "I have completely deflected
the gamma rays of actinium." Rutherford glanced at the plate. "Do
it again", he said with a smile. "Certainly, I will do it at once", replied
Godlewski; but he tried week after week without a shadow of success,
and it may well be wondered what malignant sprite had placed the
flaw in just the very place to delude the enthusiastic Godlewski. Alas!
he, a physical chemist, died in Lemberg the victim of a slow and
undetected escape of gas containing carbon monoxide.

By this time Rutherford had become a good lecturer and Prof.
A. Norman Shaw has done good service in preserving for us a vivid
picture (*McGill News*, Winter Number, 1937).

Witness the reactions of the eloquent John MacNaughton, Professor
of Classics, who had shot many a satiric dart at scientists and their
doings—'plumbers' and 'destroyers of art' he has sometimes called
them—but Rutherford seems to have captured him in one lecture. He
wrote as follows in the *McGill University Magazine* for April 1904:
"...We paid our visit to the Physical Society. Fortune favoured us
beyond our deserts. We found that we had stumbled in upon one of
Dr Rutherford's brilliant demonstrations of radium. It was indeed an
eye-opener. The lecturer seemed himself like a large piece of the expensive
and marvellous substance he was describing. Radioactive is the one
sufficient term to characterize the total impression made upon us by his
personality. Emanations of light and energy, swift and penetrating,
cathode-rays strong enough to pierce a brick wall, or the head of a
professor of Literature, appeared to sparkle and coruscate from him all
over in sheaves. Here was the rarest and most refreshing spectacle—the
pure ardour of the chase, a man quite possessed by a noble work and
altogether happy in it."

In May 1904, Rutherford gave the Bakerian Lecture to the Royal
Society in which he not only stated the theory of radioactivity but
applied it with convincing evidence to the chain of radioactive products

beginning with radium. He laid down those fundamental principles which have since been found to hold good for over forty elements of the radioactive families, nor has it been necessary to add, alter or detract anything relating to the precision of his main statements. For this great achievement he was awarded the Rumford Medal in 1904.

In order to understand the principles involved in a chain of radioactive changes, it is necessary to remember that at every instant each number of the series is obeying the elementary rule that the loss of its population of atoms per second is proportional to the number present. A similar rule would hold for human beings, if all died from violence, and natural death were eliminated.

Evidently there is a proportionality factor involved which is called the 'radioactive constant'. Each sort of radio-element has a different constant of this kind, whose reciprocal is the 'average life', which is about 44 per cent higher than the 'period to half-value'. The *loss* of an atom of any one group in the chain is the *gain* to the atoms of the next group and the rate of change is the same all the way down the line. Every time a baby dies, an angel is born! Hence in a very old radioactive ore body, which has had time to produce a steady or equilibrium state in the chain of radio-elements, there will be a large population of slow changing atoms of any one kind, and a smaller population of faster changing atoms of another kind. If it takes one minute to glide down a hill, and thirty minutes to walk up, there will be thirty times as many people walking up as gliding down, provided there are very many of them. So too if there are three million atoms of uranium to every one atom of radium in an old mineral, the uranium atoms on the average last three million times as long as the radium atoms. If the half-period of radium is 2000 years, then the half-period of uranium is three million times as great or 6000 million years—so uranium is quite a Methuselah among the atoms, but it decays away, none the less. The fact that it has not all gone means that we live in a universe of very considerable but not unlimited antiquity.

Atoms do not grow old, not the radioactive ones. They are as likely to explode in youth as in old age. All that can be stated is that they have an average period of gestation and then an alpha or beta is born and the old mother becomes instantly an entirely new woman with distinctly new characteristics. Consider radium, each of whose atoms, after an average life of 2300 years, gives birth to a helium atom and the remainder is the radioactive gas, the emanation, which after five-and-a-half days' labour, produces another helium atom and the residue is yet a new atom, radium A, with but a few minutes' life! And so the extraordinary story continues until radium G is reached, which from any point of view seems to be a truly stable atom. Perhaps it is possible to venerate the deliberate uranium atom which after a quiescent life, averaging 6000 million years, decides to burst into activity and produce a genealogical tree with fourteen successive heirs. And when the word 'decides' is written, comes the question, Who decides? or What decides?

Rutherford also gave a lecture to a full audience at the Royal Institution dealing with the energy of radium, the heat from which had been determined by Curie and Laborde (1903) and, for different products and radiations, by Rutherford and Barnes (*Phil. Mag.* Feb. 1904). He was prepared to state at this lecture the effect which radium had in prolonging the heat of the earth, and thus vitiating the forecast of Lord Kelvin, which limited the past and future age of the earth to a few hundred million years.

Sir Archibald Geikie has written of Lord Kelvin (*Proc. Roy. Soc. A.*, 1908, 81, p. 70):

In 1862 he published *The Doctrine of Uniformity in Geology briefly rejected.* Again in 1868 he returned to the attack and brought forward additional lines of argument in support of his charge that British popular geology stood 'in direct opposition to the principles of natural philosophy' and required 'a great reform'....In his latest writings he restricted the time [age of the earth] to between twenty and forty million years. His papers have given rise to a long controversy and no final agreement has been reached.

Kelvin's conclusions were based on the laws of cooling of a body losing heat by radiation. Rutherford pointed out that there was an *internal* source supplying heat; and therefore the earth was not simply a cooling mass, and Kelvin's results did not hold.

The actual words written by Lord Kelvin about the duration of the heat of the *sun* are given in Appendix E of Thomson and Tait's *Natural Philosophy*:

It seems on the whole probable that the sun has not illuminated the earth for 100,000,000 years and almost certain that he has not done so for 500,000,000 years. As for the future, we may say, with equal certainty, that inhabitants of the earth cannot continue to enjoy the light and heat essential to their life for many years longer, unless sources now no longer known to us are prepared in the great storehouse of creation.

A certain journal welcomed with large headlines the announcement by Rutherford that the earth would endure many millions of years longer than Kelvin had estimated, and wrote "DOOMSDAY POSTPONED".

As for Rutherford, he used to tell humorous stories about this lecture long afterwards:

I came into the room, which was half dark, and presently spotted Lord Kelvin in the audience and realised that I was in for trouble at the last part of my speech dealing with the age of the earth, where my views conflicted with his. To my relief, Kelvin fell fast asleep, but as I came to the important point, I saw the old bird sit up, open an eye and cock a baleful glance at me! Then a sudden inspiration came, and I said Lord Kelvin had limited the age of the earth, *provided no new source was discovered*. That prophetic utterance refers to what we are now considering tonight, radium! Behold! the old boy beamed upon me.

About this time Rutherford, walking in the Campus with a small black rock in his hand, met the Professor of Geology; "Adams", he said, "how old is the earth supposed to be?" The answer was that various methods lead to an estimate of 100 million years. "I *know*", said Rutherford quietly, "that this piece of pitchblende is 700 million years old."

This was the first occasion when so large a value was given, based too on evidence of a reliable character; for Rutherford had determined the amount of uranium and radium in the rock, calculated the annual output of alpha particles, was confident that these were helium, measured the amount of helium in the rock and by simple division found the period during which the rock had existed in a compacted form. He was the pioneer in this method and his large value surprised and delighted both geologists and biologists.

RUTHERFORD TO HIS WIFE

Cambridge: 4 *May* 1904. I have had a very busy week trying to get my papers in order and preparing for my lectures. I came up to Cambridge yesterday and stay till Tuesday. The Thomsons are well but the two kiddies have whooping cough. . . .They are giving a Cavendish Dinner for me here Monday night. . . .I have a round of entertainments next week. Tonight, Monday, I go to dine at the Fellows Table at Trinity. Tuesday with Lockyer, Wednesday smoker with Lockyer. Thursday, Bakerian and dinner at the Royal Society Club with J. J. afterwards; Friday, Royal Institution and reception, I expect, afterwards at Dewar's house for half an hour. Then up to Terling Place on Saturday. I got your letter re convocation and am glad you are having a gay time. Charlie says he is coming down for the lecture at the Royal Institution on Friday and will return Sunday. I am in such an awful rush that I will be glad when the lectures are over.

Terling Place, Essex: 22 *May.* . . .I came down here yesterday with Mr and Mrs Kempe and found Professor Schuster and his wife on the same train. Lord and Lady Kelvin came down in the morning and there are in addition Campbell Swinton and Mallock. . . .Lord Rayleigh is here, Lady Rayleigh and young Strutt. Lady Rayleigh is a sister of Balfour. I played a game of bridge last night but have not yet had any opportunity to talk to her. Kelvin hankers about information on every subject which it would take days to go over the elements of. . . .I am feeling comparatively jubilant as I have done my two lectures. The Royal Society was crowded and fearfully hot, but I got through all right. My throat got bad for five minutes early in the lecture but recovered later. J. J. came down, Lodge and Kelvin, and generally

there was a very good audience. J. J. and Lodge made very complimentary remarks about it and I think people were satisfied enough.
The Royal Institution lecture went off pretty well I think, at least people
say so. I had about 800 people—the biggest audience I believe for this
year if not for five years. The experiment went off flying without a
break and I got through exactly in the hour....I am continuing this
later in the evening. I think most people were pretty pleased and the
newspapers gave me a great send off. Charlie came down for the
lecture....The Bakerian Lecture was I think a success at any rate J. J.
seemed to think so. Lord Kelvin has talked radium most of the day,
and I admire his confidence in talking about a subject of which he has
taken the trouble to learn so little. I showed him and the ladies some
experiments this evening, and he was tremendously delighted and has
gone to bed happy with a few small phosphorescent things I gave him.
He won't listen to my views on radium, but Strutt gives him a year to
change his mind. In fact they placed a bet to that effect last evening.
Lord Rayleigh believes, I think, in my views as far as he knows the
details. I did not mention that J. J. and Mrs J. J. looked after me very
well and gave me a dinner with a few Cavendish people on Monday
night. I have decided to return to New York by the *Cedric* sailing
June 1. She will be an eight or nine days boat, so don't expect me very
early.

In 1903, Lord Kelvin clearly belonged to the school that believed
that the energy expelled by radium was received from some external
source. With considerable courage, he abandoned his theory publicly
at the 1904 British Association Meeting. He also paid a bet of five
shillings to R. J. Strutt, now Lord Rayleigh, who had a few months
previously backed the view that Kelvin would shortly fall in line with
Rutherford's ideas.

In June 1904, Rutherford returned from England in the *Cedric* and
landed in New York with a very small box, encased in lead, which
caused great consternation to the customs official, for it contained about
30 milligrammes of radium bromide. There was nothing in the customs
laws relating to radium and so it was proposed to send the box to the
appraisement office for a ruling by the Treasury. But the "doctor was

by no means willing to part with his treasure" and finally he bonded it through to Canada where no difficulties occurred. There was an added insult because the New York paper described Rutherford as a Professor of Physics at McGill University, Toronto, which is much like saying that the Cavendish Laboratory is in Oxford.

By this time, Rutherford had become, as it were, the clearing house for all discoveries in radioactivity and letters reached him from professors and research students from many parts of the world. He influenced their work and his mind reacted to their discoveries and ideas. For this reason an adequate account of Rutherford must include some letters written to him as well as those by him.

Prof. B. B. Boltwood of Yale wrote many letters of the greatest interest to Rutherford. Some day these should all be printed in full, with Rutherford's replies. All that can now be done is to give a few short extracts from them:

BOLTWOOD TO RUTHERFORD

11 *May* 1904. You will perhaps recall meeting me when you were here in New Haven and talking over the question of the determination of the relative proportions of uranium and radium in natural minerals, a subject on which I was at that time working. Within the last ten days I have succeeded in obtaining a series of results which seems to throw considerable light on that question and leads to conclusions which seem to me of very great importance.

He was in fact finding, with six different minerals from various countries, that the ratio of uranium to radium was the same (about three million to one). He expressed high hopes of growing radium from uranium, but like Soddy he was disappointed, because it was later found that there was an intermediate product of a fairly long life.

8 *Aug.* 1904. The receipt of your letter of June 20 afforded me very great pleasure and I was very glad to learn that you found the results obtained with the uranium mineral interesting.... The thing which started me on this matter was something you said while in my laboratory about the very possible interrelationship between all the different radioactive

substances, and I have been thinking how it would simplify matters if only thorium, which seems to occur so persistently with uranium minerals, could be fitted in with the uranium and radium....

This idea proved erroneous however.

Next Prof. H. A. Bumstead of Yale University made an interesting discovery. Elster and Geitel had shown that an insulated wire with a negative charge, exposed in the atmosphere, collected the active deposit of radium due to the radium emanation (radon), present in minute quantities in the air, which leaks out of the ground from the radium present in all sedimentary and igneous rocks to an amount of perhaps one part by weight in a million-million, or thereabouts.

Now Bumstead exposed such a charged wire for about half-a-day and collected not only the active deposit of radium, which decays away in a few hours, but also the active deposit of thorium. This substance decays to half-value in about $11\frac{1}{2}$ hours, so that the thorium active deposit, collected from the air and due to the thoron in it (leaking from the thorium in the ground) can be observed slowly decreasing in value for two or three days.

Bumstead was a fine physicist and was particularly successful in suggesting fruitful research work to graduate students.

BUMSTEAD TO RUTHERFORD

5 June. I got hold of your new book yesterday and have read most of it, last night and this morning. I want to offer you my warm congratulations on the admirable clearness of the language, the good arrangement and the great logical force of your presentation of the theory. The collection of the evidence in this way is very impressive and unless I am prejudiced, the future historian of physics is going to think a lot of this book as a most brilliant example of the application of the true scientific method to a perplexing problem. It is going to be one of the classics.

For that reason and also on personal grounds, I was immensely pleased at the kind way in which you spoke of Wheeler's and my little contribution; it showed a most Christian spirit on your part too, and I hereby resolve never in the future to accuse you of having leaky gas holders or any such thing!

My experiments on atmospheric radioactivity were finished a month or so ago and since then I have been getting ready to go away. The paper will be published in Am. Journal for July and I shall have a copy sent you. It turned out very much as you prophesy in the book; and the second constituent (besides the radium excited activity) on a negative charged wire is due to our old friend thorium. With a three hour exposure about 3 or 4 per cent of the initial activity is due to it and with a long wire it is easy to get its rate of decay, etc., after the radium effect has disappeared. With a twelve hour exposure it sometimes reaches about 15%. The combination gives a nearly exponential fall for the first two hours with the exception of the initial rapid drop due to radium A which I can always get if I am quick enough. There is some evidence that there is a small quantity of something else present but it will have to decay quicker than Debierne's results for actinium excited activity.

Boltwood tells me that he wrote you about his uranium-radium results. I notice in a recent *Nature* that Soddy is inclined to give up uranium as the parent of radium, but in spite of this and Strutt's results, I think Boltwood's results are correct. This, because I know he is very skilful and careful, and because I have seen a good many of the experiments done. If I am not greatly mistaken, his results will stand and will thus confirm another one of your predictions.

It was on Bumstead's suggestion that H. L. Bronson went as a Demonstrator to McGill where he proved himself conscientious and successful. He was quickly caught in the Rutherford vortex and busied himself with an improved form of electrometer, which took so much time to make that Rutherford became impatient and said that Bronson thought more about apparatus than about physics. When Bronson achieved his purpose and measured the half-periods of several radio-elements with accuracy, Rutherford withdrew this, and was full of praise for his work. Bronson also made standard high resistances, using the ionisation of air by minute traces of radium bromide as his low conductor. He became Professor of Physics at Dalhousie University.

On 13 Sept. 1904 A. J. Balfour expressed to Rutherford "the very great pleasure with which I have received your letter of September 1st. Your own name will for all time be associated with growth in our conceptions of the physical universe, and it is with the utmost satisfaction that I learn of your approval of my attempt to deal from the outside, with the problems to whose solution you have so greatly contributed." It is noteworthy that this philosopher quickly understood the importance of the work of the physicist.

Between the first (1904) and second (1905) editions of *Radio-activity* Rutherford had worked out the interesting and difficult problem of the later products of radium—those following radium C. He took thin foils of six different kinds of metal and exposed them for a week to a large amount of radium emanation, placing all together in an air-tight tube. After two days, when the short-lived products had decayed away, he tested his foils for active deposits with an electroscope sensitive to their alpha rays. The activity steadily *increased* for 200 days! On the other hand the beta rays remained constant after the active deposit was one month old. Here was a mystery which at the time seemed to me insoluble. Rutherford followed every clue and never lost the scent, and his conclusions (2nd ed., p. 403) were these: Radium C turns into radium D, a rayless product which is half-transformed in 40 years; next comes radium E with beta rays, and a half-period of six days; whence radium F, which emits only alpha rays, is half-transformed in 143 days, and is identical with Mme Curie's first discovery—polonium. These were Rutherford's findings more than thirty years ago, and the current values today after so many subsequent investigations, are these:

> radium D, about 25 years, beta rays,
> radium E, 5 days,   beta rays,
> radium F, 136·3 days,   alpha rays.

The substance radium D, with a life of many years' duration, must be fairly abundant in radioactive ores and indeed it has been discovered and named 'radio-lead'. This explains the letter which Rutherford wrote to Professors Meyer and Schweidler of the University of Vienna:

5 *Nov.* 1904. I have read with the greatest interest the reprints of your valuable papers which you have so kindly forwarded to me. The results in your last paper on radio-lead have been of especial interest to me and I congratulate you on your proofs of the identity of the products of radio-lead with those of the active deposit of radium. There can be no doubt of their connection. I was glad to see you had found the six day period (rad. E) for your palladium wire. I have much pleasure in forwarding you some copies of my recent publications.

Next month there came a cheery letter from his old friend G. Elliot Smith from the Turf Club at Cairo:

6 *Dec.* Many thanks for the newspapers containing the eulogy of your noble self. I had seen the announcement some weeks ago in *Nature* and had the noble resolve of writing to congratulate you on this latest emphasising of your great fame.

It is a great record for a babe like you—you are I believe two weeks younger than I—to achieve such greatness.

May you continue to thrive and do great things. But I hope that in your upward path to greatness and glory you will not forget those of your old pals who are condemned to struggle through life as mere nonentities.

We wish you and Mrs Rutherford the jolliest of Christmases.

One day there came from Brazil to see Rutherford a huge man with a vast voice, so that Rutherford looked a dwarf and was almost inaudible beside him. The man had a curious story to boom. German ships for some time past had been landing goods in Brazil and asking permission to sail home in ballast taken from a certain cove. It turned out that the ballast that they thus got free of charge was monazite-sand, containing thorium needed for the mantles of Auer or Welsbach lights. It was worth at least a pound a ton! The Brazilian Government had sent the big man to McGill to find out all about thorium. He stayed with Rutherford, who used to come quite tired to the laboratory, saying, "I can't hear myself speak in my own house!"

The International Congress of Arts and Science at the Universal Exposition, St Louis, lasted 19–25 Sept. It was truly a gigantic affair organised by men like Nicholas Murray Butler, Simon Newcomb and Hugo Muensterberg. There were two eminent lecturers speaking on every subject under the sun from Religious Education to Demography, and Rutherford's share was a lecture on the Physics of the Electron, a title which was changed to Radioactive Transmutations.

Rutherford also participated in the International Electrical Congress at St Louis organised by Elihu Thomson, President, and A. E. Kennelly, General Secretary. He told the Congress the remarkable story of radium D, E, F.

In 1904 Mrs Newton travelled from England to Montreal and later went home to New Zealand, accompanied by Mrs Rutherford and Eileen. During a busy session Rutherford spent the winter with M. Morin, Professor of French, and his wife. It will be seen elsewhere that Rutherford finished his work with difficulty so as to go to New Zealand (1905) and return with his family in time for the session which then began in September.

RUTHERFORD TO HIS WIFE

Montreal: 2 Nov. 1904. Still not a word from you, but I suppose the next mail will bring something. I have bought a card catalogue from Morgan's, so as to keep a record and references of scientific papers. I will start, next month, preparing a new edition of my book and the catalogue will be invaluable. I heard from the Press that they are going to charge £40 for a German translation of my book, £20 to me and £20 to themselves. They think the new edition will interfere with their German sale to a large extent. I had a visit the other day from Professor Fischer of the University of Munich who came up specially to see me. He spent a year in Cambridge and knew a good many of my friends. I had him to dinner and lunch at the Morins. He arrived in time to attend a meeting of the Physical Society and I had Eve and McClung and Barnes in during the evening. He is a very decent fellow and speaks very good English. He tells me all the German Universities take my book as a standard, and thinks as a whole they prefer the English rather

than the translated edition....My other piece of news is that my Dr
Godlewski has arrived....I am glad he talks enough English to make
himself understood and with a good accent. He is quite a handsome
young fellow, about twenty-five I should think, and well dressed and
looks distinguished. Altogether he has made a good impression and
I think will prove an acquisition. He is a pure Pole and has no Jew in his
composition. We had John Morley here last week. He stayed with the
Petersons. Peterson asked me to lunch and I had quite a chat with him.
Morley is looking pretty old and tired, but bucks up in conversation.
I have not dared mention the fact about, as some of the Faculty are very
jealous of such things. He gave a good speech when he got his degree
...and there was a reception afterwards. My tenant has not yet paid
up. Walter tells me the piano is used very seldom. Our old garden is,
I believe, a wilderness. Give Eileen a kiss from me and tell her I will
write her a letter.

A new honour now came to Rutherford, the award of the Rumford
Medal by the Royal Society; his colleague at McGill, John Macnaughton,
Professor of Classics, wrote his congratulations:

14 *Nov.* 1904. I saw with great delight that you had been awarded the
Rumford Medal for research. It has been real research in your case and
the brilliant prize splendidly earned. In my department what they call
'research' especially in the American Universities is often no more than
mechanical totting up of the number of times something occurs—a
thing which can be done with one eye open and both feet up against
the chimney-piece. Yours has required a brain of most unusual penetrating power all focussed to one point and brought to bear like a
Bunsen-burner on the hard material to be fuzed. Accept my most
hearty congratulations. I told them to send you my inaugural address.
I don't think you will disagree with it, though it is not very much in
your line.

The Rumford Medal of the Royal Society, consisting of a gold medal
with a silver replica struck in the same die, is awarded once every second
year "to the author of the most important discovery or useful improvement which shall be made and published by printing or in any other
way made known to the public in any part of Europe during the pre-

ceding two years on Heat or on Light, the preference always being given to such discoveries as, in the opinion of the President and Council of the Royal Society, tend most to promote the good of mankind".

RUTHERFORD TO HIS WIFE

Montreal: 5 Nov. 1904. ...I really have quite a big piece of news to tell you, but have been working off the local gossip before I start it. I received a cable from Lord Strathcona marked private as follows. "I congratulate you on the great distinction of being awarded the Rumford Medal for researches in radioactivity." I have not heard anything further yet and do not expect to for a week, but I presume it refers to the Rumford Medal of the Royal Society. On looking up a book on the Royal Society, I find that the Rumford Medal is a great distinction indeed. On reading the account given in the Yearbook of the Royal Society I made the agreeable discovery that the sum of £70 usually accompanies the medal. I think that this portion as well as the award itself is correct, but nothing public will be announced till I hear definitely. The Rumford Medal was instituted on account of some money given to the Royal Society by Count Rumford in 1796, and he at the same time gave an equal sum for a medal to the American Academy of Sciences. Rumford himself, whom you remember I often talked about last year in connection with the army of beggars at Munich, and so on, was the first recipient of the medal. On looking up the list of recipients I found myself in the company of notables of which I am not in the least degree ashamed. The following is a partial list: Fresnel, Faraday, Sir G. Stokes, Pasteur, Clerk Maxwell, Kirchhoff, Tyndall, Sir N. Lockyer, Sir W. Huggins, Hertz, Dewar, Lenard and Röntgen, Sir O. Lodge, Parsons (steam turbine man). I am not sure of Lodge but think so. The above is only a selection of those you know of, so you see if I have that medal I am in company better, I am afraid, than I deserve. I trust it is all right and that it has not been divided as in the case of Lenard and Röntgen for X rays. I will forward you a photo of the plate in the book as soon as I get it done, just to see what it is like. I believe there is a big dinner given to the medallist this month by the Royal Society but I of course will not be able to be present. I don't suppose I shall get hold of the medals until about Xmas.

I got an account from the Cambridge Press about the sales of books. I find that up to the end of last July £41 was due to me in royalties. This will not be paid to me till January. About twice that amount is probably due to me altogether for sales up to date, so my banking account will have a chance of growing again.

RUTHERFORD TO HIS MOTHER

27 June 1904. I have arranged to write an article for Harper's Magazine on "Radium, the source of the Earth's heat", for which I will be paid 350 dollars—pretty good pay....

10 *Aug.* 1904. The reviews of my book have been very satisfactory but I will not know much about the sales till next year. I am sending you a copy of *Technics* with an article of mine on Radium with a photograph in the beginning. If you get the August number of Harper's magazine you will see a photo of my noble self in an article by Sir Oliver Lodge on "A New Kind of Matter" so that I am getting as much advertisement as is good for me. These things, however, don't count scientifically, for it is work that tells. Things scientific are very quiet just now, but I am working steadily at the lab.

27 *Sept.* 1904. I have just returned from my trip to St Louis. My main object in visiting St Louis was to attend the Scientific Conference—one of Electrical Engineers, before whom I read a paper, and the other an International Congress of Arts and Sciences which was a very big affair. A large number of the most prominent representatives of the different branches of learning were collected from all over the world and gave addresses on their respective subjects. I had rather hard work attending meetings and trying to view the 'Exposition'. There were a number of interesting side-shows of which the Boer War was one of the most interesting. Several hundred Boers, including Cronje and Viljoen in person took part and the whole thing was very well done and attracted great crowds. The illumination at night was very fine—all by electricity of course.

16 *Oct.* 1904. The Frisco mail came in this morning and I received a letter from you and Florrie. I was very glad to hear you are yourself again after your illness. May has been gone about a month and I am now getting used to my new quarters with the Morins. I am very comfortable and find them very agreeable people. I had a bad cold

after my trip to St Louis but managed to shake it off and am now feeling in good form again.

I got a letter this morning from Dr Godlewski—a graduate of the University of Cracow, Austria. He is coming at the end of this month to work under my direction at research. He is my first foreign student and it is a compliment to the University, not to speak of myself. I am rather interested to see what kind of man he is. I have working with me a graduate of Yale University, so I have two strangers in the laboratory this year.

People have been very kind to me in asking me to suppers and dinners as they consider a grass-widower requires looking after. I have been out several times last week and have as many invitations as are good for me for the ensuing week. I go out golfing about twice a week in order to get a little exercise before the coming winter. We have had beautiful sunny weather the last few days and it seems a pity to spend it indoors. I am told that I am considered the sort of 'show' man of this University. It rather startles some of my colleagues to find that the University is known wherever they go in Europe—chiefly in connection with radio-active research. However don't think my head is swelling for I still take the same size hat.

I got a letter from a distinguished chemist Baskerville in the States who wishes to dedicate a book to me. I think I am rather youthful for inscriptions of that character as they are the special perquisites—like honorary degrees—of Scientific Octogenarians. However, it is a compliment and I suppose libel actions are not in order. A conservatoire of music has just been opened in connection with the University in which there are several hundred musical students, and which promises to be a great success. I expect my article will appear in Harper's within the next month or so. I am just finishing writing up my article for publication which I delivered at the International Congress. I hope father is well. Give my best wishes to Art. and his fiancée.

RUTHERFORD TO HIS WIFE

Montreal: 9 Nov. 1904. I wrote you in my last letter about Strathcona's cable re Rumford Medal, I have not had time to receive official word from England.

A week later. I received the following cable from Cambridge, "Cavendish Professor and Research Students send heartiest congratula-

tions Rumford Medal." It is very good of the Cavendish people to
remember me this way. I am expecting to hear from the Royal Society
during next week. The office on Friday communicated the news to the
Gazette and it appeared in all the papers. The account isn't up to much,
but is correct as far as it goes....It never rains but it pours. On Saturday
morning I received a letter from Professor Hadley of Yale asking me if
I would deliver the Silliman Lectures at Yale this year. You remember
J. J. delivered them two years ago and came over for the purpose, and
Professor Sherrington (the physiologist) of Liverpool, last year. It is a
great distinction, more especially as the sum paid to the lecturer is
2500 dollars. There are supposed to be from eight to twelve original
lectures which are to be published in book form by Yale University,
and the proceeds, if any, of the sale go to them. The fee you will agree
with me is not to be sneezed at. It is not often one can earn a year's
salary for ten lectures. Now Hadley tells me that the lectures should be
delivered in April and May, if possible, so if I deliver them I shall not be
able to start for New Zealand quite so early as I anticipated. I now put
the conundrum to you. Would you prefer 2500 dollars or three weeks
more of my company? It reminds me of the story of the lady and the
lion. I don't think however that I shall miss such a unique opportunity
of increasing our funds and of gaining scientific distinction at the same
time. The difficulty will be to get sufficiently new material for the
lectures as I will by that time have all the MSS through for my new
book. Hadley said at the end of the letter that the proposal was quite
independent of the talk re the Chair of Physics at Yale, but he thought
it would be a good opportunity to renew our discussion. I told Cox
about it and I think he is going to tell Peterson that they will have to get
a move on over the salary question pretty soon, unless they want to come
in after the event. I have an idea that the question will ultimately work
out to our financial advantage. Peterson was very pleased over the
Rumford Medal, and they have plenty of reason for raising me specially,
without raising a revolt among the others. I am sure you would like
to be here at these times in order to keep in touch with what is going on,
but I will keep you as well informed as I can. My tenant paid up two
months rent all right, I wondered whether he was going to forget much
longer. Harkness is coming along this evening to discuss the Yale
business. I went to see Dr Birkett about my throat which was worrying
me a good deal. He said there was not much the matter, but that I had

had a granulated throat for many years, but nothing serious enough to trouble about. He gave me some tabloids to take which put me pretty right, but I find I have a sore throat again this morning.

Montreal: 18 *Nov.* 1904. ...I have written down to Yale accepting the Silliman Lectures and hope to get them through by the end of April, so my stay in New Zealand will not be much shortened, and if necessary we could lengthen it at the other end. You may be interested to hear the history of the Silliman Lectures. In 1883, 80,000 dollars was left in trust, in honour of the mother of the donors Mrs Epsa Ely Silliman (what a name!), for the purpose of memorial lectures "designed to illustrate the presence and providence, the wisdom and goodness of God, as manifested in the natural and moral world. It was the belief of the testator that any orderly presentation of the facts of Nature or History contributed to the end of this Foundation more effectively than any attempt to emphasize the elements of doctrine or of creed" etc. My last week has been a series of congratulations over Rumford and Silliman. The Medical Faculty conveyed their congratulations, while Dr Harrington made a complimentary speech at the Chemical Society. My hat, however, still remains of the same dimensions. The medals are given at the Royal Society meeting on November 30, but of course I cannot be there to receive it in person. Cox tells me a dinner is to be given me somewhere about the time of the presentation of the medal in London. It cannot be on the actual day, as Peterson is to be away in New York at that time. Everybody seems pleased and to agree that I deserved it, which is not generally the case with one's colleagues.... I have been pretty busy during the week as I have got a good deal of research in hand. I am feeling in pretty good form and have almost got rid of my cold.

11 *Dec.* 1904. The past week has been rather a trying one for me as the dinner in my honour came off on Wednesday in the Windsor Hotel. I have already forwarded to you a copy of the menu and a paper with a brief account of the same. Clay also sent you a cable telling you of the fact, while the dinner was in progress. I hope you got it safely and if you hadn't heard of the Rumford Medal by that time, you must have been considerably mystified. The dinner was a great success,...about 130 were present with Peterson in the Chair...at 8 o'clock Peterson accompanied me to the dining room while appropriate music heralded our

arrival. As I told you Macdonald took over the expenses of the dinner, which was given regardless of expense. Loads of flowers were scattered over the table and music was played at intervals. I naturally did not enjoy myself as much as if I had been a guest instead of a principal. After the usual toasts, Cox got up and made the speech of the evening. It was really very clever and not unduly buttery and with a good many jokes. Everyone considers that Cox excelled himself on that occasion. He was followed by Tory who said many kind things and then I had to get up and reply. I talked about twenty minutes and got along better than I thought I would. I think they all considered I made a fair speech. I took the historical order and gave credit to all the people who had worked with me and worked off a few villainous jokes. They received me extremely well and my health was drunk amid much enthusiasm (or at any rate well simulated, if it wasn't real). Walton followed with a humorous and capital speech in which he referred to a side of the question which he stated all the previous speakers had overlooked. He drew a harrowing picture of the state of mind of radium—a shy shrinking thing which had been dragged into the glare of publicity and all its secrets laid bare and instanced it as an example of scientific oppression. He was followed by Bovey, Dean Johnston, Barclay and after that Lash Millar who came down to represent Toronto University and Dr Goodwin who represented Queen's. We got away about 2 a.m. MacBride carried home an armful of flowers, as did a good many others. Everybody thought the dinner a great affair, but I was mighty glad to get it over. It is not altogether pleasant to be talked at, for four solid hours in succession. I don't suppose however I shall have to endure the like again for a long time to come. The next night I was asked to dinner at the Drummonds....I found I was the guest of honour, and took in Lady Drummond. After the ladies had left Drummond suddenly proposed my health which was followed by a spasmodic three cheers. After thanking them in the usual hackneyed terms, I said a great deal had been said about the medal, but I had doubts if I would get it, as Lord Strathcona had it! Also that Lord Strathcona was a most faithful deputy, for he had not only received my medal but had eaten my dinner....I am really very comfortable with the Morins, who do everything they can for me. They are quite a charming couple.

25 *Dec.* 1904. Xmas night, and I am staying in my rooms working and

reading. I worked on my book in the morning and went to the Walkers for midday dinner. Miss Cameron and Godlewski were there. Dr Godlewski is a very able and fine looking fellow and has made a favourable impression on everybody. He is a very hard worker and has already done a very good piece of work on actinium. He is very pleased with the Lab. and I think does not in any way regret his visit to Montreal. He now speaks English well and has no difficulty in following rapid conversations. His father is Professor of Botany at Cracow, so he is pretty well up in University matters. He told me today that he had seen that Rayleigh had got the Nobel Prize for Physics and Ramsay for Chemistry. I think they are a very good selection. I may have a chance if I keep going, in another ten years, as there are a good many prominent physicists like J. J. and others to have their turn of spending the money. It is just as well too that I have got something worth having to look forward to....I must now come to an interesting piece of news. I got my gold medal and it is a stunner. It weighs 14 ozs., and probably has £50 worth of gold in it. I haven't got the silver one and do not know if it is forthcoming. It will be a good way of saving money as it can always be melted down into dollars when required! My name is on the edge, but otherwise everything is as in the photograph I sent you. On Xmas Eve I got a letter from the Royal Society enclosing a cheque of £63. 7. 6. or 308 dollars! I felt it was a Xmas present worth having, so that the whole thing altogether is worth over 500 dollars. The money will be handy for future contingencies....I got letters from you last Monday and am glad you are both well. You may tell Farr I will, provisionally, promise to give a lecture when in Christchurch. He wrote to me about it. I am enclosing another draft for £20, don't stint yourself too much.

CHAPTER V

LAST YEARS AT McGILL

RUTHERFORD TO HIS WIFE

Montreal: 1 *Jan.* 1905. This is the first day of the New Year and I trust we will both have as happy a time as in the old year. I don't suppose this year will be quite so full of important happenings as last year for I have been amazed when I reflect on the number of things I got through. There was my visit to St Louis, followed by the lectures in the States, the publication of my book, the voyage to England and my lecture there, the publication of the Bakerian Lecture and my paper to the Electrical Congress, also the address at St Louis, and finally the Rumford Medal, and the award of the Silliman Lectures. It would be difficult to do more in the course of one year. I have had a pretty busy week of work preparing my book.

11 *Feb.* 1905. We are having a very hard winter...fortunately the house is warm. My work is progressing rapidly, both as regards my book and research. I have recently found another product in radium, and in addition have found the charge carried by the alpha rays, which J. J. has been trying to get for the last six months. I am sending a letter to *Nature* about it tomorrow. Mrs Morin now does my typewriting. It is very handy as I give her the papers, and get them back as I require them. She is more careful than Miss H. and makes very few mistakes. I of course pay her at a fair rate. She enjoys typewriting and is tickled to death at earning a little money for herself. She has already typed about 150 pages for me. Work in the lab. is getting on very well indeed. Godlewski is getting on in first rate style, also Eve and Bronson. By the way I have a most startling piece of news for you. Eve is engaged to the younger Miss Brooks of Dunham College. Eve told Harkness on Saturday. I don't know how he has managed to see much of her and have not yet seen him to gain particulars. Nobody had the slightest suspicion of the coming event. I feel we are both responsible for the event, as he would not have known the Brooks without our inter-mediacy....I am intending to insure my life for 4000 or 5000 dollars in the New York Life on the 20 year basis. I think it is a good way to

make one save money. If my salary should happen to be increased I would take out another 5000 on the same basis. By the way if I should in any way come to grief my papers will be found ticketed in my office desk. I have no premonition of any disaster but it is as well you should be aware of such details.

19 Feb. 1905. . . . I have just written a congratulatory letter to Miss Brooks at Dunham. . . . I saw Peterson on Friday and told him about the Yale offer. I think he will precipitate, but don't know how much. I will probably not know for some time until he has seen the Governors about an addition.

New Haven, Conn.: 20 March 1905. As you see by the above address I am now at Yale and start the first of my Silliman lectures at four this afternoon. I left Montreal on Saturday accompanied by Bronson who went to see his people for a fortnight, and also to help me a little with my experiments, if necessary. Had a series of mishaps to the train on the way down—a hot box—then a freight engine carried us slowly along, and finally a special to Springfield at which we arrived four hours late, just late enough to miss all the trains to New Haven for that evening. We put up at the hotel, had some supper and then to bed, getting up at six Sunday morning and taking the first train to New Haven minus breakfast. I arrived at the Clubhouse, where they are putting me up, at 8.45 and had breakfast. . . . I had a succession of visitors, Hastings, Wheeler, Boltwood, President Hadley, etc. . . . I am going in a few minutes to the lab. to arrange for my experiments. I already have invitations to dinner from Hadley and Chittenden, and other general invitations all round, so I expect to have a busy time socially. Feeling in pretty good form, and hope to pull off the lecture all right. Now in regard to Montreal. I had a terribly busy time getting my last edition of my book ready before I left, but fortunately got it all through. I have had to work like a Trojan, and never intend to work as hard again in the future, if I can help it. Now for an interesting piece of news. I told Peterson about a fortnight ago about the Yale offer and he said he would arrange matters to keep me. I believe the Governors have agreed I am to get 4000 dating from May. It is pretty satisfactory and I don't think I will seriously consider the Yale offer when it comes, as Montreal suits me. I understand from Clay that he doubts whether anybody else will ever rise to it, however much they try. It is very decent of them considering the straitened state of finances, and it will make us moderately comfortable and give something to put by for the future. I believe Strathcona has brought out my silver medal (Rumford). I called to see him with Peterson, but he had a bad cold and could not see us. . . . I doubt whether I can get away by the steamer on April 27th but will have to wait for May 15th. It will make me late in New Zealand but I want to see the proofs of my book through, and publish the work I have been doing during the year, otherwise my vacation will be full of work. I need not publish the Silliman Lectures for some time. I will cable if I can't get away from Frisco on 28th. I must not sacrifice my work for an extra three weeks in New Zealand. I have had a look round the University at odd times. . . . In our walk Dana opened up on the subject of the Yale appointment. They are very anxious I should come to Yale, but I doubt whether they can make it worth my while to come, especially as McGill has raised my salary. I am to see President Hadley tomorrow to discuss business and I will tell you about it later.

A week has passed away and I have delivered six out of ten lectures, and I think pretty successfully. I have had a very good audience all through and they keep up very well.

2 *April* 1905. My stay in New Haven is nearing its close, as I give my last lecture tomorrow. Bronson is returning with me so I shall have company on my way back. The weather has been very bright and sunny, during the past week the snow has disappeared from the ground and the grass is getting green again. I have had a very pleasant time altogether and, as far as I can gather, I think my lectures were a great success. My attendance keeps up steadily and all sorts of people turn up to hear them. In my lecture tomorrow I am expecting a large audience as I am dealing with questions of the effect of radioactivity on the age of the sun and earth. I have had a round of visits during the week—dinner, lunches, teas. Tonight I am spending at the Club, glad of a little rest. The people have all been very kind and all seem to know I am a prospective Yale Professor. They are all very anxious I should come down here. I had a long talk to Hadley and he outlined the position. The salary offered is $4000, but he led me to infer that the actual question of salary would not stand in my way if I thought seriously of coming. I imagine I could raise $4500 if I asked for it. The appointment is not to become vacant till September twelvemonth, and

I have some time to think over the matter before I give my decision finally. The place is very tempting in some respects as I would have a larger sphere of influence here than in Montreal. At the same time I am not inclined to give up a good position at Montreal to accept a salary which is about equal in buying power. I don't think I shall definitely decide until I see you in New Zealand. New Haven is a very pretty place and they all seem to consider it is a very desirable place to live in compared with most American cities. The environs are pretty and there are lots of nice walks in the neighbourhood. The houses of the professors are mostly on broad avenues and seem to have gardens and generally more country-like than Montreal. Professor Dana, who is Director of Mineralogy and also instructs in Physics, a man whose opinion counts for a good deal here, is very anxious for me to come...the men are much the same as the general university type and I should think would be easy to get along with....They are all prepared to welcome me very cordially. I am very sorry you have not been here to see things for yourself. If I seriously thought of it we might run down to New Haven on our return from New Zealand.

Montreal: 9 April 1905. I have finally decided it is impossible for me to get away from Montreal in time to catch the boat leaving Frisco on April 27th as I had hoped. I sent a cable to Rutherford, Christchurch, saying leave Frisco May 15 which I hope you received safely. My great trouble has been to get the proofs of my book through. Less than half have so far arrived but I sent a cable to them to hurry up. I cannot leave the book half finished, as I would have to re-write a part again if I left it to the autumn and it must be put through before I leave. In addition it is impossible for me to publish my own work before I leave, unless I take the extra three weeks. It is very important I should write it up as they are all following my trail, and if I am to have a chance for a Nobel Prize in the next few years I must keep my work moving. I am afraid my delay in getting out will upset your plans but two months in New Zealand will I think suffice to see most of the people I wish to.... I think we had better return 2nd class from Suva as you suggest, it will not be particularly pleasant but if there are several of us it will not be so bad. I hope Floss [his sister Florence] will come with us. I shall be quite willing to finance the journey one way in any case and give her the £25 for the outfit as you suggest. I think it will be an excellent chance for her to see a little of the world and I am sure she will appreciate it....I brought back a cheque for $2500 from New Haven. We of course don't want to spend it but the proceeds of my book for this year will I think pay her trip. My Silliman lectures have to be ready for publication by January 1st so I will have to bustle along when I come back. That is one of the reasons why I must get the second edition off before I leave for New Zealand for there will be a fearful muddle and I could not enjoy my trip with that hanging on my mind....I would like to go to the Hot Lakes, so we will make a trip there for a few days after my arrival. I came back from New Haven on Tuesday and have been very busy since correcting the proofs of my book and arranging for the numerous papers which have been turned out by the men in the Lab. We have got through a lot of good work this year, about thirteen papers in all. Dr Godlewski left last night, I am sorry he is going as he is a very good man.

Lord Strathcona invited me to dinner the other evening, about forty people there, Miss Lichtenstein and Miss Oakley were the only college people there. It was pleasant enough. He gave me the silver copy of the Rumford Medal which the Royal Society had given him to forward to me. He was very complimentary in private and seemed much interested.

Champaign, Illinois: 30 April 1905. ...Journey to Chicago fairly pleasant. Professor Townsend met me and we went to the Hotel Beardsley—comfortable enough, though at first sight not prepossessing. First lecture booked for 4 p.m., so I did not have much time to spare before going on the spout. The lecture room was crowded (over 500) and the lecture went off all right. I lecture tomorrow night at 8. Afterwards the students and instructors, or a good many of them, filed in to shake hands with the visitor. It was rather comic, and I relieved it by attempts at a few jokes. Afterwards — and I went to Townsend's house, and after about half an hour, we went down to the Hotel where a meeting of the philosophers had a dinner during which a philosopher discoursed on the 'God-Problem'....They have apparently been advertising me extensively beforehand. Tomorrow I attend an informal meeting of the Physics people and discourse to them a little, and in the evening to Carman's to dinner and lecture afterwards....The town of Champaign is the usual ramshackle township of 5000 inhabitants or more. Everybody is very kind and the audiences appreciative.

About 1905 the world caught fire and radium was the vogue. An eminent physicist was warned by his doctor that he had not long to live; he replied: "I can't die yet, I want to know more about radium!" He recovered.

A great number of Universities and Societies poured in appeals to Rutherford to come and lecture to them about radium. He did what he could. He also received offers, between 1903 and 1908, of a Professorship from Yale, Columbia, King's College, London, Leland Stanford, the National Physical Laboratory, and of the Secretaryship of the Smithsonian Institute.

About this time Rutherford began to make a confidant of Boltwood of Yale and wrote to him on scientific matters with a freedom quite in contrast with his published letters and articles.

RUTHERFORD TO BOLTWOOD

28 Jan. 1905. I have practically finished radium A, C, F, and thorium B all having e/m about $2 \cdot 5 \times 10^3$. I am convinced that the alpha particle is a helium atom with two charges.

18 April 1905. I have been working a lot over that question as to whether lead is a decomposition product of radium (radium G in fact) and am extremely impressed by the data which I find in support of this hypothesis.

He proceeded to discuss the ratio of lead to helium (at first thought to be nitrogen) in Hillebrand's careful analysis of twelve radioactive minerals and concluded,

if lead can be shown to be a disintegration product of uranium, will it not necessarily follow that all the lead existing on the globe originated in this way? I think that the deductions which can be made from this assumption will make even the metaphysicians dizzy!

Miss Harriet Brooks, working at McGill with Rutherford, obtained a remarkable result. On her exposing a metal plate to radium emanation it received a coating of induced radioactivity or active deposit. On her placing this plate inside a closed vessel, without touching the sides, it was found that the walls of the chamber became radioactive, and this effect persisted after the metal plate was removed, but gradually decayed away.

Rutherford at once grasped the situation and wrote: "Since radium A breaks up with the expulsion of an alpha particle, some of the residual atoms, constituting radium B, may acquire sufficient velocity to escape from the plate into the gas, and are then transferred by diffusion to the walls of the vessel." The analogy of the recoil of a gun firing a shot is obvious. The recoil method was rediscovered at a later date by Hahn and Meitner and used by them, as well as by Russ and Makower, in the discovery of new radio-elements.

There were many reviews of the second edition of Rutherford's *Radio-activity* which appeared in 1905 and was half as large again as the first edition. R. J. Strutt, now Lord Rayleigh, wrote in *Nature*:

Rutherford's book has no rival as an authoritative exposition of what is known of the properties of radioactive bodies. A very large share of this knowledge is due to the author himself. His amazing activity in this field has excited universal admiration. Scarcely a month has passed for several years without some important contribution from him, or from the pupils he has inspired, on this subject, and, what is more wonderful still, there has been in all this vast mass of work scarcely a single conclusion which has since been shown to be ill-founded.

The review went on to weigh the probabilities, as they then stood, that the alpha particles were truly and really helium, a view which rested partly upon the dubious value of the charge per unit mass of the alpha particle. It was criticism of this kind which later induced Rutherford and Royds to collect alpha particles, apart from all else, and to prove, beyond possibility of question, that they were indeed helium.

In 1905, Rutherford ingeniously measured the number of beta rays ejected per second from radium in equilibrium with its subsequent products, and he found 4×10^{10} per gramme. This was a first effort and not far away from a more accurate measurement made by Makower at Manchester, 5×10^{10}. The idea of the experiment is not difficult.

Measure the total amount of electricity carried by all the beta particles in a second, and divide by the charge carried by one beta particle (the electronic charge). The result is of course the number of beta rays per second.

Rutherford had also determined the number of alpha particles released from a gramme of radium every second, and from each of its successive alpha products. The value he found was of the right order but somewhat high, namely, $6 \cdot 2 \times 10^{10}$. Hence he could find the amount of emanation in equilibrium with the radium and could deduce the average life of radium, about 2000 years. Lastly, he had determined the energy released by each alpha particle and calculated the heat generated by a gramme of radium in calories per hour. His figures may not have been highly accurate; they were not, but they were of the right order of magnitude. This remarkable work showed that he was master of the game! It was only a matter of time, patience, and improved technique to obtain more accurate results. The work thus begun at McGill was brought to a fuller fruition at Manchester.

On 20 May 1905, Sir William Ramsay wrote from University College, London, supporting Dr Otto Hahn's application to work with Rutherford at McGill. Ramsay stated: "Hahn is a capital fellow, and has done his work admirably. I am sure that you would enjoy having him to work with you."

The arrival of Otto Hahn at McGill was a noteworthy event. He had been working with Ramsay, who had given him some thorianite with the object of extracting some *radium* from that ore, using Mme Curie's method of fractional crystallisation. To the surprise of both Ramsay and Hahn, the residue became more and more powerfully radioactive, while the production of radium was small. The concentration of the residues led to the discovery of a material many thousands of times more radioactive, weight by weight, than the parent thorium. This was an experience similar to that of Mme Curie when she extracted radium from uraninite. The question was: What was the nature of this new substance which Hahn had discovered? He came to Rutherford at McGill to find out. Now, on arrival, Hahn was naturally excited and enthusiastic, and his English at that time was not altogether adequate, so that at first Rutherford seemed somewhat puzzled and sceptical, but when Hahn showed him the presence of the emanation of thorium (thoron), with a period of 53 seconds, Rutherford was enthusiastic over this discovery of radiothorium, an important and powerful member of the thorium family, which decays to half-value in 1·9 years. Hahn continued to work with Rutherford at McGill for a year or so, discovering radioactinium, and carrying out further investigations on the thorium family, which he has continued with brilliancy in Berlin, and which have justly brought him fame.

Boltwood was at first very sceptical about the discovery of radio-thorium and wrote to Rutherford (22 Sept. 1905) that "the substance appears to be a new compound of thorium X and stupidity", a view which he soon recanted.

An important further step was made at this time. In a specimen of any ore containing uranium and radium which was formed a great many years ago and has remained compacted ever since, there must be a great accumulation both of helium and the end product, lead. Since the rate of production of these materials had been determined, it was a simple matter to find the age of the given specimen by dividing the amount of the material present by the rate at which it was produced. Thus Rutherford wrote in his second edition of *Radio-activity*, p. 484: "I have, for some time, considered it probable that lead is the end or final product of radium. The same suggestion has recently been made by Boltwood (*Phil. Mag.*, April 1905)." The reason is weighty. Uranium has an atomic weight 238·5 and the ejection of three alpha particles, each helium of atomic weight 4, would leave radium with an atomic weight 226, the value found by Mme Curie. It was known that the transformations of radium involve the loss of five more alpha particles, or a reduction of 20 more units, and this would leave 206, which is nearly the atomic weight of lead, 207.

Boltwood wrote Rutherford a letter (18 Nov. 1905) stating that he

Rutherford at McGill, 1905

was using this method, suggested by the latter, to calculate the age of a mineral "from the relative proportion of lead and uranium in it, using the rate of disintegration as determined by Rutherford".

The ages of about 27 ores were determined, beginning with a specimen from Glastonbury, Conn., aged 92,000,000 years, and ending with thorianite from Ceylon, aged 500,000,000 years; the others were fairly equally spaced between these values. According to geological friends of Boltwood these results were not contradicted by the geological data on the relative ages of the different deposits.

RUTHERFORD TO BOLTWOOD

5 *Dec.* 1905. ...You appear to have been diving deeply into the mysteries of matter and certainly manage to obtain very plausible results. The way the ages of the radioactive minerals work out is certainly striking and I am glad to find that a professional chemist when properly infected is quite as rash in theorizing as a physicist. I have been much amused at various articles the last six months by writers in Engineering Journals, who hold up their hands at the audacity of the experimentations of the workers and sagely reflect how Newton would have sat down and worked out the whole subject and then given a theory. It never occurs to them that it would have wanted half a dozen Newtons to accomplish the experimental work in a lifetime and even they could not have put forward any more plausible theory than we work on today. These dam'd fools—whom I think must once have been chemists— (excuse me—no personal reference) haven't the faintest notion that the disintegration theory has as much evidence in support of it as the Kinetic Theory of Gauss and a jolly sight more than the electromagnetic theory, which they all swallow as the eternal verities. Apart from this miniature outburst, I quite agree with you that the only way to get any idea of what are the products of the radio-elements is, to examine carefully every available mineral. If we don't find it that way, I think it is extremely improbable we shall get any further at all. I feel sure helium is the alpha particle of radium and uranium products but it is going to be a terrible thing to prove definitely the truth of this statement for I feel confident e/m will come out to be 5×10^3 instead of $2 \cdot 5 \times 10^3$. It may conceivably be a hydrogen molecule—half atom of helium, or helium atom with two charges, and nothing but a pure

scientific nose can say with certainty that one is more profitable than the others. My nose (which may be prejudiced) leads me to avoid the H molecule like the devil. It is too plebeian in character to be sired by such blue-blooded stock like radium whose ancestors certainly existed before the flood. However, for plebeian thorium, hydrogen seems to be very well fitted. I really see no reason why all the active bodies should emit helium—actinium and radium certainly do—but H is the most likely material to be turned off. However, it will be mighty difficult to prove. Until you furnish me with more evidence, I shall still cleave to helium if only for Galileo's doctrine of simplicity. You remind me of Japhet in search of a father, for you seem bent on finding a mother at any rate for the two waifs cerium and lanthanum. Your proofs seem to me too convincing to be true, but all things are possible if thorium B breaks up into two fragments of about equal weights. I have long thought such an effect must occur among some of the products and I feel confident actinium owes its origin to some product which has two distinct forms of equilibrium—the smaller percentage part yielding actinium. I thought I had a fair amount of scientific nerve, but I am left far behind in your last essay. This search for a father is becoming positively indecent. Why not accept the chemist's view of separate creation and rest happy?

Boltwood was unfortunate enough to follow a false trail, for on 10 Dec. 1905 he wrote to Rutherford:

As to my attempts to trace the lineal descendents of thorium, my efforts may be misdirected but they are none the less earnest. I am beginning to believe that thorium may be the mother of that most abominable family of rare-earth elements, and if I can lay the crime at her door I shall make efforts to have her apprehended as an immoral person guilty of lascivious carriage. In point of respectability your radium family will be a Sunday school compared with the thorium children, whose chemical behaviour is simply outrageous. It is absolutely demoralizing to have anything to do with them.

Rutherford sailed for Auckland from San Francisco on 18 May 1905. It was 6 June when he joined his wife at Auckland for a happy tour of the Thermal district with some of his family. Then after a stay with his people at Pungarehu, they visited Christchurch and Rutherford lectured on radium at Canterbury College. Everywhere they received a warm-hearted welcome from their relations, friends and the kind folk of New Zealand, and they all returned in time for the opening of the session at McGill. There were four of them, Rutherford, his wife, daughter and sister Florence.

There are many letters preserved which passed between W. H. Bragg, then at the University of Adelaide, and Rutherford at McGill; and arrangements were made to prevent overlap in their research work. As an example the following letter may be quoted:

BRAGG TO RUTHERFORD

16 *July* 1905. You have indeed done a lot of work; I shall look forward to the description of it with the greatest interest. Your result about the velocity of the alpha rays, when they cease to ionize, is most surprising, as you say. I thought the critical velocity would be about 10^5, as for beta rays....It is a very generous thing to offer to keep away from the line of research which I propose to take; and I want to thank you for your courtesy. Of course I would be glad to go on with the work I am doing now, and to do so without fear of clashing with you. But I am sure you will understand when I say that I could not let myself be in the way. I am very happy that I have been able to help a little bit....

Bragg had discovered that when gamma rays from radium struck a thin plate of metal, the radiation forwards was greater than the back radiation. If the gamma rays were waves, or pulses, the radiations should be of equal intensity fore and aft. This result made a strong case for the viewpoint that gamma rays, and therefore X-rays, were corpuscular in nature. It is probable that most physicists today would admit that radiations combine the qualities of waves and of corpuscles, but to explain how that can happen is quite another thing.

RUTHERFORD TO HAHN

25 *Sept.* 1905. I have spent some time over the MSS on radioactinium and forward you the corrected copy. I have not made alterations except to reconstruct some of your sentences in more idiomatic English. I think the paper is clearly written and should be easily understood. Your

treatment of the mathematical theory, while correct in the main essentials, was not very rigorous, so I wrote it out in the MSS enclosed with your books. I hope that you will agree with the emendations. I tested some of the numbers on your theoretical curves and found them substantially correct, but it would probably be advisable for you to test a few points again with the theoretical formulae. Eve kindly reconstructed your final sentence which I did not feel equal to altering. I have to thank you for your kind expression of good-will but doubt if I deserve it.

I am prepared to communicate the paper to the Phil. Mag., if you decide to send it to them, you might signify the same to me. I would fix up the curves well and refer to them when you are speaking of the active curves earlier in the paper.

I have been back here three weeks, and just started lectures. Bronson and Eve have started work and I am doing a few things.

I am expecting to get the papers on the mass of the alpha particles of radium, actinium and thorium published in the October Phil. Mag.

The plates I forwarded of our photographs were all smashed to pieces in transit, so we had to prepare fresh ones.

We have had a very warm summer but the weather is now very cool and pleasant. I intend to pursue an easier life this year as I find that I feel lazy. You saw no doubt that Soddy says the alpha particle is initially uncharged. I will believe it entirely when I have seen it for myself.

You were quite right about the radium bottle. It has all precipitated again. Don't worry about it, however, you did your best to put it into shape and I think I can manage all right, if it is lead, by doing a little chemical work myself.

I was surprised to hear that Boltzmann had taken his own life. He is one of the last of the big men of the old school of Maxwell, Helmholtz and Kelvin.

I got a letter from Levin from Paris, before his visit to Cambridge. He feigns happiness even if he has it not. Dr Rümelin will be over next month to represent your native land.

I am glad you are getting things in shape in Berlin. I trust that you will be able to convince the 'savants' that there is some method in our madness.

On 1 Oct. J. J. Thomson wrote a long and careful letter from Cambridge giving Rutherford an unbiased opinion as to the advisability of his applying for the Chair of Physics at King's College. The authorities were anxious to make the post as attractive as possible and to give as many opportunities for research as they were able. The laboratory was not at that time well provided with apparatus and there was every prospect of having to secure a new staff. He added:

the most serious drawback is the interruption of your work, and this must occur to some extent whenever you leave McGill. I do not like to take the responsibility of advising you strongly one way or the other and I am afraid my judgment might be warped by the strong desire to have you back in England.

The letter concludes with a statement of the difficulty of finding room at the Cavendish, "but we shall have more room soon, as Lord Rayleigh has given £5000 to build a new research laboratory".

In April, there was a large gathering at the Franklin Institute, Philadelphia, to commemorate the Bicentenary of the Birth of Benjamin Franklin. Rutherford received an honorary degree and on 18 April lectured to the Institute on "The Modern Theories of Electricity and their Relation to the Franklinian Theory". So much attention is paid to Franklin's work in connection with drawing electricity from the clouds by the moist string of a kite, thus identifying lightning with the more innocuous electric sparks in the laboratory, that one is apt to forget his careful experiments on insulated bodies (generally human) with a frictional machine given to the Library Company of Philadelphia by Peter Collinson of Edinburgh.

In his lecture Rutherford quoted the actual words used by Franklin in connection with his experiments and pointed out that, although the one fluid theory of Franklin scarcely survived today, yet the very terms of Franklin were still in common use, for example:

At the same time that the wire and the inside of the bottle [Leyden jar] is electrified *positively* or plus, the outside of the bottle is electrified *negatively* or minus, in exact proportion; that is, whatever quantity of electrical fire is thrown into the inside, an equal quantity goes out at the bottom.

Rutherford pointed out that "positive electricity is always found associated with bodies atomic in size", a statement which is no longer true since the discovery of the positive electron, or positron, by Anderson. He continued:

With regard to the question 'What is Electricity?' so often asked the scientist by the layman, science cannot at present venture an adequate answer. Nor is this surprising when we consider what a fundamental part electricity plays in nature. We have seen that electricity is a constituent of all matter, and, indeed, that what we call matter is electricity in motion. Attempts have been made to explain electricity as a manifestation of the universal medium or ether, but until we know more of the properties of the ether, such theories must of necessity lack physical definiteness. Even if we may ultimately explain electricity in terms of ether, there remains the still more fundamental problem, 'What is the ether?' An attempt to explain such fundamental conceptions seems of necessity to end in metaphysical subtilties.

Rutherford was President of Section III (Mathematics, Physics and Chemistry) of the Royal Society of Canada in the year 1906. He had been elected a Fellow of the Society six years earlier. He went to Ottawa to lecture on radium and on the way he lost a small tube of that element in the car. In his address he stated what had happened and added that "the radium will be giving off emanation for several thousand years". It was suggested that the railway carriage should be called 'Rutherford' in accordance with the Canadian custom of giving a name to every passenger car.

In 1906, Rutherford asked me to discover whether it was possible to measure the amount of radium in ore, such as pitchblende, by measuring the gamma radiation which came from it as compared with that from a standard. The result, and the ore specimen, were sent to Boltwood for confirmation. He wrote: "My chemical nose would tell

me in two minutes that Eve's uraninite contained over 40 per cent of
uranium just from smelling it, so that I am quite sure that his 27 per cent
is quite out of the question!" It was finally found that the radium solution standard was potentially half-strength. "The water has attacked
the glass and the silica and other things have precipitated the radium."
A new standard solution was prepared from Giesel's radium bromide
to which some acid (HCl) was added. Then Boltwood wrote: "The
factor obtained by the use of this new standard gives a very good agreement for the amount of radium in Eve's uraninite as determined by
him by the gamma ray method. He finds 0·32 mg. radium per
kilogramme, while the present standard gives 0·31 mg." It was thus
clear that gamma rays would reveal the true quantity of radium in an
ore body, or in a radium preparation in radioactive equilibrium.

In *Nature* (19 July 1906) there appeared a review of the work of
Rutherford at McGill:

So much work and such novel theories have naturally called forth
criticism, but the discussions have always been chivalrous, buttons have
been on the foils, and Rutherford's extreme care of verifying every
step by thorough experimental evidence has saved him from error to an
extent quite exceptional. It is fortunate that so much of the development
centred in a man to whom the remarkable instinct is possessed of rarely
following side issues.

His own successes as an investigator may be traced to a few well-
marked characteristics. The first is his pertinacious and reiterated assault
at the particular point which he wishes to attack. He has also an
instinctive insight which often makes his initial point of view more
trustworthy than the deliberate conclusion of some befogged experimenter. Most noteworthy of all is the extreme simplicity and directness
of his experimental methods. Some observers seem to grow happier as
their apparatus becomes more complex. Rutherford selects some
ingenious, straightforward attack, but the simplicity is supplemented
by the genius which has enabled him to make such great contributions
to our knowledge of the mutability of matter and of the atom in
evolution.

Professor Rutherford inspires research students with some of his own enthusiasm and energy. He follows their results closely and is as delighted with any of their discoveries as with his own. He is generosity itself in giving a full measure of credit to those who do research work under his guidance.

It has been justly said that many a stream of research, bearing the name of another, may be traced back to a Rutherford source.

After the Annual Meeting of the British Association at York, 1906, a famous controversy was started in *The Times* by a letter (9 Aug.), from Lord Kelvin, who assailed the idea of the evolution of elements; declaring that the heat of the sun was due to contraction under gravity and that to assign it to radioactivity was a hypothesis without foundation; and maintained that radium was a molecular compound consisting of lead(?) and five helium atoms.

Kelvin had to fight his battle almost single-handed, except for the embarrassing assistance of Prof. H. E. Armstrong, who criticised physicists in general as "strangely innocent workers under the all potent influence of formula and fashion" (*Nature*, 1906, p. 516); and claimed that 'radium' was so scarce that no one knew anything about it!

Sir Oliver Lodge on 15 Aug. wrote a firm but rash letter in rebuttal. He declared, writing of Kelvin, that "his brilliantly original mind has not always submitted patiently to the task of assimilating the work of others by the process of reading, and our hope has been that before long he would find time and inclination to look into the evidence more fully". This statement roused the ire of *The Times* which wrote an ill-informed leader with an indignant and just repudiation of the slur cast on Kelvin.

It is, perhaps, hardly respectful to suggest that Lord Kelvin publicly opposes opinions with the grounds of which he is not acquainted. There are, no doubt, many things which Lord Kelvin does not read and upon which he offers no opinion. But that he offers a considered opinion upon a subject which lies in his own province, without knowing what has been done by others, is more than the public will readily believe even on the authority of Sir Oliver Lodge....

As for Lord Kelvin, he protested on the 20th:

I am quite sure my old friend Lodge could not wilfully be unjust to me, but I do not think he knows how carefully and appreciatively I have done all I could by reading, and by personal intercourse with many of the chief workers in the field, to learn experimental results and theoretical deductions regarding radioactivity, ever since its discovery ten years ago by Henri Becquerel. I scarcely think any other person has spent more hours in reading the first and second editions of Rutherford's *Radio-activity* than I have.

R. J. Strutt inquired in a letter (15 Aug. 1906): "What then becomes of the heat generated by the radium admitted to be present in the earth?"

As the correspondence continued, it became clear that Kelvin admitted the production of helium from radium, which he therefore considered to be a molecular compound; hence his contention that there was no evidence for the evolution of atoms.

On 22 Aug., Sir Oliver Lodge took the opportunity to "assure him [Kelvin] of my profound and affectionate regard".

In the meantime the man most concerned had been silent, but on 25 Oct. Rutherford, who had been away from Montreal, wrote to *Nature*, and quoted pp. 482–3 of the second edition of his *Radio-activity*, where he had already answered most of the points raised by Lord Kelvin. What are these answers?

If radium is a compound, it is of a "character entirely different from that of any other compound previously observed in chemistry. Weight for weight, it emits during its change an amount of energy at least one million times greater than any chemical compound known." Moreover,

the rate of breaking up of this compound is independent of great ranges of temperature—a result never before observed in any molecular change. ...Radium, as far as it has been examined, has fulfilled every test required for an element. It has a well-marked and characteristic spectrum, and there is no reason to suppose that it is not an element in the ordinarily accepted sense of the term.... The radium atom is built up of parts, one of which, at least, is the atom of helium.... We must regard the atoms of the radio-elements as compounds of some known or unknown substances with helium....I have, for some time, considered that lead is the end or final product of radium. The same suggestion has recently been made by Boltwood....

The controversy died down; the victory has rested with Rutherford, who is credited with saying that "when a single experimental fact is established which does not conform with the disintegration theory it will be time to abandon it".

Soddy, who had opened the discussion at the British Association, and taken part in the correspondence, gave a summary of the whole case in *Nature* (1906, p. 516). There was something in Sir Robert Ball's remark that "radium was not a mystery but a miracle".

How amazed Kelvin would have been, had he known, as we now know, that these remarkable feats under discussion were the properties, not of the atom as a whole, but of that minute central citadel, or nucleus, whose dimensions are but about a hundred-thousandth part of the linear dimensions of the atom.

Mme Curie ably summed up the case thus:

I see no use in combating the theory (that radium can no longer be regarded as a simple element) enunciated by Lord Kelvin. There is no disadvantage in occasionally stirring up scientific ideas and in discussing researches from different points of view.

In every respect radium is a distinct chemical element in the sense which chemistry attaches to this word. It is highly improbable that Lord Kelvin considers radium to be a compound analogous to other molecular combinations. And it is thus possible that the discussion may have turned rather upon words than upon definite ideas, since it is apparent that all atoms are complex entities and are formed originally out of more simple elements the nature of which still remains almost unknown to us.

Long afterward Hahn wrote about his old days at McGill:

Rutherford was so sincere and unassuming in his dealings with his students and with everyday things of life, that we two Germans [Hahn and Levin] in particular were constantly filled with surprise and admiration. We had no doubt imagined that such a distinguished professor would be an unapproachable person, conscious of his dignity. Nothing could have been further from the truth. I still possess a small photograph which shows him clearing away the snow from the entrance to his house. In this house were often evening guests, listening in rapt attention to the intimate piano-playing of Mrs Rutherford or to the spirited narrative of the Professor.

Early in the year 1906, a photographer came to the Macdonald Physics Building to take a photograph of Rutherford working in his laboratory, for publication in the columns of *Nature*, with an article by Dr A. S. Eve on the Macdonald Physics Building. Rutherford was at first reluctant, but later granted the photographer permission to take a few flashlight photographs showing him seated at his alpha-ray apparatus. The photographs were duly taken and were quite good. In the opinion of the photographer, however, the already famous professor was not dressed elegantly enough for the readers of *Nature*. Not even cuffs were to be seen peeping from the sleeves of his coat! But the photographer found a way out; I was to lend Rutherford my loose cuffs. They were so arranged that they protruded well beyond the end of the sleeves. The photographer expressed satisfaction with the new photograph. As a result, in one of the volumes of *Nature* for the year 1906 (*Nature*, 74, p. 273), we see not only Professor Rutherford seated alongside the apparatus with which he carried out his epoch-making experiments on the alpha-rays, but also one of the cuffs of a young research student, who treasures his sojourn with one of the greatest masters of physical research as one of the most beautiful memories of his life.

In the summer Rutherford wrote to Hahn who had then left McGill and returned to Germany:

20 *Aug.* 1906. I am writing at present from a place called Lost River in the country about 70 miles from Montreal where my family are vegetating for three weeks. I arrived back from California about a week ago and left next day for the country, so I had not much time to attend to correspondence. I had a very good time in California and enjoyed my lectures. I had a class from 30 to 40, mostly professors and instructors, and managed to keep them interested. I found it very

difficult to get through much work, but finished the indexing of my book and the writing of the papers on the mass of the alpha particles. I sent off our joint paper at the same time as my own. The proofs will be returned to me at McGill for correcting, so it will be some time before they appear. I saw in California that your two papers appeared in the Phil. Mag. They both read very well.

I had an interesting time in California and wandered about a good deal. The weather was rather chilly most of the time. I spent two days at the Arizona Canyon on the way back and enjoyed myself thoroughly. I was glad to hear from Levin that your arrangements in Berlin were satisfactory. I hope to have the pleasure of renewing our acquaintance next summer as I hope to visit Germany. I shall miss both you and Dr Levin very much next year. The latter has sent a brother-in-law in his place.

About 1906 Rutherford bought a piece of land in Montreal on the north-west heights of the West Mountain, with a fine outlook towards the Lake of Two Mountains. Plans for a house were prepared and there was every intention to build, but his appointment at Manchester upset the scheme.

The preliminary steps had been taken by Prof. Schuster, who was retiring, to secure Rutherford as his successor to the Langworthy Chair of Physics at the University of Manchester. Rutherford wrote in reply:

26 *Sept.* 1906. I was very glad to receive your kind letter in reference to the Chair of Physics at Manchester, as it came at a time when I was seriously considering my future plans. I have had to decide during the past year between the attractions of McGill and Yale University and finally decided to remain here. My chief reason for this step was my hope to return ultimately to England to a position where I would not have to sacrifice laboratory facilities by so doing. The position at King's College seemed to me to invite the probability of the latter.

I very much appreciate your kind and cordial letter and am inclined to consider very favourably the suggestion of becoming a candidate for the position you propose. The fine laboratory you have built up is a great attraction to me as well as the opportunity of more scientific intercourse than occurs here.

It may possibly save time if I briefly state my views on the suggestions made in your letter. I should be prepared to give five lectures a week but no more, as otherwise there would be a serious curtailment of time and energy for research. At the same time, I should like to feel that there was sufficient lecturing staff in the department to give over one or more of these lectures occasionally in order to substitute a course on some special subject in which I was interested. I have had a good deal of committee work here as I belong to two faculties. I would be quite prepared to do my duty in that respect but would like to escape, as far as possible, from too much committee routine not involving my own or allied departments.

There is one point of importance on which I should be glad of further information—namely the state of the funds for the working of the department, including the amount for general teaching, research, etc., and how far the powers of decision in regard to expenditure are in the hands of the Director.

I need hardly tell you how much I appreciate the suggestion coming in the way it has. I fully recognise the spirit of self-abnegation displayed by you in the letter. Nothing could give me greater pleasure than to have you a member of the department to add your strength in the branch—Mathematical Physics—in which I should most value assistance. It is hardly necessary for me to say that I should be only too delighted to have your assistance and advice and I am sure that, as far as I am personally concerned, you would never regret the arrangement.

I should be glad to hear when you expect a decision to be made. Should I be appointed, I presume that I would not have to take up my teaching duties until next October.

Classes are now in full swing and I have settled down to another year of busy work. Kindly remember me to Mrs Schuster.

BRAGG TO RUTHERFORD

Adelaide: 10 *Dec.* 1906. Thank you very much for your paper on the retardation of the alpha particle in going through matter. It was splendid to get your results so clear. I want to congratulate you also on your success on measuring the ratio of charge to mass, and the velocity, in so many different cases. It is a great achievement for it must have made the most exacting demands on your skill and patience....

Rutherford, following some important work by Dewar, passed air and radium emanation through a tube containing coconut charcoal. He found that the emanation was wholly absorbed by the charcoal; the emanation could be released at will, by raising the temperature of the tube below a red heat.

He at once requested me to determine the amount of radium emanation per cubic metre present in the atmosphere, which was duly carried out. The absorption of gases by charcoal has not only played a successful part in the use of gas masks as protection from poison gases in war time, but its use has greatly improved the production of high vacua in electric lamps, etc. This method is also employed in making Dewar flasks, which, placed in a metal holder, are better known as Thermos flasks. Our debts to Dewar are great.

Prof. O. W. Richardson wrote to Rutherford from Princeton University, New Jersey, raising a question of interest and importance:

4 Nov. 1906. There is a point in connection with the measurement of the electric charge per unit mass for the alpha rays which seems just worth while considering, particularly in view of the idea, which Soddy claims to have established, that the alpha rays start by being initially uncharged. The point is this, that it is possible that the rays keep getting charged and discharged in succession during their flight, these changes taking place presumably during collisions. The discharge may not necessarily be complete; the particle may simply have a greater charge at one period than another....

I expect the alpha particles are helium with the double average charge as you suppose but I am inclined to think there is something in this charging, discharging idea. I shall be glad to know if you have ever considered it, and if you have not if you think there is anything in it. In any case it is interesting to know that even with such a queerly constituted alpha ray the velocity would come out right and the e/m only altered by the factor specified.

At a much later date Henderson of Dalhousie proved this surmise to be correct, and Rutherford, taking a great interest in the subject, also verified the results experimentally.

Max Levin wrote from Göttingen (18 Nov. 1906) to Rutherford on the "strange fact that potassium is radioactive"; he added that his brother-in-law Dr G. Rümelin of Freiburg was writing "delighted letters from Montreal", where he had joined Rutherford to do research work with him. Rümelin, a gentle, kindly soul, a good physicist and musician, was killed in the War. Levin translated Rutherford's *Radioactive Transformations* (Constable & Co.)—the Silliman Lectures—into German. He also wrote to Rutherford: "It is extraordinary how well the period of radium deduced by Boltwood agrees with your calculations."

RUTHERFORD TO HIS MOTHER

Montreal: 11 *April* 1906. I go to Philadelphia next Monday to attend the bicentenary of Franklin. I am giving an address there. I hear this morning that the University of Philadelphia intend to confer on me the honorary degree of Doctor of Laws (LL.D.). I am rather youthful for such honours, as they are usually the special perquisite of septuagenarians. This is my first honorary degree. They don't worry me much, I can assure you, but one is supposed to value them very highly—I imagine the esteem is largely dependent on whether you feel you deserve them or not. It is possible I may go to California to lecture in the summer and will spend a week at the University of Illinois to give six lectures. It will be an interesting experience and will add a little to my depleted banking account. (Later.) I had a very interesting time and the lectures were a success in every way. I leave on Monday next for the University of California stopping en route with Professor Trowbridge, of Madison, for the University of Wisconsin is giving me an honorary LL.D. degree there on June 20.

University of California, Berkeley: 10 *July* 1906. I am taking part in the summer session—at which about 750 students attend. I have passed through Frisco twice and seen the ruins of the fire [after the earthquake]. It is certainly a most depressing sight. For miles there is nothing but heaps of bricks and tangled ironwork. Wooden buildings are going up everywhere for temporary use.

4 *Aug.* 1906. I am staying at the El Tovar Hotel right at the edge of the Grand Canyon of Arizona. The hotel is located right on the edge and

you look twelve miles across a mass of peaks of sandstone and rocks of various colours left like battlements and castles 3000 to 6000 feet above the level of the Colorado ravine, which is out of sight from here and a mile directly below the hotel. The colours are very beautiful. I went down by mule today to the Colorado River. It was a very steep trail in parts and I walked and rode alternately. We got back after nine hours pretty steady work and I now feel pleasantly tired. There is an Indian house nearby and the Indians are dancing for the benefit of the guests of the hotel. The main point seems to be the music, which reminds me of catcalls. I visited the Lick Observatory and stayed with Professor Campbell, the director, for two days. I had a very pleasant time and saw through the big telescope. I also visited the University of Leland Stanford at Palo Alto, and saw the way the earthquake had knocked it about.

17 *Dec.* 1906. I have received the offer of the Physics Chair at Manchester. I think it quite likely I shall accept. I think it is a wise move for a variety of reasons. I shall receive a better salary and be director of the laboratory and what is most important to me, will be nearer the centre of things scientifically.

RUTHERFORD TO HIS WIFE

Berkeley, Cal.: 25 June 1906. Arrived at Oakland on Sunday afternoon after a pretty pleasant journey. The weather was a little hot at times.... Met at the station by Prof. Lewis...took a car up to the house where supper was ready....The household is very simple and things are run in a quiet way. The house is built on the side of a hill and is of the type of summer cottage you are so enamoured of. This morning I went to the University and met a few of my class, but I start regular work tomorrow. I have arranged for two rooms with a Prof. Wrinch and his wife—not with them but in their house. I shall take my meals at the Faculty Club. I move in tomorrow....Today has been foggy and distinctly chilly and we had a fire in the sitting room all the afternoon....I did not tell you about my concluding day at Madison. It was stormy but we went for a drive round the lake front where it is very pretty.

Berkeley, Cal.: 27 June 1906. I went into my new quarters on Tuesday. I have a small sitting room...and a small bedroom...I think I will be very comfortable. I go over to the Faculty Club for meals and do not lack for people to talk to. I have an advanced audience of twenty-five to attend my lectures, mostly lecturers and professors, so I cannot complain. Everybody is very kind and I think I shall enjoy my stay here....Loeb has invited me to go from Friday to Monday to see his seaside lab. at Pacific Grove about 100 miles from here. Mrs Loeb will be there and we are to make a twenty mile driving expedition on Saturday. I go with Lewis tomorrow to look at the ruins of Frisco. Last night I heard a lecture on the Frisco earthquake by a Jap, and Prof. Lawson. The Jap's voice was weak and I did not hear a word, but Lawson was interesting and showed exactly what happened, by photographs.

29 *June* 1906. ...In order to get to Pacific Grove this afternoon I have to pass through the ruins of Frisco and I will give you a report thereof. I expect to be back Monday afternoon. I am to be given a free pass over the Californian Railways, so can career round with little expense. I am very busy preparing the index of my book. The proofs were sent to me direct, and I have forwarded them back again. There is no sign of the effects of the earthquake in Berkeley. There was a camp of refugees in the University until a fortnight ago but they have all gone. Oakland is filled with people, but a great number are now located in shacks in Frisco itself.

3 *July* 1906. I left here at 2 to go to the Loebs....We first took the electric car to the ferry at Oakland and then to San Francisco. We passed through the city by car to the Southern Pacific Station right through the burnt part of the city. There was not a single house standing unburnt the whole route, but the streets on either side are covered with masses of fallen brick and twisted iron. Occasionally there is the shell of a building which still remains erect though burnt out. It is a pretty desolate looking spot, and dust is flying everywhere. The car lines are all running and are filled to bulging with people, but there is nothing doing in the burnt part, except here and there a wooden shack has been built for business temporarily. They are awaiting the money from the insurance people before rebuilding. We left by train at 3 and had a very pleasant, though warm, journey through beautiful country to Pacific Grove, arriving there about 7. We saw the ruins of a number of buildings en route and could just see the shell of the ruined chapel at Leland Stanford University. Loeb met us, and had taken rooms for us at the local hotel. We then

took a car to his lab. and bungalow, about a mile away, on the sea shore between Pacific Grove and Monterey. His house is a small one of wood of four rooms, including a small kitchen. Water and light is laid on. Mrs Loeb who had gone down a day before had dinner ready for us which we enjoyed thoroughly. She was main cook, an art which she acquired a year ago. A big salmon was the centrepiece of the meal. They are catching these by the thousand in the bay.... Spent the morning in Loeb's Lab., he is very busy there doing experiments in parthenogenesis and showed me all his methods of developing the sea urchin's eggs without fertilization. They were extraordinarily interesting. I first tried the normal method with the spermatozoon, and saw the egg develop in the course of a few hours. I then developed the beggar from the egg till he swam, by adding appropriate chemicals, without calling in the aid of the male. It took only about a minute to form the membrane and about three hours to get division of the cell and twenty-four hours to have them divided into sixty-four and swimming round. They are really wonderful experiments and he appears to me to be on the right track for great discoveries. Loeb is a short dark man, German Jew by origin, but not Semitic-looking, about forty-five and getting grey. I like him very much and I find we have a lot in common. He is very modest but a terrible enthusiast, worse than I am. Mrs Loeb is a very pleasant woman about his age. She was a Ph.D. of Zürich in Philology. He is rather helpless outside his lab. and she appears to look after him like a child. She is a pretty fiery character when she is roused and gives Jacques Loeb, as she calls him, a hot time when he gives vent to views she doesn't approve of. As she has no maid I helped wash up the breakfast things while Robertson swept out the dining room. We went to lunch at a restaurant. On Sunday we took a horse shay and went the seventeen mile drive through a natural park and reservation a good deal of the way close to the sea shore. The day was ideal and we all enjoyed it thoroughly. On Monday morning Mrs Loeb was to accompany us back to Frisco, but her husband forgot her ticket so she waited for a later train. We passed through Frisco again, had lunch at an impromptu Italian restaurant, and we got back here at three.

Berkeley, Cal.: 6 July 1906. ...Not much to record except an attack of diarrhoea on my part. The meals at the Club are not up to standard, and I am sure I got a dose of ptomaine poisoning from some soup there on Wednesday evening. I was feeling sick and miserable all Thursday, but managed to get through my lecture. Am pretty well right today. ...My lectures are going very well, and I have quite a fair class....Tea with Mrs Loeb, and the Taylors were there. He is a physiological chemist and a good fellow. She is a well known concert player here and certainly plays well even to my ignorant judgment. They were living in Frisco at the time of the earthquake, and she was in the midst of an attack of diphtheria. He had to secure a waggon and carry her out of town. The first night he intended to leave her in a house unhurt by the earthquake, but her nerves gave way and she refused to stay there. He then had to get another waggon and drive her into the country in the middle of the night and was held up by hoodlums, but managed to calm them with a rifle. Finally he placed her in a country sanatorium but in the mean time she had got reinfection. Altogether he had the devil of a time for three days....I am worried with reporters, but trust I shall escape without much damage. We have had beautiful summer weather for the past week—an absence of daytime fog, which is so common here. A fog however always comes down at sundown.

9 *July.* It strikes me that I am writing more letters than I receive....I lectured today on the condensation of the radium emanation and had the president for one of my audience. He told me that there was one thing he understood and that was the use of the sealing wax for charging the electroscope....I feel that I am getting a little lazy in this fine climate. I got through the index of my book, however, which I am very pleased to get off my hands....I find that if Montreal is expensive, Frisco is far more so. I understand a large part of the professors are perennially in debt.

15 *July* 1906. ...I don't know when I will be able to get away. In any case I ought, now I have the opportunity, to spend a couple of days at the Arizona Canyon and see something of the greatest wonder of the world. I forget how far I told you of my doings this week....Yesterday I went with the Taylors and Robertsons for a trip to Mt Tamalpais about 2000 ft. high. We first took car and ferry to Frisco, then ferry to Sausalito and then the mountain railway to the top. Started at 8.40, and got to the top at 11.50. After lunch, we started to walk down. It was a beautiful sunny day and we got a splendid view of the surrounding country. The trail was rather rough through chaparral, and a path

washed out by the rains. Got back to Frisco at six and then went down to a restaurant, had dinner and got back about ten. This morning the president called and wanted me to lunch at his house....I am trying to work as well as lecture, but find the combination difficult in this climate, with so many distractions around. I hope to visit the Lick Observatory next Saturday. Make any arrangements you like for your country trip....

While at McGill, and indeed all through his life, Rutherford was an omnivorous reader with a retentive memory. At the end of a hard day of work and thought, it was necessary to switch off the resultant nervous strain by turning to light reading. Novels and detective stories often served this useful purpose and the librarians at McGill sometimes had difficulty in meeting his insistent demands. His wife says that four library subscriptions did not supply quickly enough the stream of biographies, history in popular form, novels and books of general interest, which Rutherford read in the evening after he had stopped work. It was hard to meet the demand at McGill and it became much more difficult in Cambridge. In the country, after lunch, he could take a short sleep and wake up perfectly refreshed, without the unpleasant interregnum that some day-sleepers experience. Bridge would also serve as a recreation in the evening, but a day in the open air, preferably at golf, was his most invigorating tonic.

Rutherford insisted that many scientific men, particularly in the United States and Canada, spent too great a part of their time at their offices. He recommended that professors should spend more time in thinking and less in doing. "Go home and think, my boy", he used to say.

The hunt for the elusive parent of radium was continued with much zeal and little success. In November, 1906, Boltwood had written to Rutherford: "I have found the long-sought intermediate product between uranium and radium. It's actinium. You will remember that I have suspected for a long time that actinium was the father of that radium infant, and, as I wrote you last winter, I just put the suspected gentleman under observation." This hasty conclusion of Boltwood's turned out to be erroneous. He had prepared a solution of actinium and it was true that radium continued to increase in that solution. He sent some of it to Rutherford who verified that it grew radium—but did the radium come from the actinium?

The difficulties of this period are set forth in a letter from Rutherford to Hahn:

6 *Jan.* 1907. I was very glad to hear from you at Xmas time and to know that all was going well with you. You will be interested to know (I expect Levin will have informed you too) that I am to leave McGill having been appointed to the Professorship and Directorship of the Laboratory at Manchester (Eng.). I will leave here in June and start work in October.

The Laboratory is a very good one and also the salary, so I expect to have a good time there. I shall be glad too to be nearer the scientific centre as I always feel America as well as Canada is on the periphery of the circle.

I went down to New York for a day to the Physical Society at Xmas and gave a paper on the "production of radium from actinium".

My results will appear in *Nature*, I trust, in a week or so. It is very difficult to prove that actinium is *for certain* the parent of radium. The points you raised in your previous letter also occurred to me and I had them well in hand experimentally when you wrote.

Boltwood came up for a couple of days to stay with me before New Year. He is a very fine fellow and good company and is very well fitted for his special line of work. It rained hard all the time he was here. There was a scare the other day as Professor Porter's rooms in the Chemistry Building took fire. Not much damage was done. The fire was due to a high tension A.C. wire striking the clock circuit outside the building. It is well it was discovered.

In regard to the alpha particle and helium, etc., and the fitting in of atomic weights, I am not worrying about it at present. I think the density of helium determined by Ramsay is alright. It may yet turn out that the alpha particle is hydrogen and that helium comes from a rayless product. Boltwood swears that it is impossible that helium comes from thorium—so you will have a chance of improving this guess. The whole problem is very mixed. The difficulty lies in the fact that the alpha particle moves "in a peculiar way its wonders to perform".

Richardson has suggested to me it might be a hydrogen atom which gains and loses a charge alternately at each collision with a gas molecule. Under some conditions, the measured value e/m would then be 5000, agreeing with our value. The whole question is still sub-judice. We want a new method of attack.

Rutherford wrote again to Hahn in March congratulating him on his new discovery of a radioactive substance intermediate between thorium and radiothorium, which was fittingly named mesothorium.

It is clear that Rutherford's suspicions were now thoroughly aroused, for radium grew independently of the amount of actinium in his solution.

RUTHERFORD TO HAHN

March 1907. I was delighted to get your letter and to hear of the great things you are doing with your thorium family. It sounds very interesting and I hope you will get a good commercial method of separation of the members of the family. If you do so, kindly remember me when the time comes for the first distribution of the commercial article.

I see you think two years is about the half period of radiothorium. It is about what you supposed from your measurements here. The name mesothorium seemed good enough for the new product but I really think we shall have to get to work soon to revise the nomenclature. I expect to add another one or two soon to complete the parentage of radium but of that more anon. I will not say anything of the results in your letter till you give me permission.

I was very glad to hear you and Levin were taking part in the Radio-active meeting of the Physical Chemical Society. It will be a good advertisement for both of you and legitimate ones of that character are not to be despised by the embryo professor. I hope you have a successful meeting and would like to be there—in the audience!

By the way, do not pay too much stress to Boltwood's views of the position of actinium in the uranium-radium series. I think it not unlikely I shall put it out of court soon—at any rate experiments look that way. I can grow radium quite plentifully but I find I can grow it pretty well independently of the amount of actinium—at least my experiments indicate this as far as time has allowed. I had Boltwood up at Xmas and had a couple of pleasant days with him. I will have to be far reduced before I start tampering with the number of alpha particles expelled from an atom. If we once do that, all calculations are reduced to chaos.

My wife and Eileen are both flourishing and the former sends her kind regards and best wishes. I leave for Manchester May 17th by the *Empress of Ireland* and will spend some time in Manchester making arrangements for the season in October.

RUTHERFORD TO HIS MOTHER

Montreal: 18 *April* 1907. The time is rapidly approaching when we leave Montreal. I understand the university is giving me the LL.D. degree (honorary, my third one). A fortnight ago I went to Toronto and gave two lectures there and then went on to New York. While I was there I heard of the destruction of the McGill engineering building by fire and came back to find very little except the walls standing. The building was valued with its contents at about a million dollars, for it was one of the finest, if not the finest, engineering laboratories in America. Last Tuesday (ten days after the previous fire) the medical building was also destroyed by fire and incendiarism is suspected. The university buildings are now patrolled by night-watchmen to prevent a repetition. The university has more or less got the jumps and a telephone ring late at night gives me the 'jumps' expecting to hear of another fire.

A resolution of the Faculty of Applied Science, expressing regret at Rutherford's departure from McGill, was forwarded by Dean H. T. Bovey:

In the course of nine years, crowded with epoch-making researches, Professor Rutherford has permanently associated the Macdonald Physics Laboratory with discoveries of such significance that their ultimate effect on the conception of the Physical Universe cannot yet be foretold; has thereby extended the fame of McGill University to all parts of the world, and has attracted to her laboratories distinguished men from Europe and the United States.

His energy, directness and independence of view, combined with University experience in another Colony, as well as in the Old Country, made his opinion on University Politics one of the first and most valuable to be obtained; witness the leading share he took in organising the courses for higher degrees. In all such business his sincerity and geniality enabled him to take a strong line without risk to the warm personal friendship with every member of the staff, which would make the occasion of his departure one of pure regret, did not his colleagues rejoice in his advancement to a sphere where he may render still more eminent service to the cause of Science at large.....

In his letter to Principal Peterson, Rutherford explained that "the determining factor in deciding to go to Manchester was my feeling that it is necessary to be in closer contact with European science than is possible on this side of the Atlantic...".

In 1907 Cox and Rutherford were trying to arrange with the University authorities for the transfer of W. H. Bragg from Adelaide to McGill as Professor of Theoretical Physics.

Unfortunately, the Engineering Building and a large part of the Medical Building were burnt in two separate fires within an interval of less than a fortnight. In the crippled state of finance the appointment of an extra Professorship had to be dropped. This was a great misfortune to McGill, but a gain to Leeds, where Bragg went in 1908.

BRAGG TO RUTHERFORD

Adelaide: 17 *Dec.* 1907. I am sorry the Montreal proposal has fallen through, for the time at least. It was an attractive idea that I should be asked to do post-graduate and research work only; in such a splendid laboratory and in such good company: and in many ways I should have liked an experience of Canada. But finances have a way of ruling other things, and it is easy to understand that money is hardly forthcoming for new ventures where there have been such heavy losses by fire. It has been a great compliment to me that I should have been asked; and I do not know but that I value even more the kind way in which you did the asking and the goodwill with which Cox has worked for my transfer. You must not be troubled in the slightest because you had to tell me the proposals were to be dropped. I am no worse off than before: in fact I am better off. I have had my compliment, and found some new interests....

In the collection of letters to Rutherford from W. H. Bragg there are several dealing with the reflection of X-rays from crystals; beautiful work, for which W. H. and his son W. L. Bragg obtained a Nobel Prize in 1915. This great result rather emphasised the wave theory of X-rays although obtained by the advocate of the corpuscular aspect.

J. J. Thomson has given an authoritative summary of the magnitude of Rutherford's scientific achievements during the years he spent at McGill University:

Rutherford's scientific activity was never greater than when he was at Montreal. In the years between coming to Cambridge and leaving Montreal to be Professor of Physics at the University of Manchester he had published between forty and fifty papers; a few of these were joint papers, but the great majority were about researches of his own which led to results of first-rate importance and which could not have been obtained by anyone who was not an experimentalist of the very first order.

On 24 May the Rutherfords arrived in England from Canada and spent a few weeks in Manchester looking for a house and meeting new colleagues. The summer was spent at Mullion and Mortehoe, where they were joined by Mrs Newton and her son Charlie, then a medical student at Edinburgh.

RUTHERFORD TO HIS MOTHER

The University, Manchester: 30 June 1907. You will have heard details of our trip to England. At present I am staying here to make arrangement for work and to get things well in hand for the beginning of the session in October. The laboratory is very good, although not built so regardless of expense as the laboratory at Montreal. I went down to London a week ago and attended a meeting of the Royal Society, the Royal Society soiree, and a meeting of the Chemical Society. The soiree at Burlington House was a big affair. Lord and Lady Rayleigh, President of the Royal Society, received, and Balfour was present as well as most of the scientists and their wives. On Friday night we were guests of the staff of the Physics Laboratory and had a pleasant dinner. My predecessor, Professor Schuster, was present and also Dr and Mrs Hale of

California—a noted astronomer who is in command of the Carnegie Solar Observatory—by far the greatest and most important in the world, on the summit of Mt Wilson in California. On Saturday morning, the degrees were given in Whitworth Hall before a large audience and the Vice Chancellor, Dr Hopkinson, was quite a picture in his knee breeches and his University rig-out. After a short holiday in Cornwall I give a talk to the British Association meeting at Leicester on the "Constitution of the Atom".

P.S. I have a story to tell you which I heard yesterday and which my modesty almost forbids me to relate: Baron Kikuchi, Japanese Minister of Education, was here yesterday and was introduced to me by Schuster. Later he said to Schuster, "I suppose the Rutherford you introduced me to is a son of the celebrated Professor Rutherford!!!"

Sir William Ramsay made incursions into radioactivity which were singularly unfortunate. He was a great chemist, foremost among the men of his day. He was the discoverer of the presence on the earth of helium—an element whose existence in the sun had been revealed to Norman Lockyer by the spectroscope. To Lord Rayleigh belonged the credit of the discovery of argon, but Ramsay had then collected large quantities and explored the properties of that gas. Ramsay had three more rare and noble gases to his credit—neon, xenon, krypton. The first-named is familiar to town-dwellers today in frequent illuminating signs.

Imagine then the excitement and surprise when Ramsay announced in *Nature* (18 July 1907) that if some emanation of radium was mixed with *water* there appeared *neon*, with only a trace of helium; but that if the emanation was mixed with a solution of *copper sulphate* there appeared no helium at all but only argon, while the copper gave rise to *lithium*.

Here at last was the transmutation of elements with a vengeance! These statements, backed by Ramsay's great and deserved prestige, supported by his known skill in handling small quantities of gas and in using the spectroscope, produced a mixture of admiration, astonishment and bewilderment. The whole affair has since proved to be a mare's nest, all except the first observation of helium from radium by Ramsay and Soddy. Boltwood was scornful and wrote to Rutherford: "I wonder why it hasn't occurred to him [Ramsay] that radium emanation and kerosene [paraffin oil] form lobster salad." Perhaps the atmosphere can be conveyed by an extract from a letter from Boltwood to Rutherford. After considerable invective of a humorous type, he settles down to this:

I write with some feeling on this matter because I had a devil of a time trying to persuade some of my chemical friends last summer that Ramsay was not the whole show in radioactivity. We had a general meeting of the American Chemical Society here in New Haven at the end of June and practically every mother's son that I met was firmly convinced that Ramsay was the biggest thing that could be seen on the horizon. Practically not a single one was willing to concede that he even 'might be wrong'. I did not attempt to persuade them that he *was* wrong, but only attempted to prepare their minds for the denials of his conclusions which I felt sure would be forthcoming. I think that most of them felt the same sort of pity for me that a good catholic feels for one who is not a true believer. For the Pope can do no wrong.

I had an interesting talk with Ames [Professor of Physics at Johns Hopkins University] last winter at the time of the Physical Society Meeting. He got me to one side and asked me very cautiously what I thought about the 'lithium, argon, neon' business. I asked him whether he had read the Cameron and Ramsay paper in the Journal of the Chemical Society. He said that he had, and then I asked him whether he found anything to show that the lithium had not come from the glass or the reagents. He said "that was just what bothered me", and I told him that I thought the paper was all rot and rubbish. He replied "Th-Th-That was just what I t-t-told Remsen and he didn't like it. It was through Remsen that the first news reached the American press, he having been informed of the 'great discoveries' in a private letter from Ramsay."

It must in justice be remembered that Boltwood himself had advanced and advocated the view that the alpha rays from thorium were hydrogen and not helium, and this in spite of the large quantity of helium found in thorianite (several c.c. per gramme). Well, everyone makes mistakes,

except those who do nothing; but it was hard for Rutherford to hold
the scent when so many of the hounds were starting and chasing hares.
Some questions remain. Who discovered that alpha particles were
helium? The man who first saw the spectrum? or the man who found
that the charge per unit mass was half that of hydrogen and concluded
mass 4 and charge 2 was the true deduction, considering the large amount
of helium in radioactive rocks? or the men who collected alpha particles
and proved them helium?

When Rutherford gave his Nobel Prize address in December 1908,
he referred to these early discoveries in these words:

Using 30 milligrammes of Giesel's preparation Sir William Ramsay and
Professor Soddy in 1903 were able to show conclusively that helium
was present in radium some months old and that the emanation produced
helium. This discovery was of the greatest interest and importance, for it
brought to light that in addition to a series of transient elements, radium
also gave rise in its transformation to a stable form of matter.

A fundamental question immediately arose as to the position of
helium in the transformation of radium. Was the helium the end or
final product of the transformation of radium, or did it arise at some
other stage or stages? In a letter to *Nature* (20 Aug. 1903) I pointed out
that helium was derived from the alpha particles fired out by the alpha
ray products of radium, and I made an approximate estimate of the
rate of production of helium by radium. It was calculated that the
amount of helium produced by a gramme of radium should be between
20 and 200 cubic millimetres per year, and probably nearer the large
estimate....In 1908, Sir James Dewar found that the rate of production
of helium was 154 cubic millimetres per gramme per year, not far
from the value calculated as most probable five years previously.

There was again a spirited atmosphere of discussion at the Leicester
Meeting of the British Association (1907):

Professor Rutherford championed the electrical theory of matter....
The effective actions of the electron all seemed to be confined to a small
volume, and the electron itself was uniformly small in dimensions
compared with the atom of ordinary matter....Certain it was that the

existence of the electrons had been definitely proved, for they now knew of a variety of methods by which they could produce them in enormous quantity. The electron, he submitted, had come to stay [cheers] and the view that it was a constituent of all matter had been greatly strengthened by recent investigations....

Sir Oliver Lodge complimented the opener [Rutherford] on the agility with which he had skated over this tremendous subject....In some respects the electrical theory of matter did not receive the full approbation of Lord Kelvin. It was always interesting and often hazardous to differ from Lord Kelvin, although the latter had taken of late to the slaying of his own children [laughter] and sometimes they had to appeal from Lord Kelvin to Thompson and Tait....

At the same meeting Sir William Ramsay announced another remarkable and startling discovery. He declared that, by placing radium hermetically closed in a glass vessel, the electrons emanating from the radium through the glass, and falling on a nickel bar placed in juxtaposition, had the effect, after a certain time, of covering the bar with a film of radioactive matter, which could be separated by chemical treatment. His conclusion was that some sort of transmutation took place converting the nickel into some other substance, this being characterised by its radioactivity.

This very erroneous statement, coming from so eminent a chemist, was believed for a time by many. The experiments have been repeated and the effect may be attributed to some escape of radon leading to a radium active deposit. With proper care in avoiding leaks no such effect can be produced, no, not with five grammes of radium, as McLennan and Grimmett showed at the Radium Beam Therapy Research at a later date.

RUTHERFORD TO HIS MOTHER

18 *Aug.* 1907. The British Association Meeting at Leicester was a great success—about 200 present. I started a discussion on the "Constitution of the Atom". Lord Kelvin, who is still as lively as ever, Sir Oliver Lodge, Sir William Ramsay, Larmor, and others took part. It was the best discussion at the meeting. I also gave a couple of other papers.

Everybody is very kind and apparently glad to see me back for good in England. I return to Manchester in a week's time to get things going for the beginning of term. I shall have a number of researchers over from Germany—probably fifteen in all. My two German researchers (at McGill) have turned out very good men and have got through a good deal of work.

About this time Rutherford was elected to the Athenæum under the rule which empowers the annual election of a certain number of persons of "distinguished eminence in science, literature, the arts or for public service".

Whenever Rutherford appeared at the Club a knot of members tended to gather round him and he became the centre of animated conversation. The lunch at the Athenæum made a break in a day of meetings and was a very welcome rest to him.

CHAPTER VI

THE NOBEL LAUREATE

Rutherford left McGill with regret and went to Manchester with high hopes, but he could scarcely have imagined how great and important an acceleration was to occur in his scientific achievements and professional career. He was glad to be nearer the centre of gravity of scientific thought and research. Thirty years ago, American and Canadian physics were far behind the standard they have reached today. Publication of his papers would now be more prompt, and conferences with other savants more frequent.

Arriving in England in June, Rutherford attended on 16 June a reception at the Cavendish Laboratory where he showed some experiments with radium emanation in the large Lecture Theatre. He took up his duties as Langworthy Professor at Manchester in October, and found that his predecessor, Prof. Arthur Schuster, had left his department in a high state of efficiency. The laboratory, which had been opened in 1900, was particularly well equipped with electrical apparatus and there was a liquid air machine. The junior laboratory steward, Mr William Kay, was "young, energetic and exceptionally capable", and above all there was Schuster's young assistant, Hans Geiger—who could have wished for a better?

Rutherford settled with his wife and daughter in a comfortable house with a garden at 17 Wilmslow Road, Withington, nearly two miles from the laboratory.

Soon after his arrival at Manchester there was an interesting *fracas* at the first Faculty Meeting he attended. Just before he arrived, Chemistry had annexed certain rooms that had previously belonged to Physics. This action was condoned, though there was sympathy with the newcomer. When the time came for Rutherford to speak he brought down his fist on the desk with a resounding bang and exclaimed "By

Thunder!" He followed this up with a vigorous speech and finally pursued the Professor of Chemistry to his study protesting that he was a nightmare—"like the fag-end of a bad dream". When in his more sober days the story was recalled to him, he was quite unrepentant and hailed it with delight.

On 6 Sept. 1907, Hahn wrote his latest views to Rutherford: "As a matter of fact, I believe that the father of radium has just the same chemical qualities as ordinary thorium." He gave his reasons from experience gained in dealing with *old* thorium preparations which contained radium. He wrote, "either (1) thorium disintegrates into radium; very unlikely (but Boltwood's idea), or (2) the parent of radium is mixed with the thorium. The radium appears gradually with time. As radio-thorium follows the reaction of thorium, the father of radium has to follow the radio-thorium and I must find it after some time...."

"All these discoveries of Sir William Ramsay's are still so bewildering to me that I do not know what to make of it...."

It will be seen that Hahn was in full cry for the parent of radium, but Boltwood was able to find it first.

On the other hand, when Hahn discovered mesothorium, Boltwood at once accepted it and said "I was almost there myself". Hahn suspected (13 June 1908) that there were two mesothoriums, and afterwards this view was found to be correct. The order of succession is therefore thorium, mesothorium 1, mesothorium 2, radiothorium, thorium X, thorium emanation (thoron), thorium A, B, C, D.

Rutherford's letters to Hahn throw much light on the type of difficulties which were puzzling all the workers at radioactivity during this period.

24 *Sept.* 1907. I was glad to get your letter the other day and to hear how your work was progressing. It is very satisfactory to have the mesothorium settled. You say it emits beta rays; I suppose of only very weak penetration like radium B. If not, Levin's result of the absence of beta rays from thorium freed from Th X cannot be right, unless he

separated the mesothorium. Have you got your methods of separation on a commercial scale? I am hoping to see radiothorium and mesothorium soon on the market.

We are now comfortably settled in our house in a suburb of Manchester. We have had very fine weather the past month, but I presume foggy weather will soon be with us to show the approach of winter. I have got my work going in the Laboratory and have continued my observations on my actinium solutions. All goes well and I hope to settle the question definitely before long. I don't know that you would expect radiothorium to grow radium unless there are two methods of production of it; but there is no harm in trying. I shall have a good number of researchers to look after. H. W. Schmidt is coming, also Retzel from Leipzig, and Brill as soon as he finishes his work with Ramsay. I also have Geiger who was working here last year, one of your countrymen, and also a representative of Japan. In addition I have a contingent of local workers. I expect to be very busy as I have promised to give a number of addresses and lectures. I am in good form and my wife declares I am getting stouter—but slowly in any case.

Is there any chance of your coming over to visit us? We should be delighted to have you stay with us if you come over. Let me know any time you could run over. It would be a good way of using up a vacation.

I am writing up 'production of radium' paper. The results go well together.

P.S. I have been reading your letter again. I think your idea that the parent of radium has the properties of thorium or radiothorium is very likely correct. In fact, I find its properties to go with radioactinium and Boltwood finds thorium precipitates it. I do not think thorium changes into radium. The small amount of radium in orangite, etc., is against it.

I quite agree with you about the life of radium. I am sure Ramsay is a long way out. In fact, I believe his volume of emanation is miles too big. I have several reasons for thinking so.

Many thanks for your congratulations on the honorary degree at Giessen. I thoroughly appreciated it.

I don't know what to think of Ramsay's results; I must confess to being a little sceptical.

Boltwood's energy and skill were at length rewarded by his discovery of the parent of radium—a great feather in his cap. He wrote to Rutherford on 23 September:

You will have seen my letter in *Nature* about the 'radium parent' before this reaches you. It is a thoroughbred radioactive element all right, with all the essential attachments. It gives out nice slow alpha-particles with a range of less than three centimetres....I have sent off a second communication to *Nature* giving the above facts and proposing the name 'Ionium' from ions, etc. It is a curious and interesting fact that ionium was the chief, if not the only, radioactive constituent of the radioactive substances that I separated from pitchblende in 1899, and which I have always supposed to be actinium!

Rutherford was not at all taken with the name 'ionium' because certainly all radio-elements ionise gases. He playfully suggested PaRa, or paradium, but Hahn pointed out that this smacked of goosestep.

Boltwood wrote in reply to this criticism:

I am interested in the communications by Campbell and yourself re 'ionium' which have appeared in the columns of *Nature*. I cannot help recalling the conversation of the White Knight with Alice in the last part of *Through the Looking-glass* which refers to a song which he is going to sing her. In this case, the name is 'ionium', the thing is called 'paradium', and the name of the thing is called 'uranium A'! If later we can succeed in getting the lady (Hahn calls it 'die Muttersubstanz', so *it* must be a she) respectably tied up to Uranium, then I will not have the least objection to her changing her name. But for the present my motto is 'fest sitzen'.

Rutherford was always anxious to co-operate with chemists and mathematicians with a view to further insight into physics. He sent an invitation to Boltwood, really a physical chemist of high ability, to work with him for a year at Manchester; to which Boltwood replied:

30 *Oct.* 1907. Your suggestion about my going over to Manchester to work is certainly an attractive one. I know of nothing that appeals to

me more strongly just at present. What I would like above everything else is to be with someone who is energetic and enthusiastic, for now I sometimes feel doubts as to whether it is really worth while working nights and Sundays. One does get so lonesome all by oneself at such times. But it really isn't feasible for me to think of going over at present.

RUTHERFORD TO BOLTWOOD

20 Oct. 1907. This is a pretty active place and, but for its climate, has a number of advantages—a good set of colleagues, a hospitable and kindly people and no side anywhere....I find the students here regard a full professor as little short of Lord God Almighty. It is quite refreshing after the critical attitude of Canadian students. It is always a good thing to feel you are appreciated.

In 1907 Rutherford published his third book called *Radio-active Transformations*, which was the official publication of his Silliman Lectures given at Yale University. There was indeed little in this book which had not already been given in the second edition of *Radio-activity*. A German translation was made by Max Levin, who had been a research student with Rutherford at McGill University, and the book was most favourably received in Germany.

RUTHERFORD TO HIS MOTHER

29 Oct. 1907. I have now been lecturing a month and getting things into shape. I am naturally very busy and as newcomers we shall probably have to do a good deal of dining out. I go to two big dinners this week, one to Mr Donner—a wealthy merchant here—and one to Professor Schuster, my predecessor who, unlike most professors, is a wealthy man. Everyone is very kind and I am enjoying my life thoroughly. I have a good many outside lectures in hand and give one today at the Manchester Literary and Philosophical Society. I am lecturing later in London at the Royal Institution, at Dublin and Liverpool, and so will be kept busy. I am giving a special series of lectures on 'radioactivity', which are well attended.

Rutherford began his work at Manchester with a great shortage of radioactive material of all sorts. He had hardly settled in the laboratory before he approached the Vienna Academy for a loan of radium.

RUTHERFORD TO EXNER

5 *Oct.* 1907. I desire to make a formal application to the Kaiserliche Akademie der Wissenschaften of Vienna for the loan of some of the radium preparations in their charge. For the experiments I have in view, I shall require an amount of radium preparation corresponding to about *half a gram of pure radium.* For my purposes, I do not require pure radium bromide but preparations of about 25 per cent purity would suffice.

If the Akademie is willing to loan me the material, I should at once commence a series of experiments to investigate in particular:

(1) Physical and chemical properties of the emanation.

(2) Ionization of gases exposed to very intense radiation.

(3) Final product of transformation of radium.

I may mention that the University at Manchester possesses at present less than 20 milligrams of pure radium bromide.

I shall be glad to know as early as possible whether the Akademie will be able to grant my request.

An arrangement was made that Ramsay at University College, London, and Rutherford at Manchester should take equal shares of the 350 mg. of radium bromide lent by the Vienna Academy. Naturally, difficulties arose, and a lively correspondence followed.

RUTHERFORD TO RAMSAY

11 *Nov.* 1907. I wrote to you a fortnight ago about your views in the arrangement between ourselves for the use of the radium from the Austrian Academy, but I have not received a reply.

Today, I received the official communication from the Academy notifying me that Dr Brill is taking over to you the 350 mg. of radium bromide and that it is loaned for our common use.

I shall be glad to know when Brill is expected to arrive and also to consider some arrangement for use of the material. I presume that the Academy will expect to know the general arrangement between us, more especially as the results of the experiments are to be communicated through the Academy.

I do not know whether you are aware of my state of poverty in regard to radium. The maximum quantity available for experiment is 7 milligrams: I am unable to do any experiment that involves the use of even 30 milligrams and am consequently hampered all round. I have a number of physical experiments I wish to carry out as soon as I get a workable quantity. I presume you will continue along the lines of your last communication to the British Association. I shall naturally leave that direction free to you at any rate for some time. I shall be very pleased to consider any proposition that is mutually convenient.

RAMSAY TO RUTHERFORD

11 *Nov.* 1907. I have delayed answering yours of the 29th Oct., till I was sure of getting the radium bromide. It has now arrived.

Instead of your suggestion, I have another to make, or rather two alternative suggestions. It would be an infinite pity to divide the radium. It is so infinitely more valuable as a whole. So please agree to my retaining it for some time. I will measure the emanation volume, try its action on salts, get its spectrum (and all the machinery for this is nearly ready) and after a year, or a year and a half, hand it over to you. My second suggestion does not really interfere with the first. It is that I can occasionally send you four days growth of the emanation, sealed in a tube along with H_2 & O_2 from which you can extract it without loss. After all, it is the emanation and its product that you want, and if you leave the stuff in my hands, it will be not merely the emanation from the 0·388 gram of impure bromide, which I have received, but from 100, and later 150 mgms. additional; i.e. 0·488 and later 0·538 gram. Brill is at the atomic weight of my purest 50 mgms., which he has recrystallised; when he has finished, I shall dissolve it and place it in a bulb, and shall then again have 150 mgms., of my own, besides that of the Austrian Academy. I think you might leave any experiments for which you want solid radium bromide to the last. They are determining the heat evolution in Vienna with a much larger quantity; and I don't see how the solid can otherwise be used. It would be better to run no risk of wasting it till all experiments on emanation which suggest themselves have been made—or at least a fair number.

Have no hesitation in demanding from me frequent assignments of the emanation plus mixed gases. It will give me very little trouble. I draw every four days, and if inconvenient I will tell you frankly. You shall have it only 18 hours old; i.e. if I post it at 5 p.m. it will be in your hands at 9 a.m. next morning.

Let me hear what you think.

RUTHERFORD TO RAMSAY

13 *Nov.* 1907. I received your letter this morning; my letter of enquiry must have crossed yours.

I must frankly state that I am somewhat disappointed over the radium affair. I was given to understand some months ago that I was to be loaned a considerable quantity of radium by the Austrian Academy; then the wish to please two people leads to a joint distribution; and now I find you wish to have practically the whole use of the loaned sample for a considerable time.

I had made preparations for work in several directions which must of necessity be postponed until I get a supply of radium at my own hand. In this connection also, I recognise very clearly that there is no time like the present.

However, I am desirous of meeting your wishes as far as I can, for I quite recognise the importance of not dividing the specimen. My initial proposition appeared to me reasonable since by an occasional interchange of the emanation from a half quantity, each of us would be able to get the advantage of the whole.

You suggest that you would like the use of the 350 mgrs. radium bromide for a year or so. I think it very desirable in the interests of both of us to have a very definite agreement on this point. I am willing to forego my claim on the radium in the meantime if you would agree to hand over the whole preparation to me not later than Jan. 1st, 1909. This would allow you time to carry out a good many of the experiments you have in view, with a large quantity of radium. The date of handing over must not depend on whether any particular work is finished, or not, for I recognise that there will be plenty of work to do on this subject for a decade or more.

As you propose, I should also like to have an occasional 'draw' of the emanation from your stock. To be definite, I would suggest that I do not ask for more, on the average, than eight days collection of emanation per month. I would be willing to reciprocate when my turn comes. I told you I am practically destitute of radium and this arrangement would help me in several directions. Naturally, however, I could not hope to do much serious work under such conditions.

Before you dissolve the preparation, I think it would be very desirable to set aside about 2 mgrs. of the solid preparation for gamma ray tests.

I would like to compare this standard sometime later to see how they agree.

RAMSAY TO RUTHERFORD

15 *Nov.* 1907. I am glad you are frank; it makes things so much easier.

I don't think you realise that you will be getting emanation from the Vienna quantity and from 150 mgrs. of my own. It will be much easier to get the pure emanation from my arrangement, pure except for H_2 and O_2. I don't think there will be any difficulty in giving you as much emanation as you want. I should draw every four days. That gives the maximum yield. So you might get practically two lots a week, for a week or two, with an interval of a week, and so on.

I would rather not promise January, 1909, because I might be tempted to break my word; but I will take the date as provisional, and try to meet your wishes. Let me be as frank as you. I am used to handling this stuff in quantity; I have learned by long experience; and I am a little afraid that you or your assistants would find it difficult to get the apparatus as perfect and as workable as I. I always draw, *myself, personally. . . .*

19 *Nov.* 1907. I propose to collect the gas given off by the radium bromide on addition of water; then to analyse it, and get the spectrum of the inactive gases and see if neon and argon are present. I may have a shot, too, at the spectrum of the emanation.

This involves the construction of a good deal of apparatus, and I much doubt if I shall be ready on the 25th; but I will try.

As for the emanation: I shall send it to you mixed with Knall-gas —what it itself produces—in a little tube, locked in a small wooden box. When you receive such a tube, would you use it as soon as possible, and post it back. I shall have two made, so that we can have both on the run.

I want to do this first gas, obtained from radium bromide and water, well; so I don't want to risk an accident, or to hurry too much. After the radium bromide is sealed in position, there will be no trouble to draw off what is wanted. It will not take ten minutes. But before that, I can't promise anything.

As for sending a messenger to the train, that will be easy. It will reach you in four hours, and you can meet it at Manchester. I will always tell you the time it has taken to collect emanation so that you may know its actual value pretty accurately.

17 *Feb*. 1908. Brill told me of your present, and I congratulate you on it. If I were you I should dissolve it in acid—HCl for choice—and transfer it to a bulb of say 50 ccs. capacity. You have only 3·75 grms. of solid, and that would be ample to contain it. Seal it to a pump like mine; if you want any particulars, only write and I shall be delighted to give you all I know.

All the same if you want any specially large amount of emanation, just let me know, and we shall send a 'draw' as before. We can easily spare it, from time to time, for we have difficulty in keeping pace with its production. At present we have a bulb with silver nitrate waiting to be fed.

The last of these letters refers to a gift of 300 mg. of radium (as a chloride) from the Vienna Academy on 1 Jan. 1908, so that Rutherford was no longer dependent on Ramsay for supplies. Rutherford and Royds arranged apparatus so that radium emanation, freed from gases, could be condensed in a fine tube which could be sealed and used in place of radium. Sometimes a charged platinum wire was placed in the emanation so as to attract the active deposit. The demand for such sources was frequent by all the research workers in the laboratory. Through the President of the Royal Society, he also received from Paris the residues from pitchblende after the radium was extracted. These were rich in actinium, radio-lead and polonium. Radioactively he was now a rich man and could get on with his work. He did.

Prof. Stefan Meyer of the Radium Institute, Vienna, has explained the situation thus:

As to the radium preparation which Ramsay had in hand for several years from our Academy of Sciences it was not the same as the one Rutherford got in January 1908. The first loan of about 300 *mg. RaBr*$_2$ (equivalent to about 175 *mg. Ra*) was given to Ramsay who had to participate with Rutherford in the use of it. But there was always friction between the two and Rutherford could not work in this way as unrestrainedly and profitably in his own way as he, and with him the whole scientific world, would have liked. So the Vienna Academy resolved to give him a preparation for his exclusive use and decided the loan of 300 mg. radium mentioned in several letters.

Rutherford not only had the necessary material but he had an enthusiastic chemist to help him. Thus he wrote to his mother:

20 Nov. 1907. You will have received the newspaper about the Birmingham celebration, where I received another to my list of degrees. Boltwood, of Yale University—you remember, an old friend of mine—is a great addition to the laboratory, and I am very glad to have him here for a year. I have a very good lot of researchers here, including Florance from Christchurch, Boyle from McGill, Gray from Melbourne, and three from America, not to mention Germans, Russians and English. What with looking after them and my own work, I am kept pretty busy.

The first great success at Manchester was the counting, one by one, of the alpha particles projected from a stated amount of radium. These particles were counted as a boy counts his marbles, or as a policeman numbers people going through a door.

The answer obtained was that a gramme of radium every second ejects 34 thousand million alpha particles, which become helium atoms after they have lost their electrical charges, or rather, picked up two electrons.

Rutherford when at McGill tried to count such particles by passing them through a small hole bored in a swinging pendulum into a testing chamber of small volume and low electrical capacity. He found that the resulting ionisation was just too small to be detected. When therefore Rutherford wrote from Manchester to me that he had succeeded in detecting single alpha particles, it was a great and unsolved puzzle to guess the means that he had employed. Yet the method was simple in conception, although there were the usual experimental difficulties to be overcome.

Briefly the plan adopted by Rutherford and Geiger was to admit alpha particles through a small hole covered with a very thin sheet of mica into a cylinder, partially exhausted of air, down the axis of which was a wire at such a high voltage that it was almost sparking to the cylinder. The alpha particle on entry made some ions in its path, and each of these ions made a vast number more by colliding with molecules; these new ions did the same! This principle of ionisation by collision was a great discovery due to J. S. Townsend at the Cavendish, at the time when he and Rutherford were working as research students.

The ions formed an electrical current which caused the central wire to lose rapidly much of its charge, as shown by the electrometer to which it was attached. It was quickly recharged by a battery passing a current to it through a high resistance; so the trap was set all ready for the next arrival of an alpha particle. It was possible to sit in a comfortable chair and see a spot of light on a screen move a few inches every time an alpha particle entered. Sometimes two would enter almost together and then the jump would be nearly doubled in length. The equivalent amount of radium was known and the distance from the source down a long evacuated tube to the 'trap' was measured, so that the ensuing calculations were of an elementary character and the result was not open to doubt, provided the radium standard of comparison was correct.

The detection of a single atom was a great landmark in the progress of physics. This experiment put the famous atomic theory on a sure footing and the theory has received further remarkable support in many different ways since Rutherford's first step. It is to be noted that the same man who showed that atoms disintegrated also placed the atomic theory on a firm foundation.

Rutherford described his triumphant experiment in this letter to Hahn:

27 Jan. 1908. I was interested to hear of your new product which I have no doubt is all right. This makes mesothorium the exact analogue of actinium. You appear to have a private hunting ground in thorium and I can quite understand your difficulty of names. I think it would be a pity to concoct a new name for the new product. Your suggestion Th_1, Th_2, Th_3 seems without objection, especially as you can give the other names as alternatives.

I am sorry that I have not forwarded the two copies of my paper you asked for but shall do so today. I forget whether I told you that we have detected a single alpha particle by an electric method. I find that the number expelled from radium is of the same order as the calculated number, but it will take some time to fix it with accuracy.

I am working with Dr Geiger—one of your countrymen who was here last year. He is a very excellent experimenter and is a great assistance to me. The method is to fire an alpha particle into a small hole covered with thin mica into a cylinder about 60 cm. long where the air pressure is about 30 cm. There is a thin central wire and a voltage is applied of about 1000 until a discharge almost passes. Under such conditions, the ionization produced in the gas by the alpha particle is magnified 2000 times by collision. The effect of each alpha particle is marked enough to show an audience. I am lecturing at the Royal Institution on Friday next on 'Recent Researches in Radioactivity' and, if I can get a battery, I may try and show the experiment. You will appreciate the importance of counting the alpha particles. I can obtain the charge on each and hope to settle whether the alpha particle carries two charges or not.

In 1908, six months after his arrival at Manchester University, there came an unexpected windfall to Rutherford. The Turin Academy of Sciences awarded to him the Bressa Prize. This was an impressive testimony from an outside body of scientific men to the importance of Rutherford's discovery of the "mutability of matter and of the evolution of the atom".

RUTHERFORD TO HIS MOTHER

11 *March* 1908. I received a very agreeable surprise yesterday when a telegram came from Turin announcing that I had been awarded the Bressa prize by the Academy of Science of Turin. The award was notified in the morning papers, with the addition that the prize is worth £384. When the telegram came I did not even know there was such a prize, so after hunting up the particulars, I was very pleasantly surprised. I gather that this prize is awarded every two or three years to the man who, in the opinion of the society, has done the best work or published the best book during the preceding period. It is awarded to all branches of experimental science, including in addition archaeology, history and statistics.

The Manchester Guardian (21 March 1908), in making an announcement of this award, added a story illustrative of Rutherford's industry. The Janitor was asked what were the Professor's hours of work. He replied: "You can never tell when he'll leave his lab. and go home."

Another notable contribution of this *annus mirabilis* was the careful measurement of the electrical charge carried by a single alpha particle. Rutherford had already made a rough measurement of this quantity at McGill, but now that he knew well the number of alpha particles projected from a given quantity of radium in a stated time, it was possible to direct the particles into a testing chamber and measure the total amount of electricity that they carried. Simple division gave the amount carried by one such alpha particle. The result again confirmed the view that helium was doubly charged, or, as the chemist might say, it was He^{++}, or, as the physicist might add, it was a helium atom stripped of two electrons. The value he found for the unit of electrical charge was a great surprise, for it raised the significant figures of measurement from the old established value of 3·4 up to 4·65, an increase of 36 per cent. When I protested to Rutherford that this result must be wrong, he pointed out that eight years previously Planck had obtained a similar figure to his own. Indeed, that marvellous man Planck had studied the results of radiation distribution from hot bodies; decided that the energy must pass from an atom in quanta or bundles; that each bundle must have energy proportionate to its frequency; measured the constant now called Planck's constant h; settled that 'action' was atomic; laid the foundation of the Quantum Theory (which can never leave physics, for it is fundamental in Nature); and, as a by-product, obtained an excellent value, for that time, of the electronic charge!

Planck is a giant who has greatly modified man's outlook, or insight, and in a manner which is not artificial, but inevitable; for the basic facts are inherent in Nature, however much the interpretation may be the product of a man's mind. Indeed the man of science is driven to his conclusions by his experimental facts—he does not concoct theories to suit a fanciful imagination.

Another interesting discovery had already been made. When alpha particles strike a thin layer of zinc sulphide, spread on a piece of glass, small bright specks of light appear which may be seen well when examined with a magnifying glass in a dark room. There was a little instrument first made by Sir William Crookes to show this, which gloried in the ungainly name of spinthariscope. Rutherford and Geiger, using a microscope, counted such specks on a given area at a measured distance in a known time. They proved that each light speck was due to an individual alpha particle. The atomic theory was again upheld, and a method of counting alpha particles was established, so that for a long time many physicists spent weary but exciting hours trying their eyes in counting single particles. Happily this trouble is now removed in modern counting by the use of Rutherford and Geiger counters and suitable electric valves or thyratrons.

About this time it began to be clear that the actinium family was an offshoot of uranium, but Boltwood wrote to Rutherford on 11 October:

I hate that notion of an illegitimate branch of the uranium family like the very devil....As a matter of fact I am getting a little down in the mouth on the radioactivity matter for I am afraid that the time is rapidly passing when one can hope to accomplish very much with homeopathic doses. I see that someone has given a lot of money for a Radioactive Institute at Vienna and I am afraid that the wholesale business will drive the small dealer like me to the wall....

J. Joly of Dublin University was working at the amount of thorium in ordinary sedimentary rocks and finding it about equal to that of uranium. He was especially anxious that Rutherford should find out the period to half-value of thorium, for obviously this is a factor in the heat produced within the earth by radioactive matter.

In particular, Joly was making a radioactive survey of the St Gothard Tunnel and it seemed as if places of high temperature in the tunnel coincided with regions where radioactive contents were high.

In Rutherford's laboratory there were experiments in progress by Geiger, who was firing alpha particles through thin metal foils on the fluorescent screen and seeing how much the particles swerved aside from the straight line motion. This was the beginning of a great matter which came to a head two years later.

Sir James Dewar of the Royal Institution (*Proc. Roy. Soc.*, A, 1908, 81, p. 285) wrote on the rate of production of helium from radium:

Professor Rutherford in his work entitled *Radioactive Transformations*, 1906, p. 186 on the theoretical assumption that the alpha particle is an atom of helium carrying twice the ionic charge, deduced from electrical measurements that the number of particles expelled per year per gramme of radium would reach 4×10^{10} and as 1 c.c. of a gas at standard temperature and pressure contains $3 \cdot 6 \times 10^{19}$ molecules the volume of helium produced per year would amount to 0·11 c.c., which is equivalent to about 0·3 of a cubic millimetre per day. Considering I have found a rate of helium products of the order of 0·37 cubic mm., the agreement between the experiment and the theoretical prophecy of Rutherford is almost too wonderful, substantiating as it does the accuracy of the theory of radioactive changes he has done so much to initiate and develop.

No doubt the agreement was better than the experiments justified, but Rutherford had an uncanny way of approximating to the correct value even before more exact experiments could possibly be carried out.

LARMOR TO RUTHERFORD

22 *July* 1908. I have been fascinated into reading up your white magic, which I hope is going to press tomorrow to appear in the next number of the Proceedings of the Royal Society.

I understand that radium and its products are in equilibrium when each of the latter is destroyed as fast as it is produced—and that each of them gives out the same number of alpha particles. In deducing their atomic weights you assume that they give out nothing material in addition to the alpha particles. I suppose the experimental proof you refer to is that the radium in equilibrium is four times as active as radium freed from its products. You find the heat is substantially the energy of the alpha particles stopped. The recoil velocity of the molecules counts in any case for practically no energy, only 1/2500 of the whole. I suppose you say that this energy of its internal shake up goes away in gamma radiation....

The varied activities of the year 1908 are brought out in two letters to Hahn:

24 *Feb.* 1908. It was very good of you to tell me of Regener's paper. I read the first one of his on the probability estimate of the number of alpha particles. The general idea was not new to me as Geiger—one of my men—had finished a paper on the determination of the experimental error in comparison with the calculated one.

I am rather surprised that the method of counting scintillations gives somewhere about the right result. I should have anticipated that the chances were that the number of scintillations would be smaller than the number of alpha particles. It will be interesting to see how closely Regener can determine the number.

Our counting experiments are going all right and I hope in a few days to be able to say pretty definitely how the theoretical and experimental numbers agree. The method is going fine.

By using a sensitive electrometer I could easily make an alpha particle swing the spot of light off the scale. An account of my Royal Institution lecture will probably appear in this week's *Nature*. A reference to my counting experiments will be found there. I am enclosing here an abstract of my paper before the Manchester Literary and Philosophical Society. Regener would probably be interested to see it.

14 *July* 1908. I am afraid there is no hope of getting over to Germany this year unless I manage to get away to the association at Köln where I have been invited by Professor Kayser. I shall do so if I can manage it. I have a lot of university business on hand which will keep me here to the end of the month. I hope to see you soon but I must delay that.

Really you have a remarkably keen nose for unconsidered products. Congratulations on your latest additions. There is fortunately no difficulty about names. I am interested to see it makes the agreement of the changes of actinium with thorium still more complete.

In *Nature* which you mentioned I gave a statement of my volume results but also the main lines of the emanation spectrum. We have done it up in good style I think and will publish in full soon. Ramsay is doing his best to claim priority, etc. His work makes me feel tired—to use an Americanism.

I am now quite an expert in gas purification and in spectroscopy. It is all very interesting and amusing. My two main papers with Geiger, of great solidity and I think of the first magnitude, should appear in a month or so. I go to Dublin to the British Association this year.

RUTHERFORD TO BUMSTEAD

11 *July* 1908. I have now got through the stress of my first year and have got fat on it, at least my wife says so. At the same time I have never worked so hard in my life and have got off four papers to the Phil. Mag. (August) on the volume and spectrum of the radium emanation and two to the Royal Society on the counting of alpha particles and the measurement of the electric charge carried by each. First as to the counting experiments. Geiger is a good man and worked like a slave. I could never have found time for the drudgery before we got things going in good style. Finally all went well, but the scattering is the devil. Our tube worked like a charm and we could easily get a throw of 50 mm. for each particle.

He then proudly gives his results:

Charge per alpha particle	..	..	$4 \cdot 65 \times 10^{-10}$ e.s.u.
Number of alpha particles	..	..	$3 \cdot 4 \times 10^{10}$/gm./sec.
Heating effect	..	..	113 cal./hour.
Volume of emanation/gm. radium	..		$\cdot 585$ cubic mm.
Life of radium (half-period)	..	..	1760 years.
Production of helium ..	..	..	158 c.mm./year.

He continued:

I am sorry to reduce the life of radium for Boltwood's sake, but his result must be a maximum estimate. . . . Geiger is a demon at the work of counting scintillations and could count at intervals for a whole night without disturbing his equanimity. I damned vigorously and retired after two minutes.

This referred to the verification that each alpha particle gives one sparkle on a fluorescent screen. The letter went on:

I had a laugh at Crookes when I gave my paper at the Royal Society. He hadn't the faintest notion of how the things were done, but remarked "What a pity it was that so few people had the quantity of radium required to repeat the experiment". I rose up and gave the kind of remark which I generally let loose on Ramsay when I want to disagree with him without saying so. I said with an infantile smile that any one with a hundredth part of a milligramme of radium could repeat it, if he felt inclined that way, but that his work was cut out for him. General grin of enjoyment....

Rutherford had so greatly improved his methods of purification of the emanation of radium that he was not only able to obtain a good measurement of its volume (in equilibrium with a gramme of radium) but he also succeeded in getting a good spectrum of this gas, free from impurities, and he used to recount with some glee how he had really beaten the chemists at their own game.

RUTHERFORD TO HIS MOTHER

July 1908. On September 2 I shall go over to Dublin to the British Association. Trinity College, Dublin, is giving me an honorary degree on that occasion. I have plenty of these insignia but they are gratifying to the extent that they are meant to be an appreciation of one's scientific work. I may possibly go later to Cologne for a week to attend the German Association meeting, to which I have been invited. I have published a good deal of work this year and this has kept me very busy, with the whole additional work of running a large department. I found my eyes were troubling me a little, so I may have occasionally to wear glasses when I have a good deal of reading to do. Otherwise my eyesight is as good as ever. Mr and Mrs W. H. Skinner, of New Zealand, had lunch with us the other day.

12 *Sept.* 1908. Since writing last time I have been over to Dublin to attend the meeting of the British Association. I stayed in a country house of a Mr Darley, about seven miles from Dublin, and we motored in every morning. I was given the honorary D.Sc. at Dublin. There were twelve of us with red gowns with green facings and the public orator reeled off our virtues in Latin. I am afraid that I have not yet arrived at the stage of taking these things seriously. On my way back I foregathered with Professor Coker, formerly of Montreal. There were a number of garden parties, dinners and evening parties which I attended. The Provost of Trinity, Dr Traill, invited me to dinner with a number of notables. I sat alongside Sir James Murray, the lexicographer. He is a pretty old man and a very jolly one, and we got along famously. He

informed me the Rutherford family is closely connected with his and inquired of my antecedents. By the way, I had an inquiry some time ago from a man—a Rutherford—who wished to know the names of father's Scotch antecedents. He informed me that Rutherford meant 'Cattleford'. The Ruther is the same as in Rotherhithe and other names. The 'Bosphorus' is exactly the same meaning.

I am sending by this mail a copy of the diploma of honorary doctor of philosophy (Ph.D.), just awarded me by the University of Giessen, Germany, on the occasion of their tercentenary. This is my fourth honorary degree, and the first one from Europe. A free translation of the diploma is as follows: "Under the command of his most serene highness Ernest Ludwig of Hesse and the Rhine and the most noble Rector of the Ludwig's University, we, the philosophical faculty of the etc.... University, have awarded on the occasion of the three hundredth anniversary of our University, Ernest Rutherford, Professor of Physics, University of Manchester, for his exemplary investigation and his masterly presentation of radioactive phenomena, for his bold pronouncement and consequent working out of the theory of atomic disintegration, the Honorary Degree of Doctor of Philosophy, and have announced the conclusion through our Dean, Professor Walter König, at today's celebration. For evidence of which this diploma is forwarded with the great seal of the University attached and signed by the Rector and Dean."

There were always heavy demands on Rutherford's time for public lectures, and he usually kept, in a state of readiness, diagrams, slides and a few experiments, illustrating his latest discoveries. Thus, in the year 1908, he lectured on 12 Feb. to the Royal Dublin Society about 'Radioactive Transformation'; on 5 March to the Liverpool University Chemical Society. In March he also gave a Friday evening lecture at the Royal Institution, where he summarised his discoveries in a manner which was particularly clear and helpful. On 16 June, the new addition to the Cavendish Laboratory was opened by Lord Rayleigh, this being his first appearance in Cambridge since his appointment as Chancellor of the University. There was a reception by Lady Thomson, various experiments were shown, and Rutherford gave a lecture and demonstration on radium emanation in the large Lecture Theatre.

Since the Nobel Prizes had been instituted, the successive laureates in Physics had been: Röntgen, Lorentz, Becquerel, Rayleigh, Lenard, J. J. Thomson, and Michelson. Perhaps it was not a great surprise when Rutherford received a telegram and a most welcome letter announcing that he was to receive a Nobel Prize; but he must have been amused and delighted to realise that it was to be awarded for Chemistry.

RUTHERFORD TO HAHN

29 *Nov.* 1908. I much appreciate your kind congratulations and wishes on the award. It is of course quite unofficial but between ourselves I have no reason to doubt of its correctness. I must confess it was very unexpected and I am very startled at my metamorphosis into a chemist.

I may tell you in *strict confidence* that my wife and I are going to Stockholm via Hamburg and Copenhagen to arrive there on the 10th. We shall probably return via Berlin for the express purpose of seeing you. Do not say anything about this till after the proper date. I think it possible we may arrive on the night of the 13th or 14th but will let you know later. It would be convenient for me to get my letters forwarded to Berlin. Could they be addressed to you at your private address? If so, let me know as soon as you can the exact address. I should be glad if you would take charge of them before I arrive. We should of course also be glad if you could mention a comfortable hotel handy for looking round Berlin.

Rutherford arrived at Stockholm with his wife on 10 Dec. 1908, and dined that evening with Prof. and Fru Retzius. On the following day the medals were presented by the King of Sweden at the Musical Academy. Each prizeman—Lippmann, Rutherford, Ehrlich and Metchnikoff—was presented to the King with a suitable address. That same evening there was a banquet at the Palace attended by the Crown Prince and Princess. Rutherford was said by an eye-witness to have looked ridiculously young and to have made the speech of the evening. He declared that he had dealt with many different transformations with various periods of time, but that the quickest he had met was his own transformation in one moment from a physicist to a chemist.

On the afternoon of 12 December, Rutherford gave his Nobel Lecture on "The Chemical Nature of the Alpha-particles from Radioactive Substances". He clearly took great pains to explain the chain of reasoning which led to the inevitable conclusion that

the alpha particle is a projected atom of helium, which has two unit charges of positive electricity.... The alpha particle is released at a high speed as a result of an intense atomic explosion and plunges through the molecules of matter in its path. Such conditions are exceptionally favourable to the release of loosely attached electrons from the atomic system. If the alpha particle can lose two electrons in this way the double positive charge is explained.

In order to clinch this matter, Rutherford and Royds had carried out an experiment which made a great appeal to the public (scientific and lay), and he described this in his Nobel Lecture:

A cylinder of glass was made with walls so thin that alpha particles could shoot right through them, and yet the walls remained quite gas tight. Radium emanation was introduced into this cylinder and the alpha rays were collected in an outside tube, and compressed in another fine tube through which a spark was passed. After a time the spectrum of helium appeared, proving that the alpha particles were, or became helium!

This work was done in 1908, and the results published in the *Phil. Mag.* (17, p. 281) in 1909.

The recipients of the Nobel medals were entertained at a banquet by the King and Queen of Sweden, and the mathematician Professor Mittag-Leffler and his wife also gave the prizemen a large dinner party at their country house.

RUTHERFORD TO HIS MOTHER

24 *Dec.* 1908. I am sure that you have all been very excited to hear that the Nobel Prize in Chemistry has fallen my way. It is very acceptable both as regards honour and cash—the latter over £7000. [It was £6800.] We have just returned from our journey to Stockholm, where we had a great time—in fact, the time of our lives.

Rutherford took the opportunity of visiting the unique low-temperature laboratory of Kammerlingh Onnes at Leyden. At Berlin, Dr Otto Hahn and his wife entertained and guided the Rutherfords. So this wonderful expedition passed off admirably and Rutherford looked for new fields to conquer. He had not long to wait.

RUTHERFORD TO HAHN

22 *Dec.* 1908. We arrived in Amsterdam about 8.30 Saturday morning and went on at 12.30 to Leyden where we spent till 9 o'clock in the evening with Professor Lorentz, looking round the old town and laboratories. I saw the apparatus of Kammerlingh Onnes for production of liquid helium, but the latter was not well enough to see people. We had a great passage to Harwich and arrived home Sunday at 4.30 after a long slow journey in Sunday trains over England. We arrived tired but satisfied we had had a first class time. I am now hard at work trying to get through my correspondence which rather alarms me by its dimensions.

My wife and I are very grateful to you for the way you looked after us in Berlin. You worked nobly to give us a good time which we certainly had. I am enclosing a P.O.O. for 100 marks in payment of my just debts. I really did not want it but it was just as well to have a reserve. Many thanks for your kindness in that particular.

We found Eileen just arriving at the house as we came in. Our boxes arrived safely and there is great excitement over Xmas presents. By the way, I thought I had the idea of the removal of atoms by recoil in my Radioactivity somewhere—see page 392 2nd Edition. It is given in explanation of the volatility of radium B observed by Miss Brooks. Send me a copy of your paper as soon as it is out.

With best wishes for a merry Xmas and happy New Year from my wife and myself. Hoch! Hoch!

At the dinner in Whitworth Hall given in Rutherford's honour by Manchester University on 10 Feb. 1909, J. J. Thomson paid a great tribute to the new Nobel Laureate:

Professor Rutherford has never received the credit that he should have had for his work at Cambridge in connection with radiotelegraphy in 1895. His success was so great that I have since felt some misgivings that I persuaded him to devote himself to that new department of physics that was opened by the discovery of Röntgen rays. The study of radioactivity is almost a consequence of the discovery of the Röntgen rays....The discovery at McGill of the emanation of thorium was a stroke of genius, because the new gas had to be endowed with properties not recognised as belonging to any such known substance, as it is a gas which exists for a few minutes only; half of whatever there is of it always disappears in less than a minute!

Of all the services that can be rendered to science the introduction of new ideas is the very greatest. A new idea serves not only to make many people interested, but it starts a great number of new investigations.... There is nobody who has tested his ideas with more rigour than has Professor Rutherford. There can be no man who more nearly fulfils the design of the founder of the Nobel Prize than he does.

In his reply Rutherford said that

there was nothing more interesting than to watch the progress of science. Its progress was exceedingly gradual. It was like the progress of man going through a swamp, with islands of firm earth in between. The advances would be very slow from week to week, but at the end of a year it was great, and at the end of ten years it was enormous. In physics there had been a revolution since 1896. There was nothing impressed him more than the certainty of the scientific method wherever applied.

CHAPTER VII

THE NUCLEUS

About this time Rutherford was receiving many letters concerning the thorium family from Hahn, who, with Fräulein Lise Meitner, was employing the recoil method most successfully. When an alpha particle leaves an atom with a high velocity, the residue gets an equal kick or momentum backwards, as when a gun fires a shell horizontally. The speed of the recoil is less than that of the projectile, and the ratio of these speeds is the ratio of the masses, of course inversely. The recoil atom, as it may be called, is generally freed from its moorings and may be collected on a negatively charged body such as a metal wire.

On 1 Feb. 1909, Hahn sent some radiothorium to Rutherford from Berlin. It did not arrive and there was unavoidable delay in preparing more. In his letter of regret Hahn wrote: "Ramsay got several cubic centimetres of carbon dioxide in one year from 300 grammes of ordinary thorium nitrate. Rather a rapid transformation of thorium into carbon!"

One of the chief scientific events of the year 1909 was the Meeting of the British Association at Winnipeg, under the Presidency of Sir J. J. Thomson. Rutherford was President of Section A (Mathematics and Physics). There was a good gathering from far and wide: Larmor, Hobson, Hahn, H. E. Armstrong, Bumstead, Nicholls, Hull, and others. Winnipeg is at the very centre of the North American continent, but the centre of population is far south of it. It is a city of the plains with cold winters and hot summers, but all the year round its people are kind, warm-hearted and hospitable.

Some letters, preserved by Hahn, throw light on the preparations for the Winnipeg Meeting.

RUTHERFORD TO HAHN

13 *Feb.* 1909. I received your last paper and am interested to see how far you have gone. There was no necessity to give such pre-eminence to

my ancient statement [about the recoil atom after the ejection of an alpha particle]. I have had the view prominently before me for several years and discussed the question in detail with Russ a year ago.

Your thorium compound turned up after all. It had been found in Manchester and forwarded back to London. I have not had time to test it yet.

I have found the vapour-pressure curve of the radium emanation. I think the boiling point of pure emanation at atmospheric pressure is about $-65°$ C. The experiments are a little troublesome on account of the difficulty of purifying the emanation and keeping it reasonably free from impurities in the capillary. I understand Ramsay finds the volume of the emanation even greater than his old value. It will be a great joke if he publishes it, for I can easily compress the emanation in a capillary at atmospheric pressure occupying less than 6/10 of the volume he originally claimed.

The purification of the emanation is not child's play, and requires a lot of work.

You mentioned some time ago about the possibility of going to the British Association meeting at Winnipeg. The circulars are now out. You can become an ordinary member on payment of £1 and have all the privileges. I enclose a circular with statement of routes, prices, etc. Please *send it back* when you have studied it. I hope you will be able to come along. The University gave me a dinner the other evening.

In his presidential address to Section A Rutherford showed that his ideas on the constitution of matter were beginning to clarify. He was impressed with the ease with which alpha particles would pass through thin foils of metal or very thin glass, while helium gas could not pass through those same layers of material. The alpha particle must be not only swift but small. He said: "The old dictum, no doubt true in most cases, that two bodies cannot occupy the same space, no longer holds for atoms of matter if moving at a sufficiently high speed."

He called attention to the work of his friend Boltwood, who "not only suggested the view that lead was the end or final product of radium, but collected much evidence in its favour".

After the Meeting at Winnipeg the Rutherfords visited old friends in Montreal, and found many changes there. Rutherford also went to the United States and lectured at a gathering at Clark University, together with Volterra from Italy and Michelson from Chicago, both Nobel Laureates.

In Oct. 1909, Rutherford received an honorary degree at Birmingham. The next year he opened the new laboratory at University College, Dundee, called the Peters Electrical Engineering Laboratory. In his address he pointed out that his "grandparents had sailed for New Zealand from Dundee about seventy years previously when his father was a boy of six" [more probably three].

On Sunday, 29 Nov. 1909, there was a mixed entertainment at the New Islington Hall, Ancoats, Manchester, when Rutherford lectured on 'Weighing an Atom', preceded and followed by song and music, such as Schubert's 'Who is Sylvia?' and a Haydn Quartet in G major. There is no record of the effects of this unusual treatment of the atom.

In Dec. 1909, there was a celebration of the twenty-fifth anniversary of Sir J. J. Thomson's appointment as Professor of Experimental Physics at the Cavendish Laboratory. Before his time there had been no great research school in England and it had been a period of very great development.

A *History of the Cavendish Laboratory* was written to which Rutherford contributed Chap. VI (dealing with his own period at Cambridge from 1895 to 1898), giving a full and clear account of all the discoveries made in those years.

The events of the year 1910 are brought out in some extracts from letters by Rutherford to his mother:

12 *Jan.* 1910. We did not go away at Christmas, but had a quiet time at home, and had a number of the research men in to Christmas dinner. We have decided to order a motor-car—a Wolseley-Siddeley. It is very desirable to have some means of getting fresh air rapidly. We have chosen a four-seated car, 14–16 h.p., which about suits our requirements for quiet travelling.

6 *April* 1910. We got our car on Good Friday and spent three days running around while I practised driving. We started on our tour Tuesday, and have so far gone 500 miles, first south through Herefordshire, then through Salisbury and Swanage and back northwards to Surrey. We shall probably start after lunch towards Windsor. We have enjoyed ourselves very thoroughly. I have learnt to drive fairly well without a single incident, even of running over a chicken. A car is very easy to manage and far more under control than a horse. We average about 17 miles an hour over country, and on a good road run along freely at 25. We can go 35 or 40 if we want to, but I am not keen on high speeds with motor traps along the road and a ten guinea fine if I am caught. These are the woes of motorists that I hope to avoid!

26 *May* 1910. Lectures end tomorrow and after a month of exams., the university session closes, but I shall continue to work on into August. I will attend a congress in Brussels on radioactivity on September 16, and may possibly go for a short tour in Germany. I am feeling unusually well and fit for work, due to motor exercise. We went last week for three days in Wales, and motored 330 miles in that time. There was very universal mourning here on the death of the King, and great processions on the day of the funeral.

14 *Oct.* 1910. I have been rather lax in writing to you lately, but I have been busy wandering about. I wrote to you last from the Lakes. I then went on the Continent and spent nearly a fortnight in Munich in company with Boltwood. Munich is a very pleasant city to live in, and I had a good rest. While there we met several of my old students, including Hahn, who now works in Berlin. He is doing the best work in Germany at present. We visited Professor Baeyer at his country house on the Steinberger See (Lake), about 20 miles from Munich. He has been Professor of Organic Chemistry in Munich University for over 30 years, and is one of the greatest living chemists. He obtained a Nobel Prize several years ago. We had a very pleasant day there and went out on the lake in a petrol launch.

Boltwood and I returned to Brussels in time for the Radiology Congress, in which I took an active part and spoke on nearly all the papers. There was a large number of people there and I met a number of old friends, including Arrhenius of Stockholm. Madame Curie was there. She looked very wan and tired and much older than her age.

She works much too hard for her health. Altogether she was a very pathetic figure. I had a good deal to do with starting the committee for fixing a radium standard.

I stayed in Brussels four days, talking about 18 hours out of the 24, so I was not surprised that I got a relaxed throat, and by the time I arrived home I had one side of my face swelled up, due to an ulcerated tooth and a bad cold. Altogether I was a miserable object, but fortunately I got rid of both very quickly and am in excellent form again.

I went up to Dundee on October 6 to open a new electrical engineering laboratory, and had a most pleasant time. They treated me very well and gave a big dinner in my honour. I have sent you a paper giving some of the points of my address. You will see I mentioned that father came originally from the neighbourhood of Dundee. I stayed with a member of the Council, a wealthy man of German origin, who looked after me in great style. I visited the University of St Andrews on my return journey. It is an old-world place which is to celebrate this year its 500th anniversary.

I have been busy starting my classes and people at original work, and naturally have not much time to spare. I shall be going to London in about a fortnight to help select a Professor of Mathematics and Physics for the new University of Brisbane.

4 Nov. 1910. Last week I went to London to attend the Agent-General's Office of Queensland to select a man for one of the new chairs in Brisbane. I went over to Leeds on Tuesday to interview a candidate from Egypt. I saw Professor Bickerton in London. He is trying to float his impact theory. Although he is seventy he is very active and energetic notwithstanding his trials in Christchurch. I am kept pretty busy with ordinary and advanced lectures and my book writing, for I am preparing a third edition of *Radio-activity*. I shall probably be appointed to the Council of the Royal Society. This will entail a dozen visits to London per year. I enclose some cuttings of the discussions on the rowdiness of the students when I was at Dundee. You will see Asquith suffered an even worse fate at Aberdeen. I did not mind it. It was only a frolic on their part.

16 Dec. 1910. Christmas is fast approaching, lectures have ceased and I am endeavouring to get some work done in my comparatively free time. I have to go up to London once a month to attend the Royal

Society's meetings. I am also going to London on January 5 to deliver a lecture before the Röntgen Society. We are having a small heating system put into our house to keep the lower floor a bit warmer. When it is cold here, I feel it much worse than in Canada.

After much discussion and experimental work, Rutherford was hopeful about the question of beta rays; in fact, there are difficulties not yet overcome, but in August he wrote to Hahn:

2 *Aug.* 1910. Many thanks for your batch of papers which I have just received. The whole question of beta rays is in a very interesting state and should be reasonably cleared up before long. I am hoping to see you in Brussels and I will talk to you on the matter there. What do you think of the 'nomenclature' question? The more I think of it, the more difficult it appears to change much. You will have seen Ramsay wants to call the emanation 'niton' as one of the air elements.

By the way, you told me in a recent letter mesothorium is now for sale. Could you arrange to buy £10 worth for me? I hope I can get reasonable value for the money. Let me know about it.

We are all well and happy. I am finding it difficult to get through all the work I have in hand.

At the Brussels Congress it was agreed that the amount of radium emanation in equilibrium with one gramme of radium should be called one 'Curie', in honour of Madame. It was further agreed that Mme Curie should be requested to prepare an international standard consisting of about 20 mg. of radium. Other standards would then be prepared and provided on purchase to the various countries of the world. This was a matter of great importance as different leading laboratories were not sure of the amount of any given specimen of radium that they used or possessed. Rutherford and Stefan Meyer took endless pains in the negotiations connected with this scheme. The final meeting of the Congress was humorous in the extreme, as there were several hundreds in attendance, and motions, amendments, and amendments to amendments, were rapidly proposed in a number of different languages. The Chairman lost his head and there was great confusion until a German, Hallwachs, took control and in a loud voice expressed and moved the opinion of the Congress, and it was carried! It was interesting to see one man get cosmos out of wild chaos.

One day, Rutherford, Willy Wien and I were lunching together out of doors at Brussels, when Rutherford began twitting Wien about relativity. Wien explained that Newton was wrong in the matter of relative motion, which was not the joint velocities $u+v$, but that expression, according to Einstein, must be divided by $1+uv/c^2$, where c is the velocity of light. Wien added, "But no Anglo-Saxon can understand relativity!" "No!" laughed Rutherford, "they have too much sense."

The Barnard Medal for Meritorious Service to Science given every five years had been awarded in 1895 to Lord Rayleigh and Sir William Ramsay, in 1900 to Prof. von Röntgen, and in 1905 to M. Henri Becquerel, each award signifying a landmark in scientific progress. In 1910 the National Academy of Sciences recommended to the Trustees of Columbia University that the medal should be presented to Rutherford, especially for his investigations of the phenomena of radioactive materials. The suggestion was approved and the President, Dr Nicholas Murray Butler, brought the medal to England and duly presented it. At the same time he made very generous financial offers to Rutherford to go to Columbia as Phoenix Professor and Director of Physics, but Rutherford, wishing to keep near the centre of physics, thought it wise to decline.

In Nov. 1910 the solicitors of the British Radio Telegraph and Telephone Company expressed a desire that Rutherford should serve as an 'expert witness' with reference to his magnetic detector, later improved by Marconi. In his reply Rutherford expressed a strong desire not to be called, as his published papers gave a clear history of the case. It is to be noted that he did claim that when he saw Marconi's paper of June 1902 in which he referred to the use of a moving band, he thought it only right to draw the attention of scientific men to his previous work *and his experiments with moving steel wires*. (*The Electrician*, Oct. 1902.)

The lawyer who went to Manchester reported that Rutherford was

most emphatic in his desire not to be associated with either side. His position is one of complete detachment and disinterestedness and he feels that his independence would be surrendered if he assumes the position of expert witness, a role to which he is most averse. He has never exploited any of his scientific work for commercial purposes, or with a view to patenting and thereby securing for himself alone the benefit of such research; he fully and freely publishes the results of all his scientific work, and therefore his research possesses the greater value, inasmuch it is not done behind closed doors, but is done with a view to the world's knowledge.

Rutherford was not called; his published writings were quite sufficient for the purpose of proving the historical facts, which after all was what the Company particularly desired. He never made any money for patents, inventions or discoveries, nor did he seek to make any.

The year 1911 was one of the great years of Rutherford's interesting life. It began in the ordinary way. He gave a lecture to the Röntgen Society on 5 Jan. at which he called attention to Hahn's then recent discovery of mesothorium, intermediate between thorium and radiothorium. On 2 March he reviewed in *Nature* Mme Curie's two-volume *Traité de Radioactivité*. He wrote: "There is very little to criticize and much to admire in this notable work. It is remarkable what little difference of opinion exists among radioactive workers on the interpretation of the main phenomena."

The following letter to Bragg, then at Leeds, shows how Rutherford was beginning to think that the central core of the atom was *negatively* charged, while his ultimate decision was in favour of a positively charged small nucleus in the centre of an atom.

Geiger is working out the question of large scattering [of alpha particles] and as far as he has gone results look very promising for the theory. The laws of large scattering are completely distinct from the small scattering, and there are a number of directions in which the theory can be verified. I am beginning to think that the central core is negatively charged, for otherwise the law of absorption for beta rays would be very different from that observed. The working out of the absorption of beta rays is going to be rather difficult on account of the impossibility of getting an integral equation when one takes account of radiation of energy. I have thought a good deal about the possibility of accounting for X and gamma rays, but have made no definite progress.

Did I tell you that Hahn finds that the penetrating beta rays from a simple product are not homogeneous, while the slower ones are? I think I can offer an explanation of that on my theory. I can see that the working out of a complete theory of the alpha and beta rays is going to be rather a large job; but the outlook certainly is very promising. The experimental proof of the various theoretical points will take a considerable time to clear up; it will mean a number of special investigations under rather difficult conditions.

Work goes on much as usual. The experiments of several people on recoil phenomena give rather unexpected results, and it appears doubtful whether the simple theory of the balance of momentum holds in all cases. Progress is rather slow on account of the smallness of the effects. I have sent in Gray's paper to the Royal Society. It is very interesting.

In an earlier letter to Bragg there is a clear prediction in mathematical form of what must happen if Rutherford's nuclear theory is correct. This has been precisely verified.

9 *Feb.* 1911. Your conclusions are quite right re scattering from thin plates. I have worked all that out and find the amount returned should be proportional to thickness for small thicknesses, and varies as NA where N is the number of atoms per unit volume and A the atomic weight, supposing the charge at the centre is proportional to the atomic weight. This is in agreement with Schmidt's scattering coefficient, as it should be.

I have looked into Crowther's scattering paper carefully and the more I examine it the more I marvel at the way he made it fit (or thought he made it fit) J. J.'s theory. As a matter of fact, I find I can explain the first part of his curve of scattering with thickness, in *large* scattering alone. I believe it is only the use of imagination, and failure to grasp where the theory was inapplicable, that led him to give numbers showing such an apparent agreement.

Your remark about loss of radiation being small is quite right if the centre is negative, but it is of the right order to be taken into account as a factor in reducing velocity. This is my trouble, for I want to get at the fundamental factors, if any, besides ionization, which lead to loss of energy of the beta particle.

I am pretty sure the theory is going to give a rational account of all scattering phenomena big and little for the same atom. The laws to be expected are so marked that verification should be comparatively easy. These are:

(1) The number of scintillations per second on the screen (for large deflections of alpha particles) varies as $\mathrm{cosec}^4\,\phi/2$, where ϕ is the deflection angle and varies inversely as $(\tfrac{1}{2}mv^2)^2$.

(2) The number scattered through any angle for *equal thin* thicknesses of different kinds of matter varies as A^2Nt where

$$A = \text{atomic weight},$$
$$N = \text{number of atoms per unit volume},$$
$$t = \text{thickness}.$$

(3) For thick sheets of matter of any kind, alpha radiation scattered through more than one right angle varies as $A^{3/2}$ (this accounts for Geiger's original numbers published). I am willing to bet these laws will hold pretty accurately.

Altogether, I think the outlook is distinctly promising and I am awaiting the results of Geiger's measurements on these points.

I think Schmidt's theory is excellent as far as it goes but his beta absorption coefficient may be difficult to account for *numerically* in the general theory, but there is no serious drawback. Otherwise, I think the whole scattering phenomena, qualitative and quantitative, will fall into line.

I will be in London on the 16th. Sorry I can't see you.

Glad to hear your Royal Institute lecture went off in good style. I would like to see your gross atoms.

Note too the following passages:

Consider an atom which contains a charge $\pm Ne$ at its centre surrounded by a sphere of electrification containing $\mp Ne$ supposed uniformly distributed throughout a sphere of radium R.

Consider an atom containing a positive charge Ne at its centre surrounded by a distribution of negative electrons Ne uniformly distributed within a sphere of radium R.

(*Phil. Mag.*, 1911, 21, p. 669.)

These quotations show that Rutherford was getting off the fence on to the side of the positive nucleus.

Max Born in *Atomic Physics* (p. 52) states that Lenard, who found that ordinary matter was transparent to swift electrons, and conceived a small impenetrable centre, a *dynamid*, to the atom, should "properly be credited with the first suggestion of the modern model of the atom". He goes on: "Usually however the title of the father of the atomic theory is given to Rutherford, who took up the research with more adequate instrumental resources, and carried it farther; to him we owe our concrete, quantitative ideas on atomic structure."

Rutherford stated later that he was enormously impressed with the fact that when the alpha particles are fired at a very thin gold leaf, a few, very few, return almost back towards their source, as if a bullet fired at a sheet of paper should bounce back towards the rifle! He may have thought of a fast comet swinging round the sun, controlled by gravitational attraction, and having a hyperbola as orbit; so that the negative nucleus would attract the positive alpha particle. He gradually saw that he must have a positively charged nucleus, so that the alpha particle would be repelled thereby, and the path would still be a hyperbola. He had a model made. An electromagnet stood on a table with (say) its north pole uppermost. Another electromagnet, at the lower end of thirty feet of flex wire, had its north pole downward and just over the fixed pole. These poles repelled each other, and if the long pendulum was swung to-and-fro it would swerve away from the lower pole whenever near to it. The pendulum might indeed approach and retire in the same straight line.

It is necessarily true that the results quoted in Rutherford's letter to Bragg of 9 Feb., which seem to be stated for a negative centre, are the same as those finally quoted by him for a positive nucleus.

Geiger has given a remarkably clear picture of how the light gradually dawned on Rutherford's mind, so that he achieved one of the greatest of his results, on which so much depends—that at the centre of every atom is a small nucleus, containing the main mass of that atom, and the whole of the positive electrical charge.

This is Geiger's account:

One day (in 1911) Rutherford, obviously in the best of spirits, came into my room and told me that he now knew what the atom looked like and how to explain the large deflections of the alpha particles. On the very same day I began an experiment to test the relation expected by Rutherford between the number of scattered particles and the angle of scattering.

Chadwick, who speaks with authority, very justly adds:

The genius of Rutherford had seized upon an apparently unimportant detail and transferred it into a clue to the problem of the inner structure of the atom. The nuclear theory of the atom was published in a surprisingly complete form in the same year (*Phil. Mag.*, 1911, 21, p. 669). Supported at first by the preliminary experiments of Geiger on the variations of the scattering with angle, it was in the next few months confirmed in its minutest details by a beautiful series of accurate, quantitative tests. It would be difficult to exaggerate the influence of the nuclear theory of atomic structure on the whole range of the exact natural sciences. The theory will surely rank as the greatest of all Rutherford's contributions to physics.

It is difficult to appraise justly Rutherford's mathematical powers. In the first Bakerian Lecture he handled his exponentials with quick and easy facility, and though such work and calculations are now familiar to physicists, that was not the case when he first used them for his radioactive theory. When he was discovering his small, positive nucleus in the atom, and with Geiger put the theory to a severe test, he used mathematics of a very Newtonian flavour, as though he had been studying the *Principia* or some rehash of it, such as Frost's *Newton*, which was closely studied by mathematical students in the rather stereotyped but excellent training of the famous 'coach', E. J. Routh. It is interesting to find that he flirted for a time with the idea that the central nucleus might have a negative charge so that the positive alpha particle was attracted. But then how could the electrons be held to the atom? He then changed his viewpoint, the nucleus had to have a positive charge—the alpha particle (+) had to be repelled by a positive nucleus, the path must be a hyperbola, and the mathematics continued to be Newtonian and took the final form, now exactly justified by the classical as well as by the most modern theories of wave-mechanics.

C. G. Darwin wrote:

4 *Feb*. 1938. I count it as one of the great occurrences of my life that I was actually present half-an-hour after the nucleus was born.

It was a Sunday supper at his Manchester house and I remember his saying to us, I would think before supper itself, that he had been looking into the business of big scattering of alpha particles and that there must be enormous forces in the atom to do it. Then he said that it would require a charge of something like a hundred electrons in the case of the gold atom, but that may have been added a few days later.

As to his mathematics I do more definitely remember being surprised at the way he worked the problem out.

It seems that Rutherford recollected a theorem in geometrical conics which Darwin had entirely forgotten since he was at school. Rutherford knew that the orbit was a hyperbola and he recalled a relation between the eccentricity and the angle between the asymptotes. An energy relation—loss of potential equals gain of kinetic—was also available, and so we can give to Rutherford himself credit for the primary mathematical development of his great physical idea. Darwin definitely states that he was "immediately asked to check the thing up, which I only did the next day". Darwin also worked out the results on the assumption that the law had been that of the inverse cube, instead of the inverse square; and further considered the motion of the nucleus itself, a recoil motion, obviously most important for a light nucleus like hydrogen. Rutherford took this idea up at once, and he was particularly interested in the question "How close to the nucleus can an alpha particle approach?" He got his answer from the scattering experiments. The alpha particle which came almost straight back, rebounding from a thin gold leaf, gave him the result, about 3×10^{-13} cm., or three Rutherford units, if you like, where a Rutherford unit is the radius of an electron, perhaps!

In the early days all spoke of the 'central positive charge' which was happily replaced by the 'nucleus', which included mass and electrical charge.

Henceforth every physicist pictured an atom as having a central sun of the minutest dimensions even compared with the size of the atom itself. The mass of this nucleus would be nearly the atomic weight, and the number of units of positive electrical charge about half the atomic weight numerically; for the great clarification of Moseley was still to come.

The discovery of the nucleus with the proof of the correctness of his idea was the peak of Rutherford's research, and it will remain a most prominent landmark in the history of science.

Nagaoka, the greatest of Japanese physicists, visited Manchester and afterwards wrote from Tokyo University to Rutherford:

22 *Feb.* 1911. I have to thank you for the great kindness you showed me in Manchester.

I have been struck with the simpleness of the apparatus you employ and the brilliant results you obtain. Everybody engaged with the experiments on radioactivity seems to be impressed with the same fact and expresses admiration of the splendid results which you obtain with extremely simple means.

In a colloquium in Würzburg a report was given of your paper on the calculation of alpha particles which was in progress when I visited Manchester. All the members present expressed great admiration at the splendid result obtained with such a simple device. It seems to me that it is a genius, which can work with such simple apparatus and glean a rich harvest far surpassing that attained with the most delicate and complex arrangements.

K. Fajans wrote a letter (10 April 1911) about the displacement law and the periodic table; he was one of the quickest to grasp its significance. It was known that U X (uranium X of Crookes), a beta ray product, followed uranium. Next it was found that there were two uraniums, U I and U II, each projecting alpha particles; they were non-separable. Hence, said Fajans, there cannot be only *one* U X between the two uraniums; there must be two. He tried to find the new product and succeeded. Fajans had the wits to see that it must take two beta ray changes to balance the one alpha change.

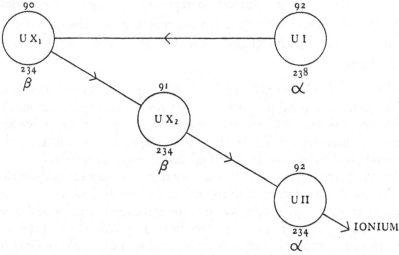

The changes of uranium, with atomic numbers above and atomic weights below

In 1907 Boltwood had discovered ionium from actinium preparations by the addition of a little thorium. "The thorium and ionium remain always together when once mixed, and no chemical process is capable of separating them again. The two are chemically identical, or as we now say, isotopes" (Soddy).

So, too, radiothorium cannot be separated from thorium, nor meso-thorium from radium. "A period of steady work followed on the chemistry of the numerous disintegration products, notably by Fleck in Soddy's laboratory at Glasgow, to ascertain which known element each most resembled, and whether or not it was chemically separable from that element...." As regards the alpha ray change the Displacement Law dates from 1911. The complete generalisation was given in Feb. 1913, when K. Fajans in Germany and A. S. Russell and Soddy in this country extended the law to the beta ray change. Briefly, after an alpha particle is ejected from an element A the residue B moves two places to the left in the periodic table, that is, the atomic number is reduced by two. But for a beta ray the number is increased by one, and the movement is one place to the right.[1]

In a letter to Hahn, Rutherford reported progress on the nuclear theory and wrote some early information on the 'branching products', indicating that radium C could pass by different routes, as it were, to radium D.

26 *April* 1911. I have not heard from you recently; but I hope that your work is making good progress, and that you have straightened out the beta ray problem. We have just returned from a good holiday; we had ten days motoring in the South of England, and a week in the Lake District, and have come back to find that spring is in full tide.

I have been working recently on scattering of alpha and beta particles and have devised a new atom to explain the results, and also a special theory of scattering. Geiger is examining this experimentally, and finds so far it is in good agreement with the facts. I am publishing a paper on the subject to appear shortly. You will have seen the results of Gray in connection with the production of gamma rays from beta rays excited in various substances?

One of my men, Fajans, an able fellow, has been investigating the short-lived product in radium C, and it will interest you to know the peculiar results he has obtained. He finds the period is 1·4 minutes, and that the product emits beta rays of about the same penetrating power as those from radium C. He has found no evidence of any other short period product. He has not found it possible to obtain by any method more than a very small quantity of the 1·4 minute product. He has come to the conclusion that this product is really a side branch of

---

[1] This statement is abridged from that given by Prof. F. Soddy in *The Interpretation of the Atom* (John Murray, 1933).

radium C, for the amount obtained by recoil is far too small to be in the direct succession. By examining rise curves, no evidence has been obtained of the existence of this product in the main series. In addition he had obtained radium D, and afterwards polonium by recoil from radium C in about the amount to be expected if it directly followed the expulsion of the swift alpha particle from radium C. The whole behaviour is certainly very curious, and is still under examination. The paper should be published in the Phys. Zeit. very soon.

I understand from Meyer that the atomic weight of radium is in progress in Vienna. The work has been undertaken by a pupil of Richards of Harvard, who has had previous experience in atomic weights of barium and calcium.

On 7 June 1911, K. Fajans wrote from Carlsruhe a very clear letter on the difficult question of the 'branching products' and their nomenclature. In the early days of radioactivity it was believed that there was direct and linear descent from father to son in all three radioactive families, namely, uranium-radium, thorium, actinium. It was gradually discovered that the C products always branched. A certain but different proportion of these elements divided into two groups; one group evicted a beta particle (electron) and then an alpha particle (helium nucleus), while the other group reversed that order. Presumably they arrived at the same ultimate goal.

The discovery of the very short-lived products referred to in the next letter to Stefan Meyer was remarkable and important. When the actinium emanation broke up, the recoil due to the ejection of the alpha particle freed the next product, A, which was a solid and could be drawn swiftly to a revolving disk. A testing chamber was placed close above the disk, and by varying the speed of rotation and the distance from the testing chamber the rate of decay to half-value of the products could be determined. In the case of actinium A, this was as small as 1/500 part of a second!

RUTHERFORD TO STEFAN MEYER

29 *June* 1911. ...I can mention to you now further details of the new products. Geiger has found that the actinium and thorium emanations

each contain two alpha ray products of different ranges. In reality the second of these products behaves as a solid and is deposited in an electric field on the cathode. The half period of the new product for actinium is about 1/500 of a second, and for thorium about 1/10 of a second. Several people are determining the periods here. The rapidity of decay, of course, makes it necessary to devise new methods for the purpose. Geiger and I have made a number of experiments to illustrate simply the presence of this product. The discovery of these products will make it necessary to change the nomenclature a little. I think that Geiger and I will propose later that the two products be called thorium A and actinium A, and the substances now called A be called B and so on. I do not think that this will lead to any confusion, and it brings out the extraordinary close analogy in the constitution of the active deposits of radium, thorium and actinium. What do you think of it? Geiger is publishing a preliminary note in the Phil. Mag. next month; but please regard the other information as confidential for the time. I am very busy here with examinations, and will continue to be so for another week.

C. T. R. Wilson's expansion chamber has revealed to us the unseen in the atomic world. The sudden expansion of humid air causes a mist to form on the charged carriers, or ions, produced in the path of swift alpha or beta particles. Their tracks are made apparent to the eye or the camera by the mist that marks their paths. It was noted as most remarkable that a beam of X-rays did not directly ionise the air, but ejected electrons (photo-electrons they are called) swift enough to produce fairly long visible tracks. The greater the frequency of the X-rays, the longer the tracks of the electrons! As Einstein would say, the change of kinetic energy is equal to the quantum which produced it, or $\frac{1}{2}mv^2 = h\nu$, where ν is the frequency and h is Planck's constant.

Rutherford wrote an enthusiastic letter to W. H. Bragg when he saw for the first time one of these wonderful photographs:

9 *May* 1911. I am intending to go up to London on Wednesday afternoon to attend the Soirée, and hope to be able to attend Thursday's meeting. I hope that you will come down too. Sorry I can't see you in Manchester.

I had heard something of Wilson's results, but did not know that he had photographed the trail. It is really a splendid piece of work, and will no doubt throw a great deal of light on some of the mechanism of ionization. I am sure you are highly delighted at the way things are turning out in favour of your views. The X ray photograph is really remarkable. I am sure that the experiments are not easy, for I spent several months' work on the same subject in 1906 and for the same object when I was in Montreal; but my apparatus was so contaminated with radium that I could get nothing definite. It is really very fine to see the things one has seen in imagination visibly demonstrated.

If I go to London I shall stay at the Kingsley Hotel if I can get in. Elliot Smith is going down too. I will tell you about our holidays when we meet.

P.S. I enclose Wilson's letter; but I hope you will allow me to keep the photographs for a day or two to show my men.

On 20 May 1911 there was a review of Rutherford's article in *The Athenæum*. It first refers to the 'Saturnian' atom of Prof. Nagaoka, consisting of a central mass surrounded by rings of electrons, and then explains Rutherford's views on "The Scattering of Alpha and Beta Particles by Matter", where he suggests that in the case of gold the central charge of such an atom might be that of 49 atoms of helium each with a double charge; this would give nearly the appropriate atomic weight 196, as against the true value 197·2 and a nuclear charge of 98, whereas it is now known to be 79. All this was before the enlightenment due to Moseley. The important point is that Rutherford stated that "it was not yet possible to determine whether the central charge was positive or negative". It is interesting to notice the whole theory in a state of rapid evolution.

Three letters follow, which were written by Rutherford during the year, and these throw light on the varied interests of the time.

RUTHERFORD TO MEYER

17 *June* 1911. I was much interested in your letter, and am pleased to hear that you have got the atomic weight practically finished. It will be very valuable to have some check in setting up the standard in order

to avoid possible accidental error. I was not quite sure from your letter whether you meant that Madame Curie is actually engaged in the preparation of the standard or whether it is to be left to the summer. It seems to me that the matter cannot be left to drag for several years.. . .

We have a number of interesting investigations going on in the Laboratory, and expect to add soon several new products of rather unusual interest; but the experiments are not yet far enough advanced to speak of the result. I have a number of experiments going to determine a possible variation of e/m of the alpha particle with speed. I think we have developed a method that promises considerable accuracy, but I anticipate the effect will be too small to measure. We have got several excellent photographs of alpha rays similar to Kaufmann's photograph of the beta ray, using combined electric and magnetic field.

I am going to London next week to attend the Coronation festivities and the Naval Review.

RUTHERFORD TO BRAGG

14 *Oct.* 1911. The arrival of your papers this morning reminds me of the fact that I have not heard from you for an age. I have, however, heard of your movements and of your visit to the British Association, and of the 'luminous and lucid' address you gave there; but I trust that the luminosity you excited was not responsible for the production of external radioactivity which I see Ramsay brought before your notice at the meeting. I thought it was long since dead, and its resurrection seems to me quite unnecessary. I understand that you turned him down with great kindness but firmness.

As to ourselves, we spent a very pleasant holiday in Scotland, and I have been back for six weeks working hard, and have nearly got a research through on the heating effects of radioactive products. I am full of work, but otherwise reasonably content. We had a call from the chemist you mentioned to us, Mr Bearder, and we are expecting him to lunch today and to take him for a run round the country.

Next, in regard to a business proposal. You know that J. J. T. has been our External Examiner for the last three years, and his time is now up and I have to look for his successor. I recall that you asked me to be External Examiner for Leeds; but I trust that you will not follow my example and turn down my offer, which is that you take the position of External Examiner in Physics for the University of Manchester.

I think the work will be reasonably light, and will involve a two or three days' visit at the time of our Honours and Pass Practical Examinations. We usually set a good proportion of the questions and examine most of the papers, but will require your aid in the more doubtful cases. If I may say so, I think it is an excellent excuse for a good holiday for two or three days, and you will be able to enlighten me at the darkest period of the year. I shall consider it a gross dereliction of your duty if you do not accept. In addition, Manchester is very convenient for Liverpool, so that you ought to be able to run both without much inconvenience in travelling. Please let me know your answer as soon as you make up your mind—as the matter has to come up before the Faculty before long.

I see by the papers that it is likely that you are taking Sadler as Vice-Chancellor.

When is Campbell going to publish his next paper on the delta rays? His last one reminded me of stories which came out in the magazines in which the hero or heroine is left in a most thrilling situation at the end of the number. I thought the paper was very Campbellesque, but we must await with trepidation the next issue.

You will observe that the last Phil. Mag. was highly radioactive, to which end we contributed materially.

RUTHERFORD TO HAHN

2 Oct. 1911. I was glad to receive the postcard from you and your friends, and I am glad that I am remembered on these occasions.

I am at present trying to write up the subject of beta rays for my new Edition, and I find it the most difficult task in the book, as there has to be a good deal of compression to bring it within reasonable compass. I should be very glad if you could forward me a photograph or two for reproduction of your deflexion of the beta rays, for example, thorium and any other that you think would show well. I would like to include an actual photograph to indicate the nature of the results. I should be glad, also, if you could give me a brief statement of exactly the results you consider definite so far.

I saw by an abstract that a Frenchman had found seven lines in the radium emanation, but I have not yet seen the actual paper.

Lectures start again this week, and a good many of the research men are back at work. Schrader has come over, and seems a nice fellow, and also A. Russell, who was in Berlin this year with Marckwald. We are expecting to have a pretty busy time. There are a great number of papers still on the stocks for publication; but I suppose they will appear before long. I have been hard at work for the last month on the heating effects of the emanation, and my experiments have gone very well. My main object was to test whether the heating effect is a direct measure of the kinetic energy of the alpha particles, and to see whether the radium B to C change emitted energy.

Kovarik is back in America, and Boyle in Canada. I shall miss both of them. Geiger is here and hard at work. We should find it very difficult to do without his services.

I shall be glad to hear how things are going with you. I am sending this letter to your home address as I imagine you have not yet got back to work in Berlin.

On 20 Dec. 1911 Rutherford wrote to W. H. Bragg stating that he had been driving at full speed at his book with little time to devote to thinking of the account of the theory which Bragg had sent him. This referred to the view that an X-ray was a neutral doublet of both kinds of electricity. The letter continues:

I went to the Cavendish Dinner which was a pleasant affair, and had a brief talk with Barkla. He seemed quite ready to believe that the energy of an X-ray was concentrated; but does not like your material doublet. He considers that an X-ray must be a type of wave motion, and is to be regarded as the simplest form of light. J. J. also expressed the latter view....

I was rather struck in Brussels by the fact that continental people do not seem to be in the least interested to form a physical idea of the basis of Planck's theory. They are quite content to explain everything on a certain assumption, and do not worry their heads about the real cause of the thing. I must, I think, say that the English point of view is much more physical and much to be preferred.

About this time Rutherford was repeating the electric and magnetic deflection of alpha particles from various radioactive elements so as to determine their velocities of egress and to make quite sure whether the ratio of charge to mass was the same for all. If so, he thereby proved that every alpha particle, whatever its source, is the nucleus of a helium atom. This was established. On relativity theory the greater the speed the greater the mass, but the change is not large enough to observe except for velocities approaching that of light.

1911 was Coronation Year and Rutherford's impressions of it are given in letters to his mother:

20 *May* 1911. I went down to London three weeks ago to attend the Soirée of the Royal Society and the meeting next day. There was an excellent show of experiments in which I was much interested. Professor Elliot Smith and I wandered round London in the morning and were highly edified at the ladies' hats and dresses especially in Hyde Park. It is as good as a pantomime to see the women in hobble skirts and big hats. I went to lunch with Professor Strutt, son of Lord Rayleigh. After attending a meeting at the Royal Society I went home by train.

We are hoping to go to London for the Coronation if we get seats at a reasonable price. I have received an invitation from the Admiralty to be their guest and go on board one of the vessels at the Coronation review at Spithead. I shall go as it might be a fine sight to see the collected fleets and the night decorations. I go down to Birmingham in three weeks' time when I act as examiner in physics for the university there. We have not yet decided on our summer holidays but may possibly go to Switzerland.

10 *June* 1911. There is a big women's suffrage demonstration in London on Saturday which will be over two miles long. I go down on Saturday to Birmingham and then down to London next day to attend the Coronation Naval Review at Spithead. We shall probably go up to the Highlands for our summer holidays and take the car with us. This will be our first holiday in Scotland. We go up to Edinburgh on July 6 to attend the celebrations connected with degree day. I am getting an honorary degree of LL.D. on that occasion. My new laboratory is being built rapidly and we shall probably open it before Christmas.

13 *July* 1911. The degrees celebrations was an interesting ceremony. There were a number of Scotch Divines who got honorary degrees and also a number of medical people and others. The most interesting person from the point of view of the crowd was the Maharajah of Gwalior

—a Coronation guest—who was resplendent in uniform and diamonds. I went to a lunch in the Town Hall given to the Colonial Premiers but Ward was not there.

13 *Oct.* 1911. We are in full swing of university work again and we have a laboratory full of people at work. I am thinking of going to Brussels at the end of the month to attend a congress in physics which will last four days—"Theories of Radiation". You will have seen about the strikes in England which have apparently ended for the time being. No doubt the wages are in many cases too low. It appears fairly certain there will have to be an increase all round in wages during the next few years.

CHAPTER VIII

THE ATOM

After the discovery of the nucleus three notable advances were made, in all of which Rutherford and the Manchester school played a leading part. The chemistry of the radio-elements was investigated and their places in the periodic table determined. The Rutherford-Bohr model of the atom was established so firmly, that it is now referred to as *the* atom. The charge on the nucleus was measured by Moseley and found to be a multiple of the positive unit electronic charge, so that the atomic number became a fundamental conception. Woven into the pattern of these three main advances were other events which invite attention.

Early in the year (22 Jan. 1912) Rutherford wrote to Stefan Meyer about radium standards and added:

No doubt you have heard that Mme Curie is far from well and will require a long rest before she will be able to get to work again....I trust that the rumours that you have heard of the cause of Mme Curie's illness are unfounded. I certainly sympathise very strongly with her in the misfortunes that have come to her in recent years.

The number of research students at Manchester had greatly increased and it was essential to secure more space. This had been duly provided, and on 1 March there was an opening ceremony and a conversazione with experiments and demonstrations. The new laboratories were opened by Professor Arthur Schuster, and an honorary D.Sc. degree was conferred upon the President of the Institution of Electrical Engineers, S. Z. de Ferranti.

The latter had solved the problem of electrical supply over large areas by using 10,000 H.P. generators transmitting current at 10,000 volts—an idea which at the time was regarded as most dangerous and revolutionary. He developed transformers to reduce the voltage for distribution to consumers. After twenty years' further experience, the bulk supply throughout the world was in accord with his original plans. Rutherford paid a tribute to Ferranti's largeness of conception, his original genius, his daring in execution, and his utter disregard for self-aggrandisement and money-making.

On 7 March Rutherford wrote again to Stefan Meyer stating that he hoped to visit Paris about the 24th and that he had arranged with Debierne to get apparatus constructed so that Rutherford could try his balance method of comparing radium standards. He arranged to get away on the 24th so as to join his wife and Bragg at Havre for a motor tour in the Pyrenees.

On 9 April Rutherford wrote to Stefan Meyer from Carcassonne and told him of a generous gift to meet the expense of the inter-national radium standard, from Dr and Mrs Beilby, the parents of Mrs Soddy.

Prof. W. H. Bragg had joined the Rutherfords in a "gorgeous trip in France. It has been splendid", wrote Bragg, "and I can never forget it and all the fun we had and the many varied experiences. I never expected to go such a trip and can hardly believe I have actually done so."

Rutherford's letters to his mother during this summer give a vivid picture of events which were described by him soon after they happened, in letters that certainly were not written with a view to publication.

22 *May* 1912. The Royal Society is celebrating its 250th anniversary about the middle of July in London. There will be delegates from the Scientific Societies all over the world and a number of garden parties and receptions. The King is to receive the Council delegates at Windsor Castle and as a member of the Council I shall probably attend. There will be a garden party at Windsor by the King and a special service at Westminster Abbey.

25 *July*. On arrival at Windsor we were shown through the state rooms of the castle and after a very considerable walk we went out on to the terrace and were marshalled for presentation to the King and Queen.

The ceremony was very simple. We went past and shook hands with the King and Queen and then made a bee line to find our women folk who were congregated at a suitable point. The day was not too hot and the King and Queen and family made a procession through the crowd of visitors, over 7000, including everyone of importance in London. I understand the man who called out the names as we were presented got our group of delegates mixed up and every man was given his neighbour's name. It did not matter, as I don't suppose the King knew the names of one of them.

After a scramble for tea we walked round and made for home about 5.30. It was rather a tiring function and I am quite content if I never go to a Windsor garden party again. I had to buy a silk hat and got quite accustomed to wearing that monstrosity. The ladies were all attired in their best and the ultra-fashionable had piebald stripes, for example, one half of the dress red and the other blue, the line of demarcation from one shoulder to the opposite foot. It was very amusing watching some of these queer get-ups.

The last month has been filled with congresses and celebrations and I am glad they are now over and I can settle down to three weeks' uninterrupted work before the vacation. There was first of all the Universities Congress in London, which I attended for three days and where I met a number of old Canadian and Australian friends. The week after was the 250th anniversary of the Royal Society. After the presentation of addresses from delegates there was a big dinner in the evening in the Guildhall, which was full of well known people. Asquith proposed the health of the Royal Society and speeches were made by the Archbishop of Canterbury and various other notables. I sat next to Ramsay Macdonald the head of the Labour M.P.s. On Wednesday we went to Syon House where the Duke of Northumberland gave a garden party and I met a number of old friends.

In the autumn of 1912 Hahn had written to Rutherford begging that due credit should be given to his collaborators von Baeyer and Lise Meitner, to which he received this reply:

17, *Wilmslow Road, Withington, Manchester:* 17 Oct. 1912. I am glad you drew my attention to my delinquencies, and I can assure you that it was quite unintentional. I have not my paper with me to look up, but I think I adopted the method to save repeating Baeyer, Hahn and

Meitner in different orders at regular intervals. The whole matter is put fully and correctly in my book: at least I think so. It is always difficult to be quoting three people at once, two are bad enough.

I am glad that you are going on with the beta ray work. I have done some here on radium, and intend to continue it. We have just had Boltwood staying with us for a few days, and he is now on his way back to America. He is in very good form after his holiday, but seems to have had a pretty hard time fixing up the new Laboratory in Yale. He left this morning to go to Rotterdam.

You will be interested to hear that W. Wilson is now over with McLennan in Toronto, and J. A. Gray in McGill, both as lecturers.

With regard to Ramsay, I noticed that he had revived the ancient notion. He told me all about it, but I gently insinuated that it was a great pity that one did not have a hundred times as much radium and could not spend a hundred years in proving such points. General radio-active analysis of minerals shows that neon, if produced at all, is in excessively small quantity.

I have got a couple of men at present examining whether gamma rays are produced from alpha rays. We have found it so already for several cases, and intend to carry it right through. You will see a reference to it in mine and Chadwick's paper. I have got a fine room myself now in the Laboratory, and am getting ready to do a number of things. My book is practically completed, and ought to be out before long. I hope that you will take an opportunity of seeing something of Geiger, whom you will find a thoroughly good fellow. He will be probably fairly lonely for a bit.

We are all very well and flourishing. I am glad to hear that you are going to work in the Kaiser Wilhelm Institute. I hope soon to hear that you will be appointed full Professor in that Institute, but I suppose that that is not settled yet.

The 'ancient notion' referred to in this letter was the production of lithium and neon by radiations from radium, which further experiments failed to substantiate. These efforts are not to be confused with modern transmutations effected by quite different means.

Later in the autumn there was held another Solvay Conference in Brussels, described by Rutherford in *Nature* (20 Nov.):

The First International Conference in Brussels on the Theory of Radiation, in 1911, owed its inception to Mr Ernest Solvay and proved a great success.

Shortly afterwards Mr Solvay generously gave the sum of one million francs to form an International Physical Institute (*Nature*, 90, p. 545), part of the proceeds to be devoted to assistance of researches in physics and chemistry, and part to defray the expenditure of an occasional scientific conference between men of all nations to discuss scientific problems of special interest. In pursuance of this aim the second International Conference or Conseil International de Physique Solvay was held in Brussels this year on October 27–31, under the able presidency of Professor Lorentz. Among those present were Lorentz, K. Onnes, J. J. Thomson, Barlow, Pope, Jeans, Bragg, Rutherford, Mme Curie, Gouy, Brillouin, Langevin, Voigt, Warburg, Nernst, Rubens, Wien, Einstein, Lawe, Sommerfeld, Grüneisen, Weiss, Knudsen, Hasenhörl, Wood, Goldschmidt, Verschaffelt, Lindemann, de Broglie....

On 13 Sept. Rutherford had written again to Stefan Meyer saying that while attending the British Association at Dundee he met Dr Beilby, who had made another generous gift to cover the expense of a radium standard for Great Britain, to be preserved at the National Physical Laboratory. The importance of correct standards cannot be over-estimated. Delivery of goods and adjustment of prices would be extremely difficult if no one knew what was the length of a yard within ten per cent. Another letter deals with the same subject:

RUTHERFORD TO MEYER

17, *Wilmslow Road, Manchester:* 16 *Nov.* 1912. I understand from Glazebrook that the communication from your Ministerium was at length located in the Home Office. As is the custom of our officials, they took four months in deciding that it did not properly belong to their Department. A formal request has been sent to the Foreign Office, and I trust that they will act on it without delay. I received safely the reprints of the paper on the 'Heating Effect'. By the way, how did Mr H. Robinson become 'Miss'? I shall be interested to know how the mistake arose. The Laboratory is enjoying the error, for Robinson is about four inches taller than myself and is not easily mistaken for a 'Miss'. I will have to correct the matter when I publish the English translation. Dr von Hevesy will be calling on you on his way back to Buda-Pesth. He is an able fellow and has done capable work on the chemical side of radioactivity. He will tell you that he has determined the valency of a great number of the radioactive substances.

In this year the third edition of Rutherford's book on radioactivity appeared from the Cambridge University Press, under a new title, *Radioactive Substances and their Radiations*, to avoid certain difficulties about rights for translations. The book embodied all that was known of radioactivity and its applications up to the time of publication. It was a great feat to write such a book in the middle of so busy a life, and with the research work of a great laboratory in full swing.

As Geiger remarked later: "His text-book on Radioactivity is for the initiated a record of his selflessness. How often does the name of a disciple accompany a discovery which was yet completely his own creation".

This invaluable book is now out of print and a quarter of a century has made it obsolete in a few matters of fact. It is the last great work on the whole subject written by one familiar with every aspect of the developments in which he had taken so leading a part. In this book is found an early, or perhaps the first, use of the word 'nucleus'—"The atom must contain a highly charged concentrated nucleus".

Seven years later the Cambridge University Press published *Radiations from Radioactive Substances*, by Rutherford, Chadwick and Ellis, a book which, as its name implies, deals largely with radiations.

About this time there was some prospect of W. H. Bragg becoming the first Principal of the new University of British Columbia, and he consulted Rutherford, who replied:

10 *Jan.* 1913. ...I can quite sympathise with you in the difficulty of making a choice....Of course the position would be ideal if there were plenty of money and one could have one's own way; but these are the very questions you cannot find out beforehand. I think that if I were

in the position that I felt tired of Physical work and had not an idea
left to work on, I should consider it an admirable position to occupy
one's declining years; but I quite agree with you that it would be
very difficult to leave the Physical world at such an interesting time,
when there is so much to do and so many interesting problems in
sight.

It seems to me that such positions should be taken by men who have
a real interest in administration for its own sake. I am sure that if you

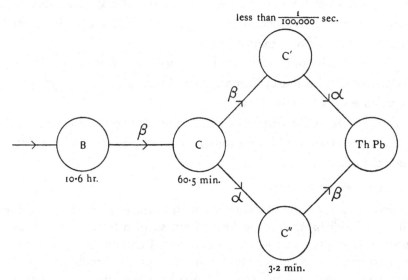

Branching products of thorium. 65 per cent go by the C′ route, 35 per
cent by the C″ route

had taken it you would have missed very much the scientific side....
I do not think that any President's job is a rosy one....I always feel that
one is very much happier in doing work in which you have a keen
intellectual interest. After all, provided one has enough to live on
comfortably and is engaged in work which you like doing, the exalted
positions of the world have very little attraction—at any rate, to me.

Darwin, Marsden and Makower at Manchester were investigating
the "branching products". Strange it is that in the case of radium C
only 1 in 2500 atoms begin their transformations with the alpha; the

great majority start with a beta ray change. In the case of actinium C, only 3 in a 1000 start with the beta; while with thorium C the betting is 2 to 1 in favour of the betas preceding the alphas. Some of the intermediate products in these changes have an extraordinarily short life, only about a millionth, or even a hundred thousandth part of a millionth of a second. The alpha particles from such quick-changing atoms have long ranges and high energies. Captives with large energy quickly burst the bonds of their prison and spread afar. But after all, these stories remind one of the Queen of Sheba and Solomon's grandeur, the half of it cannot be told.

On 18 March 1912, Rutherford had written to Boltwood: "Bohr, a Dane, has pulled out of Cambridge and turned up here to get some experience in radioactive work".

Here is one of the first letters that passed between Rutherford and Bohr, a conjunction of bright spirits which proved so fruitful in ideas and scientific progress.

BOHR TO RUTHERFORD

24 *July* 1912. Leaving Manchester I wish to thank you very much for all your kindness against me here. I am so glad for all what I have learned during my stay in your laboratory, I am only sorry that it has been so short. I am so thankful for all the time you kindly have given me; your suggestions and criticisms have made so many questions so real for me, and I am looking forward so very much to try to work upon them in the following years.

In 1913, when I was staying with the Rutherfords, there came into the room a slight-looking boy whom Rutherford at once took into his study. Mrs Rutherford explained to me that the visitor was a young Dane, and that her husband thought very highly indeed of his work. No wonder, it was Niels Bohr!

Bohr's theory of the atom was entirely complementary to Rutherford's. The positive heavy nucleus was accepted, and the satellite electrons were so constrained as to give correctly the known spectral series. It was evident to Bohr that in some way or other Planck's Quantum

Theory (1900) must be applied to the atom in order to get the correct values for the frequencies of the observed lines. Perhaps it may help some readers to regard each atom as a minute wireless station, capable, when stimulated, of sending out and receiving signals of quite definite frequencies and wave lengths. Observers had found, in the case of hydrogen, two or three simple series of such frequencies, and they were best evaluated in terms of the quantities 1, 1/4, 1/9, 1/16, ...; and their differences. There was another constant involved, named after its discoverer Rydberg.

Bohr took the mechanics of Newton and wove into them the quanta of Planck. When a signal left or entered an atom, the energy of that signal was always a 'lump', heap, bundle, or quantum of energy, proportional to the frequency of the wave. The energy change equals h times the frequency, where h is Planck's constant. Bohr created a diagram showing the orbits of the electrons round the central nucleus. In the simplest case of all, hydrogen, there would be but one encircling electron, and Bohr found it necessary to suppose that this electron could be temporarily stable only at definite distances from the Rutherford nucleus. In order to achieve this, he proposed that the angular momentum of the electron about the nucleus must also be changed by quanta. In this way, Bohr obtained a complete and successful scheme for the series of spectra of the hydrogen atom, and computed their frequencies; calculated the value of the Rydberg constant; and forecast that a series attributed to hydrogen should correctly be assigned to helium, as was afterwards proved. Rutherford was entranced with this beautiful theory which brought order into the exterior of the atom as he had done for the nucleus. Bohr's theory has had profound modification since the early days, particularly from 1924 onwards. Yet his was the key that first opened the door and without his genius the later progress, in which he was always a pioneer, could not have been achieved.

At this time, however, Bohr's theory was still in a state of evolution, and it is interesting to note the development of his ideas. Rutherford besought Bohr to be as brief as possible, consistent with clearness, for "long papers have a way of frightening readers. It is the custom in England to put things very shortly and tersely in contrast with the Germanic method, where it appears to be a virtue to be as long-winded as possible."

BOHR TO RUTHERFORD

3, *St Jacobsgade, Copenhagen*: 6 *March* 1913. I thank you very much for your letter, I was so glad to hear about all that is going on in the laboratory. I am exceedingly interested in it all, and especially in the results of your work on the gamma rays.

Enclosed I send the first chapter of my paper on the constitution of atoms. I hope that the next chapters shall follow in a few weeks. In the latest time I have had good progress with my work, and hope to have succeeded in extending the considerations used to a number of different phenomena; such as emission of line-spectra, magnetism, and possibly an indication of a theory of the constitution of crystalline structures. I have, however, some difficulties in keeping it all together at the same time, and shall be very glad to have some of it published as soon as possible on account of the accumulating literature on the subject. As besides the paper is getting rather long for publishing at one time in a periodical, I have thought it best to publish it in parts. Therefore I shall be very thankful if you kindly will communicate the present first chapter for me to the Phil. Mag.

As you will see, the first chapter is mainly dealing with the problem of emission of line-spectra, considered from the point of view sketched in my former letter to you. I have tried to show that it, from such a point of view, seems possible to give a simple interpretation of the law of the spectrum of hydrogen, and that the calculation affords a close quantitative agreement with experiments. (I have given reasons which show, that if the foundation of the theory is sound, we may assume that

$$\frac{2\pi^2 me^4}{h^3} = 3 \cdot 290 \times 10^{15} \text{ (Rydberg's number)}.$$

Putting your value $e = 4 \cdot 65 \times 10^{-10}$, I get $h = 6 \cdot 26 \times 10^{-27}$.

Putting Millikan's value $e = 4 \cdot 87 \times 10^{-10}$, I get $h = 6 \cdot 76 \times 10^{-27}$.
Unfortunately, however, Planck's constant is hardly known with any great accuracy.)

The second chapter deals with the atoms, the third with the molecules, and the last chapter with magnetism and some general considerations.

I hope that you will find that I have taken a reasonable point of view as to the delicate question of the simultaneous use of the old mechanics and of the new assumptions introduced by Planck's theory of radiation. I am very anxious to know what you may think of it all.

As you will see, I have by the considerations in the first chapter been led to an interpretation, different from that generally assumed, of the origin of some series of lines observed in stars, and also recently by Fowler in a vacuum tube filled with a mixture of hydrogen and helium. Instead of ascribing them to hydrogen, I have tried to give reasons for ascribing them to helium. This latter point might, however, be tested experimentally. In a discourse in which I prepared my point of view and tried to explain (as I have done in the paper) that the presence of hydrogen in the experiments of Fowler might indirectly be the cause of the lines considered, the chemist Dr Bjerrum suggested to me that if my point of view was right, the lines might also appear in a tube filled with a mixture of helium and chlorine (oxygen, or other electronegative substance); indeed it was suggested, that the lines might be still stronger in this case. Now, I have not in Copenhagen the opportunity to do such an experiment satisfactorily; I might therefore ask you if you would possibly let it perform in your laboratory, or if you perhaps kindly would forward the suggestion to Mr Fowler, which may have the arrangement used still standing.

When I have finished my paper, I hope to be able to come on a short visit to Manchester, and look forward to this eventuality with the greatest pleasure.

When Rutherford read Bohr's paper, he wrote:

20 *March* 1913. Your ideas as to the mode of origin of spectra in hydrogen are very ingenious and seem to work out well; but the mixture of Planck's ideas with the old mechanics makes it very difficult to form a physical idea of what is the basis of it all. There appears to me one grave difficulty in your hypothesis, which I have no doubt you fully realise, namely, how does an electron decide what frequency it is going to vibrate at when it passes from one stationary state to the other? It seems to me that you would have to assume that the electron knows beforehand where it is going to stop.

In 1913 Geiger had invented a new type of detector, which would indicate alpha or beta particles. With this instrument it was easily possible to detect a *single* electron issuing from a piece of uranium-radium ore held a few inches away. A small closed hollow cylinder had a small hole at one end covered with a thin piece of mica. The point of a fine sewing needle, just behind the hole, was almost sparking to the cylinder kept at about a thousand volts. When the electron entered there was ionisation by collision and the needle, connected to a string-electrometer, gained potential, which was quickly removed through a high resistance to earth. Every electron made the string 'kick' and this could be shown to a large audience by projection.

RUTHERFORD TO GEIGER

17, Wilmslow Road, Withington, Manchester: 10 *May* 1913. I am glad to hear that you received the proofs safely and much interested to hear of your new method of counting. It is rather surprising that the pointed conductor, which we tried to avoid so much, should prove after all the best thing.

If you can arrange to send me a working tube I should be much obliged. I do not know whether you have yet arrived at the stage of publishing. I was thinking of showing the old experiment of counting the alpha particles at the Royal Institution, but rather funked it when I thought of all the apparatus I would have to carry down. If, however, your new arrangement works well and without trouble, I might be prepared to show it. I do not know whether you will have published the method by that time. If not, I would, of course, either say nothing about the details or make any statement you thought fit. As a matter of fact, I could merely state that it was a special arrangement prepared by yourself on the lines of our old work and give no publicity to it.

I was interested to hear of the effects of pressure upon it. It certainly looks as if the effect at low pressure might prove very useful in some types of experiments. My lectures begin on May 24th, so that if you could send me a tube I would like it to arrive within the week, if possible.

I was down at London at the Royal Society Soirée where Collie was showing his production of neon and helium from vacuum tubes. I had

a good talk with him and it certainly seems that there is no doubt of
the magnitude of the effects, but he appears to have reached J. J.'s
conclusion, that the amount he obtained out of a tube diminishes with
successive heating. Fowler also showed his new hydrogen spectrum
series in helium tubes.

The Whitsuntide holidays are now in full swing, but most of the
lectures go on as usual. You will be interested to hear that Arrhenius
has sent over Dr Oholm, who is likely to be soon made Professor at
Helsingfors, to learn the elements of radioactivity, like Arrhenius in the
old days. He seemed to think that you had looked after him very well
in those days of long ago.

Lots of work is going on with beta and gamma rays, and Robinson
and I hope in a day or two to have completed the measurements of e/m.
We have now got to the stage of obtaining very definite and accurate
results.

In the summer the Rutherfords went for an expedition in the Tyrol
and were joined at Rothenburg by Geiger, who travelled with them to
Munich, where they regretfully parted company. Boltwood took his
place and they went for a three weeks' tour of the Tyrol and the
Dolomites.

The 1913 Meeting of the British Association at Birmingham was one
of remarkable interest. The quantum theory and Bohr's ideas on spectra
were in the air, and when Lord Rayleigh (Chairman of the Mathematics
and Physics Section) was pressed for an opinion, he replied that "men
over seventy should not be hasty in expressing opinions on new theories".
He was however reluctant to approve the quantum theory, for he did
not believe "that Nature behaved in that way", for, he added, "I have
difficulty in accepting it as a picture of what actually takes place".
This difficulty still remains, but when one gets used to it, the trouble
like many others is lightened.

Yet Poincaré and Jeans were crystallising public opinion by proving
that exchanges of energy *must* take place by jumps in some such manner
as that advocated by Planck, Einstein and Bohr.

Mme Curie was at the Meeting, shy, retiring, self-possessed and noble;

everyone wished to see her, but few were successful. The press were eager for interviews, and Madame skilfully parried their questions by singing the praises of Rutherford. It was not quite what they wanted, but it was all that they could get. "Dr Rutherford", she said, "is the one man living who promises to confer some inestimable boon on mankind as a result of the discovery of radium. I would advise England to watch Dr Rutherford; his work in radioactivity has surprised me greatly. Great developments are likely to transpire shortly, to which the discovery of radium is only a preliminary."

HEVESY TO RUTHERFORD

19, *Nador U., Budapest*: 14 *Oct.* 1913. I was sorry you had to leave Birmingham so soon and could not read your second paper, and also that I only saw Mrs Rutherford at the degree ceremony for a few seconds.

The meetings on Monday and Tuesday have been very interesting. It is a most remarkable fact that Aston succeeded to separate the two Neons by diffusion and gave a definite proof that elements of different atomic weights can have the same chemical properties. Thomson came in his paper on X_3 to the conclusion that the latter is a polymerized Hydrogen, a kind of H_3 (like O_3). In the following discussion Bohr—in his usual modest way—suggested the possibility of X_3 being an H atom with *one* central charge, but having a three times so heavy nucleus than Hydrogen. He suggested to let a mixture of H and X_3 diffuse through Palladium and try if it is possible to separate them for the heavier X_3 atom has to diffuse much slower.

Bohr had not been understood properly and Thomson gave a rather quick answer, saying—after a brief consultation with Ramsay—that Bohr's suggestion is useless, for not the molecules, but the atoms of H diffuse through hot palladium.—Certainly, but it was just Bohr's point. ... The general appearance was, that he told something highly ingenious and Bohr something very stupid. Just the contrary was the case. So I felt bound to stick up for Bohr and explained the meaning of Bohr's suggestion in more concrete terms, saying that Bohr's suggestion is that X_3 is possibly a chemically non-separable element from Hydrogen.... Of course it is not very probable, but still a very interesting suggestion, which should not be quickly dismissed....

Three new ideas constitute our present knowledge of Radio-chemistry and the periodic law.

(1) That elements of different atomic weights can have the same chemical properties. This was first stated by Svedborg and Strömholm (1908) but chiefly advocated by Soddy; he can claim this point undoubtedly.

(2) The fact that it is a simple connection between the occurrences inside and outside the atom.

Hevesy then points out that U II cannot possibly follow U I, as they are isotopes. Fajans saw that there must be intermediate products, and looked for and found $U X_2$.

Though Russell was already interested in the problem and I started the valency experiments when Bohr came to Manchester, no doubt he encouraged us both very much and if we trace the origin of the above ideas to their origin, we will find them in Bohr's mind, as pointed out to me by himself in his usual modest way.

(3) The third point is that after the emanations the change does not lead to the group two, but the group six of the periodic table. Which has been first pointed out by Fajans.

I spent a very pleasant time in Birmingham, it was my first visit to this part of England. On the excursion to Kenilworth Castle I met a former teacher of yours, Bickerton. He is a funny old chap and one of the comical figures you find at every congress.

Coming back from Kenilworth, we passed Leamington where the Mayor entertained us and saw us off later on. When standing on the platform he happened to start a conversation with Bickerton, and seemed to be very much struck by the versatility and nobility of the old man, and after hesitating for a few seconds he lifted his grey top hat and asked him "if he happens to be Sir Oliver?" "Not so famous, but greater" came the prompt answer and the train left the platform. I will never forget this scene.

I crossed after the Birmingham meeting the Channel and came just in time to the Vienna Congress, where I meet all the German and Austrian scientists interested in radioactivity and many others.

The very hospitable house of Prof. Meyer was the centre of the radioactive people and the few days spent there were very nice and interesting ones. It was interesting to compare the Vienna and Birmingham meetings; all together it was more knowledge in Vienna, but far more ingenuity in Birmingham.

Speaking with Einstein on different topics we came to speak on Bohr's theory, he told me that he had once similar ideas but he did not dare to publish them. "Should Bohr's theory be right, it is of the greatest importance." When I told him about the Fowler Spectrum the big eyes of Einstein looked still bigger and he told me "Then it is one of the greatest discoveries".

I felt very happy hearing Einstein saying so.

In the autumn of 1913 Rutherford attended the Solvay Conference at Brussels and he has left many notes, often difficult to decipher, made for the interesting discussions which, especially at that time, took place on the nature of things, particularly atoms and radiations; on the relation between beta and gamma rays, or as might be said today between electrons and photons; also on the number of electrons in an atom of given atomic weight, for atomic number was not yet known. Rutherford lectured for the first time in Holland at Amsterdam and visited the famous low temperature laboratory at Leyden. He was then strongly advocating improved methods of securing high voltages. "I think it is a matter of pressing importance at the present time to devise electrical machines of the highest possible voltages."

The list of New Year's Honours for 1914 included the name of Ernest Rutherford for the order of knighthood, and he was invested on 12 February by the King at Buckingham Palace.

COX TO RUTHERFORD

West Down, Hindhead, Haslemere: 1 *Jan.* 1914. ...We all send our heartiest congratulations to you and to Lady Rutherford. Of course this has 'cast its shadow before' long ago. Even in Montreal we used to try our tongues on the new style. But we are just as delighted now it has come.

But I say! You must draw the line somewhere. Remember Lamb's dreadful speculation about Lord Lamb, Prince Lamb, King Lamb and

even Pope Lamb. We will allow you the O.M. in due course. But
never think of Pope. Well, we shall drink your healths at dinner—in
lime-juice, alas!...

A distinguished research student of Rutherford's, J. Chadwick, had
gone to Berlin to work with Geiger, little foreseeing the hard trial
awaiting him—internment for the whole duration of the War. He
wrote:

Charlottenburg, Bismarckstrasse, 9 Gth III: *14 Jan.* 1914. I have wished
to write to you for some time but I was ashamed to tell you what little
progress I have made here.

First I would like to send my sincere congratulations on your knighthood. It was not a great surprise to me but Geiger was tremendously
surprised. I have had to explain to him several times what it means.
He has forgotten a fair amount of English customs but he has recovered
some since I came. From now on, no English is allowed except
dictionary English. I get along reasonably well now but have trouble
with the different accents. There are scarcely two people in the Reichsanstalt who have the same accent....

Working with Geiger is great fun and he helps me in every possible
way. I find his German very difficult to understand, and in my opinion
he has the worst accent in the Reichsanstalt. By 'worst' I mean the
farthest removed from the Berliner accent.

We go to Rubens' colloquium every Wednesday but I prefer ours.
Here there is the advantage that every subject is discussed, but the
abstracts are often carelessly done and the discussions are not at all
stimulating. The Physikalische Gesellschaft is sometimes good, but is
too long. Three hours on a hard wooden seat is too much for me, and,
if I really must keep awake and think, I prefer to have my body comfortably provided for. Hahn I find extremely nice and he gives exceedingly
clear papers. The only thing I can really complain about is that there is
no tea at half past four.

RUTHERFORD TO HEVESY

17, Wilmslow Road, Withington, Manchester: 7 *Jan.* 1914. We are very
glad to receive your kind telegram of congratulations, but I was
wondering how you heard the news unless from one of your Manchester

correspondents. It is, of course, very satisfactory to have one's work recognised by the powers that be, but the form of recognition is a little embarrassing for a relatively youthful and impecunious Professor like myself. However, I trust that it will not interfere with my future activities. . . .

I have been hard at work all the Christmas vacation determining the wave lengths of gamma rays, and have made such good progress that I see my way pretty well clear to its conclusion. The whole problem is a very difficult one, much more so than X rays, partly on account of the disturbances due to beta rays and partly on account of the great penetrating power of the radiations.

You will have seen in this month's Phil. Mag. that Soddy has found Antonoff's results. I am very glad that things have turned out all right, for I should have been very sorry to feel that the laboratory had one piece of bad work to its discredit. I think Antonoff has scored heavily in the matter, for he never wavered in his certainty that he was right. . . .

Eileen was very pleased at the news, and was greatly excited on New Year's Day with the succession of telegrams. She is of the opinion that neither of her parents has the 'swank' and natural dignity for such decorations. . . .

The reference to Antonoff is not without interest, for he had separated a branching product named uranium Y, of doubtful parentage, but a substance with an enormous progeny, for it appears to be the parent of protoactinium, also known as Pa, from which in turn descends the whole of the actinium family.

It has already been explained that when a radioactive transformation occurs, there is disintegration of the nucleus accompanied by the expulsion of an alpha or beta particle—according to the nature of the atom. What then are gamma rays? Long arguments and many experiments have been made in connection with this question. At one time the beta were said to cause the gamma rays, at another time, the exact reverse, and indeed each can incite the other in a certain sense.

Perhaps it is possible to venture on a statement, however rash such an attempt may be. After the expulsion, during the atomic disintegration, of an alpha or beta particle, the atom may be in a state of electric strain, or in an 'excited state'. A natural readjustment is effected by the emission of an electromagnetic wave, or photon, from the nucleus. This is a gamma ray. We owe this point of view to Lise Meitner, who strongly defended it against opposition.

In passing through the atom a gamma ray may give all or some of its energy to one of the satellite or Bohr electrons, thus driving it out in strict accord with Einstein's law of photoelectricity. These electrons are called beta rays, but they must not be confused with the disintegration beta rays. It may then be said, at the risk of losing every shred of scientific reputation, that a genuine gamma ray is an S.O.S. wireless signal from the nucleus of an atom in distress!

RUTHERFORD TO HAHN

17, *Wilmslow Road, Withington, Manchester: 16 March* 1914. I thank you very much for your very kind and cordial letter of congratulations on the New Year's honour. I am very glad to feel that my old students are pleased with this recognition of my labours in the past. I think you know that I do not lay much stress on such forms of decoration, for they have obvious disadvantages in the case of a scientific man like myself. However, I am getting quite used to the change of title, and trust that it will not interfere with my scientific activities. I had to go to London about a month ago for the formal investiture at Buckingham Palace by the King. I had to get a special Court uniform with a sword attached. The function passed off quite simply and pleasantly but I was glad to get it over....

You will be surprised to hear that I am paying a hurried visit to Canada to Montreal, and the U.S.A. and am leaving on March 28th. I will stay with Professor Eve in Montreal and have already arranged to give an address before the joint Chemical and Physical Societies in my old lecture room in McGill. My main object is to accept an invitation of the National Academy of Sciences in America to deliver two lectures of a somewhat popular character before them on the evolution of matter and the structure of the atom. They are starting a new series of lectures at their Annual meetings, and I am supposed to lead off. I shall have a very busy time before I go, for Bragg comes up on Wednesday to give a lecture on X ray spectra and the structure of crystals, and I go down to London on Thursday to open a discussion on the structure of the atom at the Royal Society. This is an innovation, but I hope there will be a good discussion. Our information of the atom is really growing apace, and I have a firm belief in the general correctness of my nucleus atom.

I expect to return to Manchester from America in the first week of May, and leave for Australasia on July 1st with my family. I hope to spend about two months in New Zealand visiting my parents and people. The Miss Rutherford you mention is now happily married and is the proud possessor of a son. I shall be sure to see her and will give her your remembrances.

I understand from Stefan Meyer that he has been writing to you about the Vienna Radiology Congress in 1915. I hope that all you Radioactive people will turn up in strength on that occasion to support me in my Presidential capacity. I think it ought to be a good meeting, and we shall be able to accomplish work of very definite importance.

Rutherford obtained leave of absence from Manchester University so that he might go to Washington to deliver the first "William E. Hale Lecture" to the National Academy of Sciences. He chose as subject "The Structure of the Atom and the Evolution of the Elements". He sailed to Halifax and went on to Montreal, but here is his own account of the early part of his journey, written to his wife:

Hotel Touraine, Boston: 14 *April* 1914. I have at last got a few minutes free before going in to New Haven, to tell you of my wanderings since my arrival in Halifax on Tuesday at 6 p.m. I found Mackenzie, Principal of Dalhousie, and Bronson down to meet me and we had dinner together and talked over old times....The journey on to Montreal through Maine was very tedious and uninteresting as there was deep snow everywhere. I whiled away the time with sleep and bridge with the youngsters who had the drawingroom on the car. We did not arrive at Montreal till 3 a.m. but I slept on till 6 and got to our friends at 8 for breakfast....There was snow everywhere in the streets and the weather when I was there was generally about freezing with a good deal of wind. Such conditions were of course exceptional, but I thanked God I did not have to live in Montreal permanently. The Eves' house is very comfortable and runs very happily and the children are great youngsters and the elder is remarkably intelligent and helpful for her years....

As soon as I had settled down for a smoke I had to set out for the Lab. and to make ready my lecture at four, and at tea, I met a number of old friends and Sir William Macdonald, who is now Chancellor of McGill, turned up, looking just the same as ever. After the lecture we went to dinner at the University Club, with about fifty present, with speeches by Peterson, Barnes, Eve and myself.

Rutherford spent three days in Montreal seeing old friends, particularly Harriet Brooks, then Mrs Pitcher, with three children, "all admirable specimens"; and then on Sunday night he went on to Boston and to Harvard where he dined at the Club and "gave an informal lecture which went off in great style".

He lectured at Columbia, Yale and Princeton, and there is no doubt that his wide and friendly acquaintance with so many of the leading scientific men in the United States proved of great value at a later date when the heavy strain of the War was most severe.

G. E. HALE TO RUTHERFORD

National Academy of Sciences: 1 *June* 1914. Your kind letter of May 11, telling me of your return to England, was very welcome. I assure you that we most heartily appreciate the great favour you have done the National Academy, and American science generally, by inaugurating this series of lectures. I never before saw so much enthusiasm and interest at any scientific lectures. Professor Loeb, in a great state of delight, said that Faraday never did better. The splendid thing was that you not only brought out points of the greatest general interest and the most fundamental importance, but you stirred the imagination of every investigator worthy of the name who heard you, and stimulated them to try and do likewise. I never was so stirred in my life, and I have wondered a hundred times since how I could learn to think and to do things in such a big and effective way. The rush of your advance is overpowering, and I do not wonder that Nature has retreated from trench to trench, and from height to height, until she is now capitulating in her inmost citadel....

On his return to Manchester, and in the midst of his preparations for the visit to Australia and New Zealand, in connection with the forthcoming Meeting there of the British Association, Rutherford found time to interview the press on the question of the supply of radium for medical purposes. Manchester wanted to raise £25,000 for the purchase of radium, and it is interesting to note that this city has taken a leading place in pioneer work in radiology.

Rutherford said that the price of radium was high and must fall. The two sources then were pitchblende from Bohemia and carnotite from Colorado. Two new sources have since been added—the Belgian Congo, and the Great Bear Lake district, Canada. The price was in 1914 about £16 a milligramme, and in 1938 about £4 to £5; during the War the price soared.

In 1914 the Italian Society of Sciences presented the Matteucci Medal to Rutherford.

In this same year an important discovery was made at Manchester by Marsden. He found that when alpha rays were projected into hydrogen, so that heavy projectiles, mass four, struck lighter atoms of mass one, a few of the hydrogen atoms were driven forward far beyond the range of the alpha particles. This was 'playing marbles' with atoms, the large one giving its momentum to the small, so that the latter acquired a high velocity. Perhaps this effect simmered in the mind of Rutherford and helped towards the nuclear theory, and to the greater transmutation work of 1919.

On 20 May Rutherford wrote another long letter to Stefan Meyer about the proposed Radium Congress to be held in Vienna. There were to be opening addresses by Rutherford and Mme Curie, and papers and demonstrations by Moseley, Bragg, Geiger, Hahn and others, and excursions, such as a visit to the famous mines at Joachimsthal, then the main source of radium. He further stated that Soddy had communicated a paper to the Chemical Society in which he had found "that lead from a thorium mineral was of higher atomic weight than ordinary lead".

On 29 June Rutherford wrote again to Stefan Meyer about the proposed Radium Congress with full suggestions as to speakers and subjects. He took great pains that each country should be represented by their best men. He added that he would be sailing for New Zealand on 28 Nov., returning 10 Jan. This letter ended with the words: "I am very distressed by the tragic news of the assassination of the Archduke and his wife in this morning's paper. The Hapsburghs have a very tragic family history."

Few then realised that this murder would cause the death of more than ten millions and the destruction of a world order which we believed to be stable. For a time everything continued in a normal manner. Not until August did the crash come and the world was at war.

On 27 June, Rutherford opened a discussion on the "Constitution of the Atom" at a meeting of the Royal Society. The nuclear theory of the atom was well received and Rutherford showed that an alpha particle could proceed up to a very short distance from the centre of the nucleus, amounting only to 1.3×10^{-13} cm. It has been proposed to call 10^{-13} cm.—or one ten-million-millionth of a centimetre— a Rutherford unit, suited as it is to describe the sizes of nuclei. About 100 million atoms side by side reach one inch, and about a thousand nuclei side by side would stretch across an atom; or a million-million nuclei side by side would cover one inch in length.

The next letter is one of the first written to Rutherford by H. Moseley, who flashed on the scene like a meteor, and like a meteor his course was brief, but how resplendent!

Pick's Hill, West Wellow, Romsey, Hants: 17 July 1910. Thank you for your letter informing me of my appointment. It will be a great pleasure to me to work in your laboratory, and after my failure in 'schools' I consider myself very lucky to have got the opening which I coveted....

I would like to be guided entirely by you on the subject which I attempt, since until I have had a year or two for reading of a different kind, from that useful for examinations, I cannot profitably choose for myself.

I will spend August in Oxford, and will then read up Radioactivity, in the hope that your suggestion may be in that direction. My present knowledge extends little way beyond your books.

In 1913 Moseley was still working at Manchester. Among other experiments he had imprisoned some radium emanation in a silvered glass tube just thick enough to stop the alpha particles and thin enough to let all the negative beta rays pass. This loss of negative electricity would make the inside of the tube acquire an increasing positive charge, unless allowed to discharge as in Strutt's radium clock. Moseley had a simple scheme of measuring the increasing potential which mounted to a quarter of a million volts, declining with the decay of the radon.

I had the good fortune to sit next to Moseley at lunch at the refectory and congratulated him on this interesting piece of work. He replied that he was most anxious to go forward with his next experiment which he said was full of promise, and on my inquiry as to its nature, he replied that he intended to bombard successive elements in the periodic table with cathode rays, so as to excite their natural X-rays, which would then be reflected from crystals so as to reveal their frequencies. To those who cannot readily follow that last sentence, it is sufficient to add that he was going to bombard the atoms and make them speak for themselves!

The reader may then justly inquire what part of the atom was he going to bombard? Was it the Rutherford nucleus, or those Bohr outer electrons which give rise to visible light? Neither the one, nor the other! He bombarded the electrons in the inner Bohr rings, or orbits, or levels, which are nearest the nucleus, called the K, L, ... orbits. He had already studied the reflection of X-rays from crystals, working with Darwin. With a speed of experimenting, which astounded even Rutherford, he carried out his design; putting onto a small carrier half a dozen elements at a time he moved them forward in succession, showered electrons on them and measured the frequencies of the resulting X-rays with his crystal reflector. He found that the K ring frequencies from successive elements increased by uniform steps. Something was always *one* greater for each advance of elements. *This could be nothing else*, he claimed, *than the positive charge on the Rutherford nucleus.* Much work has been carried out since that day when he completed his work in Oxford, and all that he claimed has stood, so that the atomic number is as important as the atomic weight.

This atomic number is the exact number of positive unit charges on the Rutherford nucleus. When we say uranium, atomic number 92, hydrogen, atomic number 1, that indicates that the nucleus of the uranium atom has 92 times the charge on the hydrogen nucleus. It also means that uranium is number 92 in the periodic table, and there are no fractions or decimals involved.

Towards the end of 1913 Moseley left Manchester and began to organise for a continuation of his work at Oxford, and he wrote a letter of gratitude to Rutherford.

48, *Woodstock Road, Oxford: 7 Dec.* 1913. After seeing Townsend on Saturday I took a week's holiday in Hampshire finding that there was no apparatus with which to begin work forthwith. The last week has been spent in collecting the apparatus, a task requiring patience and tact in one or two cases, but now I think I shall really be able to get started again. It is too early yet to have an opinion on the prospects of working here efficiently, but I find the professor most obliging and ready to make the way smooth. The mechanic however is likely to prove a thorn in the flesh. Things seem to move slowly here compared to Manchester. My wish to get something made in a week is thought to show unseemly impatience, while liquid air is only made apparently about once a month.

However I naturally do not expect such good conditions for doing work as obtain at Manchester; as I am sure that it is no use looking for the like in England.

I want you to know how very much I have enjoyed the three years spent in your department. When I came my brain was full of cobwebs left by reading for examinations, and even if this time has only served as an education it has been very well spent. Especially I want to thank you and Mrs Rutherford for your kindness in interesting yourselves about me, and for the debt I owe you for personally teaching me how research work ought to be done.

Moseley's progress at Oxford was phenomenal, for on 18 Jan. 1914 he wrote to Rutherford that he was obtaining spectra from elements of low atomic weight like aluminium, from the inner ring (K) of silver and tin, and from the second ring (L) of the rare earths. Here is his wonderful claim: "I do not doubt that it will be possible to put every rare earth element into its right pigeon-hole and to settle whether any of them are complex and where to look for new ones". His main difficulty was to obtain salts of pure elements but he got a number from Urbain, "who is a most friendly gentleman". By 23 April he had them all in order and wrote to Hevesy saying that he had been "searching unsuccessfully for the unknown element. Either it is very rare, or as is quite likely only occurs in a few minerals. I hardly think it does not exist". Nine years were to pass before this rare element, hafnium (72), was discovered by Coster and Hevesy.

He had already written a letter to Rutherford, congratulating him on his knighthood, in which he showed himself a stout champion of Bohr.

MOSELEY TO RUTHERFORD

48, *Woodstock Road, Oxford:* 5 *Jan.* 1914. Please accept my heartiest congratulations on your knighthood. I know well your views on the sorrows entailed by a title in the guise of blackmail levied by servants and so forth, and I am all the more glad therefore that you have sacrificed yourself for the sake of the public reputation of science....

You may have noticed that F. A. Lindemann has been going for me in *Nature*. I propose to reply and enclose a copy of my letter. I have sent it to *Nature* this evening but as it is long I hardly expect to see it this week.

I need hardly say that I should be very grateful for any criticism or advice, if you could spare time to read it. Here there is no one interested in atom building. I should be glad to do something towards knocking on the head the very prevalent view that Bohr's work is all juggling with numbers until they can be got to fit. I myself feel convinced that what I have called the 'h' hypothesis is true, that is to say one will be able to build atoms out of e, m, and h and nothing else besides. Of the three variations of this hypothesis now going, Bohr's has far and away the most to recommend it, but very likely his special mechanism of angular momentum and so forth will be superseded.

Moseley and his mother started in June for Australia to attend the meeting of the British Association. Before starting he wrote to Rutherford:

I am rather exhausted after Urbain's visit. He and his wife stayed with us for two days and proved delightful people, but unfortunately neither spoke a word of English. Ramsay shepherded them from Paris, put them up for a night and very kindly brought them on to Oxford, else they might well have been lost. The attempt to speak dog French all day, and to get definite results in two days at the same time was naturally tiring, but I thoroughly enjoyed it, since Urbain is himself an exceptionally interesting man. I gathered from him that the French point of view is essentially different from the English. Where we try to find models or analogies, they are quite content with laws. He is himself a very unusual type of chemist, since his idea of chemistry is the study of the physical properties of elements and compounds, and of the two he himself prefers to work with the elements. Hence his success in isolating these rare elements. He uses every available physical property to identify them....

This year (1914) saw the publication by Rutherford and Andrade of a piece of work of great beauty. When a beam of gamma rays from radium B and C impinged on a thin sheet of mica, those rays which struck the crystals of the mica at a suitable angle were enhanced in a certain direction and impoverished in another, so that a photographic film behind the mica appeared to be ruled with four bright lines forming a square, and with four dark lines forming another square. By measuring the deflection of the gamma rays it was possible to determine their wave lengths, which are comparable with those of X-rays, but usually shorter, though the range of values is large.

It is impossible to include here a full account of all the activities at Manchester under Rutherford's leadership. The gift of a Mathe-

matical Readership by Professor Arthur Schuster led to the appointment
in succession of Bateman, Darwin and Bohr, whose influence spread to
all the work of the laboratory. Geiger and Nuttall discovered the
law relating to the life of the atom and the velocity of the alpha particle
ejected. It is not surprising that an alpha particle with high energy
should escape more quickly from the nucleus. In fact there is a simple
linear relation between the logarithm of the life of the atom and the
logarithm of the energy of emission of the alpha particle. There is
a slightly different relation for each of the three great families, radium,
thorium, actinium. The law has received recently some explanation
on the wave-mechanical theory by Gamow and others.

Rutherford once said to Prof. E. A. Milne, "there is no page, or
paragraph, or paper in any Journal, which one cannot improve if one
tries". A true, but depressing remark, which was intended to stimulate
the writer—any writer—to take more pains!

He further urged that it was "essential for men of science to take
an interest in the administration of their own affairs or else the professional civil servant would step in—and then, the Lord help you!"

More than one of Rutherford's students at Manchester have left
a record of their impressions. Thus A. B. Wood wrote:

The laboratory housed a very happy family with Rutherford as 'father'.
Problems and difficulties were often discussed in the afternoon at tea,
and some of these occasions leave impressions, apparently trivial but
really fundamental. I well recall an occasion when Moseley raised the
question of design of large electromagnets. A serious difficulty in the
use of electromagnets was due to overheating during the long exposure
required to obtain X-ray deflection photographs. This overheating led
to breakdown of the insulation of the winding and Moseley suggested
using *bare* aluminium wires which would oxidise on heating and the
oxide coating, being an insulator, would improve with use. This topic
turned to a general talk on electromagnets, some members of the tea
party tending towards exaggeration of their merits.

Rutherford, who had remained quiet, then told, with his knowing
smile, the story of a magnet at Montreal. This magnet, he said, would

take a bunch of keys from a man's pocket as he came into the room—in fact "the magnet was strong enough to draw the iron out of a man's constitution".

At this point, of course, the tea party broke up, every man to his own job! In those days, all were keen on solving the problem on hand and many, with Rutherford in the forefront, worked very late hours in the laboratory.

I well remember Moseley making liquid air for an experiment he was doing at 2 or 3 a.m., returning after a few hours' sleep to give a lecture at 9.30 a.m. After an intensive period of research, almost night and day, Rutherford would write up the paper for publication and take a few days' holiday; then back again full of fresh ideas. His energy was infectious and almost unlimited. He made no effort to conceal his joy at the conclusion of a successful experiment, and it was his proud record that he "never put a man on a 'dud' research". When his nuclear hypothesis seemed firmly established by experiment, he would say of the 'classical atomic theorists' that "some of them would give a thousand pounds to disprove it!" The great delight in his face as he removed his pipe and made the remark was one of the characteristics of Rutherford which can never be forgotten by those who knew him.

Similarly Hans Geiger recalled "Memories of Rutherford in Manchester" in *Nature* (6 Feb. 1938):

I see his quiet research room at the top of the physics building, under the roof, where his radium was kept and in which so much well-known work on the emanation was carried out. But I also see the gloomy cellar in which he had fitted up his delicate apparatus for the study of the alpha rays. Rutherford loved this room. One went down two steps and then heard from the darkness Rutherford's voice reminding one that a hot-pipe crossed the room at head-level, and to step over two water-pipes. Then finally, in the feeble light one saw the great man himself seated at his apparatus and straightway he would recount in his own inimitable way the progress of his experiments, and point out the difficulties that he had to overcome....

In the laboratory small glass tubes were used to hold large quantities of radium emanation. To break one was a crime, for the whole laboratory would have to stop work from the spread of radioactive contamination. One day a tube was broken. Rutherford encountered Geiger and inquired after his work. Geiger was curt and said that all work was stopped and that the trouble had come from Rutherford's own room! Rutherford looked surprised and replied, "Well! There you have further proof of the power of this emanation." Geiger continued, "With this remark, he left me; but he soon returned saying that I must be somewhat upset, and that I needed a little fresh air." So Rutherford took Geiger in his car out of the turmoil of the city and he was soon discussing his experiments and all the problems that remained to be solved. Geiger concludes: "Nothing was so refreshing or so inspiring as to spend an hour in this way, alone with Rutherford. In spite of the minor provocation, I would be loth to part with the memory of such a day, spent in fellowship with a master-mind."

On one occasion a research student from another University aired a grievance with Rutherford and told him that his Professor had stolen his thunder and published in a paper a result or idea due to the student. "Well," said Rutherford, "you know it never does to quarrel with your mother's milk."

There is also a story of a breezy and friendly conversation at Manchester between the Professors of Physics and of Philosophy.

Physics. "When you come to think of it, Alexander, all that you have said and all that you have written during the last thirty years— what does it all amount to? Hot air! Hot air!"

Philosophy. "And now, Rutherford, I am quite sure that you will like me to tell you the truth about yourself. You are a savage—a noble savage, I admit—but still a savage! This reminds me of Marshal MacMahon, who was inspecting some cadets and had been asked to speak some encouraging words to one of them who was coloured. The Marshal passed down the ranks and, coming to the cadet, exclaimed—'Vous êtes un nègre?' 'Oui, mon général.' A long pause and then—'Eh bien! Continuez!'

And that's what I say to you, Rutherford, 'Continuez'."

At the Meeting of the British Association in August at Sydney Rutherford gave the popular lecture on "Atoms and Electrons". It is unusual to find criticism of his lectures and therefore it is well to quote *The Melbourne Age* (25 Aug. 1914):

Though a great scientist, Professor Rutherford is hardly an ideal lecturer, at any rate to a popular audience. He is fond of using specialised terms that convey nothing to the majority of his hearers, while he frequently drops his voice as though soliloquising in front of the screen. Nevertheless, he told a marvellous tale....

Gradually he overcame the drawback of lowering his voice at the end of his sentences, but the other point, justly raised, of using specialised terms, is one of the greatest importance and difficulty. If, every time he lectured, he had to explain the terms electron, atom, alpha particle, helium, gamma rays and so forth, he would never have got anywhere in his lectures. He had to assume a certain amount of knowledge, as well as of intelligence, on the part of his audience. Yet people who would be horrified at ignorance of Shakespeare, Beethoven, Leonardo, Julius Caesar, will not turn a hair in showing the most profound ignorance about electrons.

To return to the lecture at Sydney Rutherford said, referring to the idea of isotopes, "There may be two pieces of lead which look exactly the same and yet their physical qualities may be quite different. That may not be believed now, but it will be later."

Now is the time to consider that forecast. There are today at least seven elements all of which are properly called lead, atomic number 82, but they all have different atomic weights:

206·5, radium G, the stable end product of the radium family,

207, actinium D, the stable end product of the actinium family,

207·8, thorium D, the stable end product of the thorium family, and the following radioactive leads:

214, radium B, half-value period, 26·8 minutes,

210, radium D, half-value period, 25 years,

211, actinium B, half-value period, 36 minutes,

212, thorium B, half-value period, 10·6 hours.

The ordinary lead of commerce, when purified, has atomic weight 207·2. It is a mongrel of many isotopes.

The artificer and mathematician of the universe achieves great variety in apparent uniformity.

After the British Association Meeting, the Rutherfords went on to New Zealand, and visited his parents and other relations and friends in the North Island. They next went to Christchurch and received a civic welcome from the Mayor. At his old University, Canterbury College, Rutherford lectured on "Evolution of the Elements"—the very subject that he had chosen for his first public address to the Scientific Society before he went to Cambridge.

The choice of such a subject in his youth was something of a shock, for evolution was an unhallowed subject then. Now, what a change! He could speak with authority as the man who had detected and proved that such an evolution, or devolution, was an assured process in Nature, so that, stage by stage, the transformations of a stately series, from uranium through radium to lead, was no longer a question of speculation but of hard facts, which the most reluctant minds had been forced to assimilate and admit.

The outbreak of war caused some anxiety on the return journey, but all the members of the British Association arrived safely at their homes. The Rutherfords crossed the Pacific, Canada and the Atlantic, and returned to an England very different from the country they had left in the previous summer.

CHAPTER IX

THE WAR YEARS

The War had quickly broken up the 'family' of research students working at Manchester, leaving Rutherford and a depleted staff to carry on the teaching work of a greatly reduced laboratory as best they could. At first the impact of war was dimly understood in England and it took time to adjust everyone and everything to new conditions.

Strangely enough, it was possible to send and receive letters to and from an enemy country—witness the following letter from Stefan Meyer, sent through the American Consulate from Vienna:

MEYER TO RUTHERFORD

Wien: 3 *Sept.* 1915. It is now some time since I heard from you but I gather from a letter of Prof. Bohr to Hevesy that you received my last letter. I hope Lady Rutherford and you are in good health and I should be delighted to hear from you to this effect. I am glad to say we are all going nicely along here and work in the Radium Institute is progressing steadily forwards. Those known to you personally who are at present working in the Institute, Godlewski, Loria, Hönigschmid, Hevesy, Hess and Paneth send their heartiest greetings. It is to be hoped that you have received our publications up to No. 77, which were sent to you via Switzerland, as well as the number of the *Anzeiger* (through Holmes) in which among other things were included summaries of the papers of Loria (who without effect tried to communicate with you) and of your countryman R. W. Lawson. The latter has been working undisturbed, uninterruptedly and diligently since the outbreak of the war when he was cut off from the homeland, and I feel certain that you will be pleased to hear that our Academy of Science accepted his recent papers unhesitatingly....It is hardly necessary for me to mention that our feelings towards you and all friends are quite unchanged, and I take this opportunity of expressing the sincerest and heartiest greetings of my wife and of all of us to Lady Rutherford and to yourself.

16-2

14 Jan. 1915. We returned a week ago to Manchester after a quiet and unadventurous journey. We came back from New Zealand via Vancouver, Montreal and New York. I saw many of my old friends in Montreal, and Boltwood came to see us in New York. He desired me to send his kind remembrances to you and Mrs Schuster. We are all much better for the holiday, and I am feeling very fit for work.

I do not know if you have heard of the changes the war has made in my Department. Pring has got a Commission as 1st Lieutenant in the Royal Fusiliers, Florance, Andrade, and Walmsley in the Artillery, while Robinson expects a Commission at any time. We shall, however, with a little re-arrangement be able to carry on the work temporarily all right.

Possibly also you have heard that Marsden has been appointed Professor of Physics in Victoria College, Wellington, in succession to Laby, who gained the post in Melbourne. He is leaving here in a week's time to take up his duties. You may remember that when Laby went to Wellington, he was at my instigation given a sum to buy radium from the R.S. Fund. I gather that he will take this radium with him over to Australia, so that Marsden will have practically no radioactive material in Wellington. I was wondering, whether, under these conditions, the R.S. would be willing to make a grant of, say £100 to Marsden to acquire radium for use in his laboratory? I think you know that I have a very high opinion of the ability of Marsden, and the large amount of important work he has already done in radioactivity is the best indication that he will continue to do good research in that subject in the future.

Of course I have not been able to buy meso-thorium with the fund that was so kindly granted to me by the Royal Society, as the substance was only produced in Berlin by Messrs Knöfler & Co.

I do not know whether there is any special fund available for the case of Marsden; but I personally think it would be a very desirable thing to help out a Colonial University like Wellington, which is in a chronic state of poverty. Please let me know what you think of the matter.

I hope that you are all well. I have heard that you had rather an unpleasant time getting back from the Crimea.

P.S. I believe that Miers is certain to take the post of Vice Chancellor here (private). I think we shall all be quite contented with the appointment. Eileen leaves for Bedales to-morrow, and is quite excited at the prospect.

As soon as the British Association Meeting in Australia had ended Moseley returned to England and at once joined the Army. He was made a Signalling Officer, R.E., and trained at Brookwood, whence he wrote to Rutherford:

14 April 1915. Many thanks for your letter, which I have not had a chance to answer earlier. I am kept very busy and as we are in tents and out on horse back or foot all day long I am in splendid health. I have quite an interesting little job, as I am responsible for the communications of a Brigade, the 38th, and so I and my 26 men will be quite on our own as soon as we get to the Front. I still occasionally see the Phil. Mag. but for the rest I have dropped out entirely and never so much as hear from anyone in the game. We expect to be in France before the end of the month, and so all my affairs have to be put straight, besides a large number of details of special equipment etc., for my section. One thing lies heavy on my conscience, and that is my Sydney B. Ass. paper, for I have never published it. I must make time to get ready an abstract for the Phil. Mag., before I leave, as to chemists the reality and order of the rare earth elements is of much importance.

It is interesting to find that a letter from Geiger, serving in the German army, could find its way through some neutral country to Rutherford in Manchester:

Strassburg: 26 *March* 1915. Many thanks for your letter of Febr. 13th which I received a few days ago. Through Professor Boltwood I have had news about you already but I was very glad to hear from you directly.

I was on the Front up to middle of October but got ill and had to spend nearly ten weeks in bed. It was a rather bad rheumatism with fever and my legs were very much swollen and stiff. But now I have it over and am doing service again, but am not at the Front yet. I expect I shall be in the field again in about a fortnight.

I think nearly everyone of my colleagues whom you know is in the field. Dr Rümelin (also Reinganum and Glatzel) fell in one of the first months of the war. W. H. Schmidt is also dead; but from what I have heard he must have been in a very bad and hopeless condition all last year and so it was perhaps the best for him. I sometimes hear from Professor Hahn who seems to stand the dangers of war very well and is apparently quite happy in his military position. Dr Schrader is in the field artillery and is quite well up.

In February I had a fortnight leave of absence which I spent in Berlin. To my surprise I found that the trade in radium is flourishing more than ever and that there was plenty to do with testing. During 1914 we tested 8·2 gr. radium metal and made about £1500 with fees. I expect that the year 1915 will still be better. I noticed several papers of you in the magazines but must confess that I do not find at present the necessary concentration of mind to read anything scientific.

I am very pleased that Marsden got the professorship in Wellington and am sure that he deserves it. In case he should not have left England yet, please give him my best congratulations. I hear from Chadwick occasionally and of course we do for him what can be done under the present circumstances. But that is very little.

RUTHERFORD TO SCHUSTER

3 *May* 1915. There is some consideration here about inviting foreigners to the British Association in September. I daresay a few might come over from France or Russia, but I am a little doubtful of neutral countries, for there are special reasons why they might not feel inclined to take obvious sides. However they can only refuse.

There is to be a discussion on Friday about this matter amongst some of us. Whom would you like to suggest to invite from the point of view of Physics in general? Possibly Hale might come over; also Galitzin—possibly also Langevin, Perrin, Mme Curie, etc.

If we write at all it will be worth while seeing that Physics is not left out in the cold. Let me hear from you in the next day or so.

At a Royal Institution Lecture in June Rutherford gave an account of C. T. R. Wilson's marvellous photographs of a beam of X-rays passing through a moist expansion chamber, so that mist settled on the ions along the tracks of the electrons expelled from air molecules by the invisible X-rays.

The physicists were able to see the tracks which they had hitherto imagined and foreseen from the result of long and patient experiments.

He also referred to Joly's haloes due to specks of uranium in mica, and to the interesting relation between the length of life of the atom and the length of the alpha-ray track as discovered by Geiger and Nuttall.

In July, a friend wrote to Rutherford suggesting that Moseley would be much better employed "if he were set to solve some scientific problem presented by the war", rather than as Brigade Signalling Officer, pointing out that some particular problem should be selected, for if Moseley were brought home "simply for his own security, he would naturally be very much incensed".

Rutherford wrote at once to Sir Richard Glazebrook asking whether suitable scientific work could be found which would justify Moseley's recall from the Front. Action was taken, but it was too late! Moseley, as Signalling Officer of the 38th Brigade, had left England for the Dardanelles on 30 June 1915. In August, when twenty-seven years old, he was killed by a bullet through the head when telephoning an order. Thus passed one of the brightest spirits of that generation, who had accomplished more in his brief lifetime than most scientific men can claim in a full life of threescore years and ten.

Rutherford wrote to *Nature* (9 Sept. 1915):

Scientific men of this country have viewed with mingled feelings of pride and apprehension the enlistment in the new armies of so many of our most promising young men of science—with pride for their ready and ungrudging response to their country's call, and with apprehension of irreparable losses to science. . . .

Moseley's fame securely rests on his fine series of investigations, and his remarkable record of four brief years' investigation led those who knew him best to prophesy for him a brilliant scientific career. There can be no doubt that his proof that the properties of an element are defined by its atomic number is a discovery of great and far-reaching importance, both on the theoretical and experimental side, and is likely to stand out as one of the great landmarks in the growth of our knowledge of the constitution of atoms.

It is a national tragedy that our military organisation at the start was so inelastic as to be unable, with a few exceptions, to utilise the offers of services of our scientific men except as combatants in the firing line. Our regret for the untimely death of Moseley is all the more poignant because we recognise that his services would have been far more useful to his country in one of the numerous fields of scientific inquiry rendered necessary by the war than by the exposure to the chances of a Turkish bullet.

H. G. Moseley would have been soon elected a Fellow of the Royal Society. An obituary notice of him written by Rutherford appeared in the *Proceedings* (A, 1917, 93, p. 22).

Moseley's mother had been a widow and had married again Mr W. J. Sollas. After her son's death she wrote to Rutherford that her son had made a will on active service, leaving all to the Royal Society for the encouragement of scientific research work. This legacy was further increased by his mother's gift of an additional £10,000, so that the Moseley Fund endows two Research Studentships, each of the value of £350 a year.

Lord Fisher on his retirement from the post of First Sea Lord was appointed President of the Admiralty Board of Invention and Research. There were several departments—one dealing with submarines, mines and searchlights, on which served the Duke of Buccleuch, W. H. Bragg, Duddell, Glazebrook, Merz and Rutherford.

It soon became apparent that submarines were a deadly menace and required special methods to counteract their offensive.

Rutherford speedily turned his large basement laboratory at Manchester into a suitable research room with a large tank, where all known devices for listening to submarines were tested and new schemes quickly evolved.

These were sent to be tested at Hawkcraig where the Admiralty had a station under Captain Ryan. Co-operation proved difficult— even W. H. Bragg, when appointed there, found conditions impossible—

so a separate anti-submarine station was started at Parkeston Quay near Harwich, where Sir Reginald Tyrwhitt had his adventurous fleet of cruisers, destroyers and submarines.

It was quite important to know what kind of noise a submarine made under water and what, if any, was its fundamental note or pitch. Now Sir Richard Paget has a marvellous ear for music and can name the pitch of a note with precision. So Rutherford and Paget went out in a boat on the Firth of Forth and listened to a submarine told off for the purpose. Paget's head was under water, while Rutherford held him up by the heels. Paget emerged and named the pitch. Rutherford used to tell this story with gusto, and add: "I'm not so sure now, whether I shouldn't have let go!"

RUTHERFORD TO HIS MOTHER

18 *Aug.* 1915. I went down to London for two days to attend the meetings of the Inventions Board under Lord Fisher. We have settled down to work and I am on the committee that deals with submarine problems. We heard last night of the submarining of a troopship in the Aegean with the loss of a 1000 lives.

18 *Dec.* 1915. I am very busy with experimental work for the Admiralty. I think I told you of my recent visit to Edinburgh where I spent three days hard at work on a converted 'trawler' trying experiments and I expect to go up again shortly.

In spite of war activities there was time available for a variety of scientific interests as indicated in the following letters:

RUTHERFORD TO BOLTWOOD

14 *Sept.* 1915. The British Association at Manchester passed off very well and all the sections were as good if not better than usual. The Physics section met in my laboratory and we had very good discussions on isotopes, classification of the stars and thermionics. Weiss of magneton fame was here and gave an address and Bragg gave a report on crystal work. You will remember that Schuster was President. A small section of the Press attacked him with having a German name, but fortunately there was no disturbance in Manchester....He got news the same day that his boy was wounded in the Dardanelles....

In this second year of the War Rutherford was planning hopefully for the future:

RUTHERFORD TO SCHUSTER

4 *Oct.* 1915. I have received your letter of the 24th, and thank you for your suggestions. I enclose herewith a statement of a line of research I wish to take up, which would require more funds than the University can provide.

I have been exceedingly busy recently with my work on the Board of Inventions, and have been travelling up and down the country a good deal. I am hoping to get quite a big scheme of scientific work started under the Admiralty, and have been very busy myself making experiments. My chief business deals with the acoustic detection of submarines, on which a good deal of work has already been done, and I am very hopeful that we can push the matter further. If we prove at all successful, I think it will mean the practical elimination of the submarine as a serious factor in modern warfare.

I am going down to London to-night, and I shall be very busy this week starting up University work again. A large number of New Zealanders have recently come to Manchester from the Dardanelles, and my wife is very much occupied in looking after their comfort.

Everybody seems very pleased with the success of the B.A. Meeting here. I think, considering the conditions, it went off remarkably well.

If you have any suggestions to make about my application, please drop me a line.

One of the more important undertakings was Sound-Ranging in France, whereby enemy guns were located with remarkable precision and then quickly received the attention of counter-batteries. The sound wave in air due to the explosion of a heavy gun would cool a very fine wire and alter its resistance; a connection by wires to a distant Eindhoven string galvanometer recorded a change of electric current and this was photographed on a moving film. Similar records came to the galvanometer from other fine wire microphones, distributed in a fan-shape, and the time differences enabled the position of a gun to be fixed within a few feet. The technical head of the sound-ranging was Major W. L. Bragg, who in the previous year had shared a Nobel Prize with his father, W. H. Bragg, for their remarkable work in connection with X-rays and their reflections from crystals. W. L. Bragg was ably supported by J. A. Gray, H. L. Cooke, C. G. Darwin, H. Robinson, A. S. Russell, E. N. da C. Andrade, J. West and others.

Rutherford tried to get scientific men into such positions as would be most beneficial to the successful progress of the War. Unfortunately J. Chadwick was working with Geiger in Berlin when the War broke out and he had the misfortune to be interned at Ruhleben for the whole duration of the War. With two others he fixed up a little laboratory, with a very sensitive string galvanometer, and they conducted experiments on radioactivity as German prisoners. Chadwick wrote (14 Sept. 1915) from Ruhleben: "I was in the middle of the experiments on beta rays when the war broke out", and he reported that Geiger and Hahn were still all right.

A. B. Wood, who saw much of Rutherford in connection with the anti-submarine work, has written:

It might have been supposed that a man like Rutherford could not easily 'switch over' from atoms to submarines. No doubt such a change required a great effort on his part, but he was equal to it, and tackled the problem with his customary energy. He studied under-water acoustics, about which little was known, in a special tank he had constructed on the ground floor of his laboratory at Manchester. His early work in this laboratory with Broca tubes, diaphragms, microphones and various under-water senders and receivers demanded tests under service conditions. From November, 1915, onwards, a constant stream of ideas and instruments was sent to the sea research Station at Hawkcraig, Aberdour, to be tested on trawlers and submarines. At the same time he continued his university work and made frequent visits to London for meetings of the Board of Invention and Research. He particularly enjoyed his visits to Aberdour, where the sea air and work on ships acted like a tonic. When he took a few days' rest, after six months' work, he was followed by a deluge of letters from the Board. Professor W. H. Bragg took charge of the Station at Hawkcraig and Rutherford was a frequent and very welcome visitor. He was always brimming over with suggestions, and sometimes impatient with delays in obtaining ships for his experiments.

One of the first hydrophones evolved was a bronze plate fixed in a heavy bronze ring, and in the middle of the plate was a microphone connected by wires to head-phones, through amplifiers which had recently been evolved and improved. The noise of a submarine cruising under water could be heard from a considerable distance, because water is so good a conductor of sound. In wind and stormy weather, on a ship moving through the water, listening became difficult because of the background of other noises. The original hydrophone could receive but little sound when 'edge on', but the maximum reception was when *either* face of the hydrophone pointed to the submarine. The Admiralty declared this to be useless, because the hunting trawler could not tell if it was moving towards or away from the hidden submarine. Therefore a screen was made to cut off the sound from one face only of the hydrophone, which thus became a uni-directional instrument well suited to submarine hunting. The first and best type was a hand-made round flat piece of wood, covered with lead, called a 'baffle plate'. On production in bulk these were found not to work, and Hopwood devised a shot-filled cavity in a xylonite baffle-plate which was ultimately accepted and used in large quantities. The delay between discovery and successful manufacture in bulk was a great surprise and annoyance to Rutherford, particularly as it occurred at the time when the submarine menace was at its height, threatening the food supplies of Great Britain and the communications with the armies in France and elsewhere. As Wellington said of Waterloo: "It has been a damned nice thing—the nearest run thing you ever saw in your life."

Finally it was possible to build and buy ships faster than the enemy could sink them and, coupled with the convoy system and more rapid destruction of submarines, a solution of the problem was found, although it was a poor solution, involving great loss of life and of wealth.

W. H. BRAGG TO RUTHERFORD

17 *July* 1916. The diaphragms have come and I am very grateful to you for taking so much trouble about them. We are trying them now.

There has been a new development in connection with these things.

As you know, when a screen is placed on one side of the instrument and at the proper position, there is a very strong bias in its indications. It gives a very loud sound opposite to the unscreened side and a very little on the other. Also very little sideways. The general opinion now is that this bias instrument is the most useful of all. It reads by the maximum, although the maximum is not so sharp as the minima are on the simple instrument. Paget, Ryan and I had a trial to test this point on Saturday morning, and were exceedingly pleased with the result. We feel sure that we have now got a really satisfactory solution of the question. Ryan says that this instrument is the only good direction finder there is, and that he is sure that he could catch any submarine with it that ventured into the Firth.

There is a tremendous rush on them at the Admiralty, and we understand that an order for 1000 listening instruments is very probable. A fair proportion of these will be directional instruments.

We are being pressed to send up final specifications and have sent in something, asking only for freedom to design the screen so that it shall be more robust than it is in its present form.

Our best instrument has a diaphragm cut out of rolled sheet brass with no rim to it. There is a separate clamping rim. I do not say that this shows that sheet is better than the turned diaphragms but it shows that they can be made of sheet if required.

Paget has been here a week and we hope he can stay another. Duddell is here too for two days.

Towards the end of the War, Capt. Ryan, R.N., did indeed put out to sea with the avowed purpose of sinking a German submarine, using a uni-directional hydrophone as detector and a towed underwater mine for demolition. He used the explosive paravane invented by Capt. C. D. Burney, R.N. He was successful! Some said it was a 'fluke'; but that cannot be the case when the intention and method were both specified in advance.

On 7 Feb. 1916, Rutherford was speaking at the New Islington Public Hall on 'Radiations from Radium'. He said that scientists wanted to ascertain how they could release at will the intrinsic energy contained in radium and utilise it for our own purposes....It had to be borne in mind that in releasing such energy at such a rate as we might desire, it would be possible from one pound of the material to obtain as much energy practically as from one hundred million pounds of coal. Fortunately at the present time we had not found out a method of so dealing with these forces, and personally he was very hopeful that we should not discover it until Man was living at peace with his neighbours.

In 1916 the British Association, which was overshadowed by greater events during the War, met at Newcastle-on-Tyne. Eddington discussed Einstein's relativity theory of gravitation, while Rutherford spoke on the X-ray spectra of the elements and referred to the work of Siegbahn, who was carrying on the work of Moseley to the elements of high atomic weight from gold to uranium, using the L spectra.

It must be remembered that the Hydrophone Service had been established at Hawkcraig before the Board of Invention and Research began to do experimental research there. The transfer of the B.I.R. Station to Harwich was made after very thorough discussion between A. J. Balfour, Sir John Jellicoe and W. H. Bragg, who wished to get into close touch with the men who actually had to listen for submarines on the North Sea, then the chief seat of danger. The Senior Naval Officer at Harwich and the Captain in charge of submarines welcomed the B.I.R. Station to Harwich, and the change had the approval of Lord Fisher.

Rutherford wrote in November 1916 that he had become very interested in the question of diaphragms for hydrophones and that there was much important work to be done in clearing up various points. In particular, however much alike a man strives to make microphones they are always different, each one responding well to some band of sound frequencies. To get the best results for a hydrophone it is necessary to select or alter a diaphragm so as to be in tune with the microphone; the reverse cannot be done.

At this time too Rutherford was corresponding with C. H. Merz about a depth recorder to be fixed in a mine; how to record the flow of water at different depths; the pressure on cables exposed to tidal or other current flow. All such small but important investigations, so necessary for ultimate success, are apt to be overlooked; improvisation is impossible, they demand patient and careful work which consume time and energy.

W. H. BRAGG TO RUTHERFORD

About 1916. Thanks awfully for your offer to help. I will call on you with great willingness when the need arises, you may be sure of that.

I went up to the Enfield reservoir the other day to try the Rayleigh effect and got it all right. I had two small Brocas, one for each ear. The sound seems to shift over from one ear to the other. Sometimes it would persist in going the wrong way, and then it would right itself again. I am having a more solid set made to try at sea.

I wrote to Hall at some length after visiting Harwich. I got a long and quite friendly letter back expressing, however, the greatest surprise at our choice of Harwich, as being a place where deep waters were only to be had at the risk of Fritz's mines; and a rotten place anyway. I tried to explain that we had done enough of deep water experiments and that anyhow we could run a subsection at Blyth or Plymouth. What we want is to get in close touch with the people actually using listening apparatus. I hope we shall be able to make at Harwich the comparative experiments we want. Hopwood and Gerrard have gone down to Harwich to try.

Hall made one very interesting remark. He said "The issue is clear, we want a listening device for use up to 8 knots for a trawler or similar ship—the rest is the matter of dropping a depth charge."

I have written him to say we have practically got what he wants. We cannot make a listener which will tow at 8 knots; but we can stop a boat, listen, get the direction, and start off again. We can do this with a boat that is too shallow to torpedo, and can carry enough offence and defence to make a submarine go under. The speed might even average 8 knots; certainly it would be far higher than the average underwater speed of a submarine.

When the United States came into the War in 1917, it was decided to send a Franco-British Commission to acquaint the American scientific men with all that had been discovered and developed in the anti-submarine campaign in Europe, so as to save waste of time and useless repetition of effort. This Commission consisted of Sir Ernest Rutherford, Com. Bridge, R.N., Com. Fabry (Univ. of Marseilles), M. Henri Abraham (Univ. of Paris), Capt. de Grammant, Capt. Dupoing, and Lieut. Peterno. Rutherford went with Bridge to Paris to make preliminary arrangements and this is the account that he wrote to Lady Rutherford:

Paris: 18 *May* 1917. I sent a p.c. from Havre to Eileen Thursday morning and a telegram to you on my arrival in Paris in the afternoon. We had a great passage, docking about 6—after getting up at 5 for a light breakfast. With the aid of my Admiralty papers I got through the passports and baggage exam. quickly and found a motorcar waiting to take me from the boat to the station, driven by a soldier. We left about 7.30 and found the country very beautiful with spring foliage. I had fairly pleasant company in the train and had lunch in the restaurant car about 11, plenty of a rather dark bread, sugar in galore and some mutton, an omelette and cheese—so I did not fare so badly. I put up at the hotel de France, St Choiseul, Rue St Honoré—a quiet pleasant hotel which has few visitors at the moment. Bridge called and I saw some of the French Board of Invention and Research, including Perrin with whom I went to tea. It was Ascension Day, a religious holiday and nearly all the shops were closed. Bridge introduced me to an American, Lewis, staying at my hotel and we had dinner together here and went out to Gaumont picture palace later, a good show, and we walked back retiring about 12. This morning Bridge and I had arranged to see special experiments at 9, but it was put off as the apparatus was out of gear. I am spending the morning preparing for a lecture this afternoon before the French B.I.R.....

About 12.30 a taxi driven by a soldier turned up in which were Perrin, Langevin, Mme Curie and Debierne, who took me to lunch and treated me in Royal fashion. We then went to the Ministry of Inventions and I gave a talk on submarine matters for half an hour followed by Commander Bridge, and there was some discussion. There I met Profs. Ames and Reed both of whom I knew in America, also Profs. Abraham and Fabry, who go with us to U.S.A. I then went with Langevin and Mme Curie to her Lab. (new) which I saw and also some experiments of Langevin. Mme Curie gave me some Lab. tea. She is looking rather grey and worn and tired. She is very much occupied with radiology work, both direct and for training others. Langevin was leaving that evening for Toulon, so I came back to my hotel and labelled my luggage. I went to dinner with Ames at the Hotel Crillon where Haig, Lloyd George and Co. stay on their visits to Paris. We had a very pleasant evening and I met a well known flying officer who told us a good deal about his views of the aviation question, and the difficulties of France in this connection and their hope for American help. Tomorrow I break-fast at 6.30 and go to the train at 7.30 to meet Bridge en route for Bordeaux. Altogether I have had a very pleasant stay here and enjoyed myself. Au revoir.

RUTHERFORD TO HIS WIFE

1 June 1917. Just a short note on a very busy day to say we are now in Washington hard at work. We spent the first evening in New York and were given a dinner party in the University Club by Dr Pupin and met a number of scientific men engaged in submarine question. I travelled down to Washington with Dr Pupin and am staying with the French Mission in the Powhatan hotel, not much of an hotel, but Washington is so crowded that it is difficult to get a room. We were met by Hale, Millikan and others, and I spent the evening with them. Yesterday we paid our official call on the Secretary for War—Baker—and the Secretary for the Navy—Daniels. We then dined in a bunch at the Army and Navy Club and called during the afternoon on the French Ambassador—Jusserand, where we were received very politely and given tea. In the evening Bridge, Fabry and I went to dinner with the Hales and had a very pleasant evening turning in at 12.30 and slept like a top under a sheet for the weather here is pretty warm. I was surprised to meet Loeb on my arrival and we had a chat here yesterday before he returned to New York. Dr Taylor is also here in the Food Commission and I saw him yesterday. Bridge and I go to dinner with him this evening to meet Hoover, the Food Controller here, formerly American Representative in feeding Belgium. Today we had a great meeting, discussing phases of the submarine question with the American committees—five hours in all today, and more tomorrow. I took a prominent part in the question so have had a full and rather tiring day as the weather has been distinctly warm. Tomorrow we dine with

M. Tardieu—the French Commissioner—and on Monday there is an official dinner given by Hale.

Our plans are not yet settled but probably we shall go next week to Nahant (?) near Boston to see some American experiments. I have so far not heard from you but hope to before long. Am fit and well and having a very interesting time. I think we shall do very good work here and I am very glad I decided to come over. Have seen a good deal of Millikan and Hale and others of my friends. Love to Eileen.

4 *June* 1917. Just another line to record further movements. We had a long all day committee on Saturday and went to dinner with Hale to meet Hoover the American Food Controller, a most interesting man. We hope next Sunday, i.e. Bridge and I, to go with Taylor on a motor excursion to the site of the battle of Gettysburg and are looking forward to the trip. We had breakfast in pyjamas on Sunday morning —grapefruit and tea—and then discussed matters of organization with Millikan and Co. At 12 we called on the British Ambassador, Spring Rice of Stockholm, you remember, and talked with him for an hour. We go to luncheon there on Tuesday after a meeting with Fabry and Co. I saw Law, an English correspondent here, whom I travelled with in the *Cedric* many years ago, and then went in to Hale to discuss general matters with him. Today we pay special calls and go in the afternoon to the Bureau of Standards. Hale gives a formal dinner to the Commission in the evening, at which the French and British Ambassadors are to be present. The weather is getting pretty warm but I sleep well in pyjamas without a sheet and generally my constitution stands well the strain of so many lunches and dinners. Have not received any letters from England but hope you are all well.

5 *June* 1917. I wrote last Sunday so I feel very virtuous. The hour is 11 p.m. and I am writing in my room in pyjamas to keep cool for the weather is very steamy. After I wrote on Sunday we went to call on the British Ambassador—Spring Rice—at 12 and spent half an hour with him. We had lunch with the French party at 1 and went in the afternoon to see Hale at his house. He had been rather ill for a day or so and he received us in pyjamas lying on a sofa. At 7.30 we dined at our hotel with an English officer Commander Mock who is here on special business for the Admiralty, and afterwards went to a local music hall for an hour or so. A sailor from the Fleet sang some excellent songs; it was a good

deal like our doings during the first month of war. On Monday morning we paid further official calls and lunched at the Cosmos Club with Prof. Stratton. I met there a number of old scientific friends. Monday evening was the main official dinner given by Hale. The French and English Ambassadors were present and all the Secretaries of the Government, so it was a highly social function about forty-one in number. I had to speak last and I believe was quite effective for the job I was asked to do. We got to bed at 1.30 and I felt pretty weary this morning but had a long meeting with admirals, scientific and naval and military men and had to speak for three parts of an hour. I felt very sleepy and tired but we had three hours of it in all. I was bored stiff by the last hour which was very uninteresting. I tried to get a little rest this afternoon, but with no success, for the telephone bell summoned me to another interview and further work. This evening I had dinner quietly with Bridge and retired to my room afterwards to rest. We have many more meetings this week and dinner at the French Ambassador's on Saturday, also a quiet and informal lunch at our Embassy today—Lady Spring Rice was there and their two children....

Next week we go to the Experimental Station north of Boston for a few days and our plans after that are indefinite so far. I think our Mission is regarded as the best informed one that has come to U.S.A. We have certainly had a fine series of meetings and told all we know. Our hotel is filled with Southern Veterans of the Civil War—in uniform of blue grey—and all between 70 and 80—a rather pathetic sight, but most of them look pretty wiry and active. There is to be a big parade on Thursday and tonight there is a dance for the younger people at our hotel. Well, I feel a bit weary so au revoir with much love to you both.

Ritz Carlton Hotel, New York: 12 *June* 1917. This is a short note at 9 in the morning before leaving for Boston. Since writing to you last we were kept very busy in Washington with general interviews and meetings. On Friday night we gave a dinner to our Washington friends and included two of the French party, also Dr Taylor and Owens, formerly of McGill, also Professor Webster. We had a pleasant evening, Owens is here in uniform—a captain in the Signal Corps—and is looking very fit and well. On Saturday we dined with the French Ambassador. The English Ambassador and others were there and it was quite a pleasant function. On Sunday we went out with Dr Taylor, Captain

Lennard and Admiral Taylor on a motor trip through the Blue Mountains round by Harper's Ferry and other points of interest in the Civil
War. We went 220 miles in all and got back at 9 o'clock after a very
interesting day, but very tired. After dinner we packed and left by
midnight sleeper to New York. Yesterday I visited the Western
Electric Coy., and saw there W. Wilson, one of my Manchester research
students. Today we go on to Boston to see some experiments and return
to Washington Wednesday night where I shall again stay a few days.
Bridge returns to England on Saturday, but I shall stay a fortnight or so
more as there is still a great deal to do.

Fairly fit and well considering our late hours and continuous talking.
With much love to you and Eileen.

23 *June* 1917. I have just written to Eileen but will not repeat the news
in her letter. I have had altogether a very busy time since I arrived. I left
Washington last Monday night and proceeded to New York; on
Tuesday morning we visited the works of Sperry and dined with them.
That same afternoon I went to New Haven to stay with the Bumsteads
and to get an Honorary Degree next day. They were very good and
kind to me and I met a number of my old friends including Boltwood,
Zeleny, Taylor, Kovarik and Professor Nichols. Boltwood has changed
over his views completely and is now busy working on anti-submarine
devices. On Tuesday we motored over to New London to see the
submarine base and had dinner with the Zelenys. On Friday morning
I returned to New York and visited Edison and his works at Orange in
the afternoon where I had a most interesting time and was received very
well by the old man, who was as enthusiastic as a schoolboy over his
ideas. Last night I dined with Major Carty of the American Telephone
Coy., a very able man who has much influence and interest in scientific
work in this country. This morning I went down town to arrange
sailings. I shall be leaving about a week from now by a direct line and
you may expect to see me in Manchester not later than the 10th, if I get
off by the boat I find is likely to go.

Tonight I go by steamer to Albany and spend Sunday to Tuesday
with the people of the General Electric Coy., Schenectady. Then on to
Montreal on Tuesday and back on Thursday to New York possibly
sailing on Saturday or Sunday. It is very warm weather here and I find
it difficult to keep cool. I have had to buy a few more shirts and under-
clothes to fill out the intervals of the laundry.

RUTHERFORD TO HIS MOTHER

New York: 29 June 1917. Since I landed I have had a very busy time discussing and explaining the submarine situation to a number of naval and scientific men. We came at the psychological moment as they had already begun work but were very uncertain what had been done in England and France. We paid a visit to Boston to see some of the work in progress. I paid visits to a number of scientific works and men— including Edison. I then went to stay at Yale University. They awarded me an honorary D.Sc. and I attended for the degree function. I walked in the procession with the late President of the U.S., William Taft, who afterwards spoke at the luncheon. Other men who got the honorary degree were Paderewski, Herrick (formerly American Ambassador at Paris) and Tardieu, the French High Commissioner whom I met in Washington.

Next day we took a motor trip of over 100 miles to New London to see a submarine station. I then took a river trip up to Albany and spent a couple of days looking over the electrical works at Schenectady, the greatest in the country.

From there I went on to Montreal partly on business and pleasure and stayed at the University Club. I found that Sir William Macdonald, the benefactor of McGill, had died about a fortnight before. I returned to New York this morning and visited the Western Electrical Company. America is very anxious to get going in the war and things look well.

A letter came to Rutherford from his mother written at New Plymouth, New Zealand, on 29 July 1917:

Your Father and I were very glad to receive a letter from you yesterday written from the Ritz Carlton Hotel, New York, on June 29th. The account of your journey over, meetings with friends, introductions to other eminent scientists and discoverers including Edison, your work in Washington and the evident appreciation of your efforts to help do your bit, as shown by their awarding you an honorary D.Sc., degree, made pleasant reading for proud parents, for though your Father does not say much, he is glad and proud of the distinctions won by you.... You cannot fail to know how glad and thankful I feel that God has blessed and crowned your genius and efforts with success. That you may rise to greater heights of fame and live near to God like Lord Kelvin is my earnest wish and prayer....

When Rutherford sailed from New York to Liverpool in 1917 there were many young United States marines on board coming to join in the anti-submarine campaign. They had a goat as mascot. One night in mid-ocean Rutherford found the animal asleep under his berth and, annoyed, rang for the Steward, who said: "It's quite all right, Sir, he's slept there every night so far!"

On 5 April 1918, Rutherford gave the first annual lecture, founded by the Röntgen Society, in memory of its first President, Prof. Silvanus P. Thompson, a man who combined the best qualities of physicist and engineer, who wrote an early text-book on Electricity and Magnetism, perhaps better than any that have appeared before or since.

I have been very interested to read again Professor S. P. Thompson's first presidential address before the Röntgen Society, which gives a very realistic account of the discovery of X rays and the general idea of their nature and origin at that time. Natural prominence is given to the medical application of the X rays for the advancement of which this Society was founded. But it was difficult then to foresee the great field of usefulness to Medicine that has since been opened up by this discovery.... Undoubtedly the discovery of this new type of radiation marks the beginning of a new epoch of science which has resulted in a revolution of our ideas almost as marked as that produced by the theory of evolution in Biology. It is a period of pioneer advance over a new and fertile territory with the almost daily discovery of new and interesting facts and the gradual unfolding and development of new and bold scientific ideas. The two decades, 1895 to 1915, will always be recognised as a period of remarkable scientific activity which has no counterpart in the history of Physical Science. The writer had the good fortune to begin his researches in the Cavendish Laboratory in October, 1895, and has lived through the exciting scientific period when progress has been so rapid that it has been difficult for the scientific worker to keep abreast even of the advances in his own particular period, much less with the whole field of advance. While it has been to some extent a time of speculation, yet it has not been as a whole a time of rash speculation, for the main ideas which governed the advance have been shown to have a solid foundation in fact and the resulting theories have been exceedingly simple, though very fundamental.

Three new lines of advance were promptly opened up as an immediate consequence, namely, (1) the discovery of radioactive bodies, (2) the discovery of the electron, (3) the ionisation of gases by X-rays.

He then referred to the work of Henri Becquerel who tried the effect on a photographic plate of a special salt of uranium and potassium— a fluorescent substance.

We now know that the photographic action obtained was mainly due to beta rays and not to X-rays at all, but this observation on the radiating power of uranium has opened up the new science of radioactivity which has had such an extraordinarily rapid development. Soon after followed the brilliant discovery of polonium and radium by the Curies, in which the radioactive property was so powerful in its effects that it was difficult to explain it away.

The rest of the lecture was largely devoted to a discussion of radio-chemistry and isotopes, and to the remarkable fact that radium B, D, G, thorium B and D, actinium B and D, are all chemically identical with common lead.

CHADWICK TO RUTHERFORD

24 *May* 1918. We are now working, or rather about to work, on the formation of carbonyl chloride in light....The greatest difficulty is to obtain a reasonable source of light....The whole thing (photochemical reaction) is really extraordinarily fascinating....I have been working too much recently, so I shall take a little exercise this summer. I shall probably play tennis. A game doesn't last very long and can be made fairly vigorous. Four long and dull years have given us a decided preference for the short and merry type of life. Within the last few months I have visited Rubens, Nernst and Warburg. They were extremely willing to help and offered to lend us anything they could. In fact all kinds of people lend us apparatus.

RUTHERFORD TO HIS MOTHER

3 *Nov.* 1918. I have been on a special journey to Paris on Admiralty business. Four Admiralty representatives, including a Naval officer, travelled together. We crossed over in the regular boats used by soldiers

on leave and they were crowded with Tommies and officers. Everyone had to wear a lifebelt. Three to four fairly swift boats leave together and are escorted by destroyers and airships (dirigibles), the journey taking about 1½ hours. We held meetings on Saturday to Tuesday on a submarine problem. It was an international congress and America and Italy were each represented by four delegates while the French had eight officials of various kinds. My old friend Langevin was one of the chief scientific men on the French side while I represented the English delegation. The French were in very good spirits. The Place de la Concorde (where the old guillotine used to work) was filled with a great number of captured German guns and there was always a big crowd looking at them. The statue of the City of Lille was decorated with wreaths of flowers as I was there only a few days after its liberation.

It really looks as if the war is nearing its end. Turkey went yesterday and probably Austria to-morrow—and Germany is likely to follow before Christmas. It is a very exciting time to live in but people here are very quiet and refrain from celebrations till our main enemy goes under.

After the Armistice there followed a difficult period of reorganisation, which began to bear fruit in the following year.

BOHR TO RUTHERFORD

24 *Nov.* 1918. I need not tell you how often our thoughts in these days have gone to all our friends in England and how great a part we take in all the joy and happiness which everybody there must feel in the great result with which the enormous efforts in the critical years have been achieved. Here in Denmark we are of course most thankful for the possibility which the defeat of German militarism has opened for us to acquire the old Danish part of Slesvig, and at the same time we feel an immense relief now the war is finished. All here are convinced that there can never more be a war in Europe of such dimensions; all the people have learnt so much from this dreadful lesson, and even here in these small Scandinavian countries, where, for good reasons, there certainly was not much aggressive military spirit before the war, people have got to look quite differently than before at the military side of life. From all that we hear, we feel also quite sure that the men now in power in Germany take a real peaceful attitude, not for the occasion and not

because they have always done so, but because all liberal-minded people in the world seem to have understood the unsoundness of the principles on which international politics has hitherto been carried on. If therefore only there will not become anarchy in Germany due to the great need and poverty at the present moment, this time may certainly be looked upon as the beginning of a new era in history.

How I am longing to be in England now and to be able to talk with you about all sorts of things. I remember, as if it were yesterday, all the times I sat in your study and you developed for me your views on the different phases up and down through which the war went, and how your unflinching belief in a happy end was always able to comfort me, however downhearted I could feel myself at times. Dear Prof. Rutherford this letter is only meant as a greeting in these eventful days, but very soon I shall write and tell about my work. I feel how happy you must be now again to be able to work in the laboratory as in old days, after in these long years to have given so much of your time to the great result now achieved, and I am longing so much to hear how the important investigations you told me about are progressing. I am sorry not to have written for so long, but I have had very much to do in the last months and it has therefore taken more time than I had expected with the completing of the second part of my paper, which I had looked forward to be able to send you some time ago. Now it will be ready for the printing office in a few days, but has, I am afraid to your disapproval, become rather long (twice as long as Part I); my only excuse is, that I have really tried to put a great deal of work in it. As regard the external conditions for my work here I must still tell how very glad I am that the establishment of the small laboratory for experimental investigations in connection with modern work in theoretical physics is now secured by a permission from the Government to start the work on the building as soon as the detailed plans are ready from the hand of the architect. This splendid result is in the first place brought about through the extraordinary kindness and generosity of a local friend of mine, who by subscriptions from himself and his friends has collected a big sum of money ($£4500$) which has been put at the disposal of the University as a help in the expenditure with the building and its equipment. The laboratory will be situated at the outskirts of a beautiful park not far from the middle of the town, and we will ourselves come to live on the same ground. I look, of course, immensely forward to it, and we hope

already that it might suit you and Lady Rutherford to come to Copenhagen and stay with us at the time for the festivals for the opening of the
laboratory, which I hope will take place in about a year's time from now.

We are looking forward very much to come to England as soon as
travelling is possible again and to see you and all my friends in the lab.
and elsewhere.

CHAPTER X

CAVENDISH PROFESSOR

During his last year at Manchester, in spite of many calls on his time in connection with the University and War work, Rutherford had persevered at some experiments of a fundamental nature. He had used the active deposit of radium and employed the energetic alpha particles projected by radium C to bombard nitrogen gas. He found that something struck the fluorescent screen, which he was observing with a microscope, even when the screen was beyond the range of the alpha particles. He had carefully examined the nature of the strange particles and by long experiment proved that they were protons. Chadwick has written as follows:[1]

In 1919 he was able to publish the results of another epoch-making research, again made with alpha particles. In four papers published in the Philosophical Magazine, vol. 37, he proved conclusively that the long range particles earlier shown by Marsden to be produced when alpha particles were fired into hydrogen were in fact fast hydrogen nuclei, and he showed that identical particles were produced by the collisions of alpha particles with nitrogen. The explanation—that these protons were the result of the disruption, or 'artificial disintegration', of the normal stable nitrogen nucleus—was so revolutionary, and so pregnant with far-reaching implications, that it clearly needed to be supported by very complete experimental evidence. Rutherford obtained the necessary support by an admirably designed series of control experiments. He did the whole of the experimental work, with Kay's assistance in taking observations of scintillations. The further development of this work, and of its extension to other nuclei, belongs to the Cambridge period.

Although the War was over, the whole world was passing through a most difficult period of reconstruction. Education, as well as industry,

[1] Obituary notice of Lord Rutherford published by the Royal Society in January 1938, p. 406.

suffered from the loss of many able men and those who returned to work were conscious of a certain decrease in their powers.

It should be added that Rutherford and W. H. Bragg had taken out a patent for "Improvements in apparatus for detecting the direction of sound in Water". This was done with the avowed intention of protecting the Admiralty, and after the War the patent was relinquished. Neither of the two men made money, directly or indirectly, for their valuable War services.

The famous physical-chemist, Irving Langmuir, wrote to Rutherford (18 March 1919) advocating the efficiency of the Lewis-Langmuir theory to account for the chemical properties of atoms in a way which the Bohr dynamical theory did not seem able to do, suited as the latter was for the explanation of spectral series. While admitting certain merits, this statical atom made no great appeal to Rutherford, and on a later occasion when Langmuir lectured at the Cavendish, Rutherford, in a light vein, twitted the lecturer on a scheme which seemed to resemble that of systematic botany. Nevertheless, there still remains the task of combining the good points of both theories.

When Rutherford had first arrived at Manchester the Steward of the Physics Laboratory was Griffiths, and under him was Kay, who succeeded to the post which he still holds, always vigorous and efficient.

Kay took his part in most of the important investigations by Rutherford and his research staff and students. In 1938, I had a talk with Kay on those wonderful days and though he was enthusiastic about the research work, he laid more emphasis than others on Rutherford's remarkable powers as a lecturer to first-year students. "The Professor", he said, "loved a good experiment, no matter how often he had seen it, and this enthusiasm spread to the students. He was not good from the point of view of examinations, but he told the students they were there to learn to think, not to remember, and that he had no belief in spoon feeding. The students understood him and loved him."

When Sir J. J. Thomson, finding the double duty of directing a great laboratory and a great college too much for one man's strength, resigned the Cavendish Chair of Experimental Physics at Cambridge, Rutherford was the obvious choice for his successor.

On 7 March, he wrote to Sir J. J. Thomson: "If I decided to stand for the post, I feel that no advantages of the post could possibly compensate for any disturbance of our long continued friendship or for any possible friction, whether open or latent, that might possibly arise if we did not have a clear mutual understanding with regard to the Laboratory and Research Students...."

To this he received a cordial reply: "I am very glad to find that you are still entertaining the possibility of coming to Cambridge as Professor. If you do, you will find that I shall leave you an absolutely free hand in the management of the Laboratory...."

The negotiations for Rutherford's appointment to the Cavendish Professorship were carried out with care and zeal by Sir Joseph Larmor and it was finally decided that Rutherford should also be Director of the Cavendish Laboratory. Sir J. J. Thomson was appointed Professor of Physics without stipend, with rooms and mechanics essential to the continuance of his research work.

The Council of the University of Manchester expressed their regret at Rutherford's resignation and congratulated him on his appointment to a position "which he will worthily fill as the successor of Maxwell, Rayleigh and Thomson". Many of his old friends wrote that they would miss him in Senate, smoking room and elsewhere. His welcome to Cambridge was no less emphatic.

RUTHERFORD TO HIS MOTHER

7 *April* 1919. You will have received the news that I have been elected to the Cavendish Chair of Physics held by Sir J. J. Thomson, who is now Master of Trinity. It was a difficult question to decide whether to leave Manchester as they have been very good to me, but I felt it probably best for me to come here, for after all it is the chief physics chair in the country and has turned out most of the physics professors of the last 20 years. I was appointed on April 2 and technically take up office from that date, but as I must finish out the term's work in Manchester,

Sir J. J. will be in charge and he is very pleased to do this for me. It will of course be a wrench pulling up my roots again and starting afresh to make new friends, but fortunately I know a good few people there already and will not be a stranger in Trinity College. The latter will no doubt offer me a Fellowship which will give me the rights of the College to dine there when I please.

In April 1919, Sir J. J. Thomson wrote, as Master, to inform Rutherford that he had been unanimously elected a Fellow of Trinity and Rutherford was admitted to the Fellowship on 2 May.

On one occasion Lady Rutherford, and her daughter Eileen, were visiting Cambridge, and found a rather derelict house and large garden which charmed them both. When Rutherford was appointed Cavendish Professor, Lady Rutherford eagerly revisited the house and found it still available. It was Newnham Cottage, Queen's Road, the property of Caius College. A lease was signed, many improvements were made to the house and Lady Rutherford constantly improved the garden, which had a charming lawn under great trees and a mulberry tree at one end. There Rutherford lived from 1919 to his death in 1937.

After more than four years of war it became possible to correspond with old friends on the other side, and Geiger wrote an admirable letter from Charlottenburg:

18 *May* 1919. I take the first opportunity to write you a few lines just to say that I am well, up and at work again. I need hardly say that all that has happened these last four years has had no influence on my personal feeling to you and I hope, dear Prof. Rutherford, that you still take a little interest in your old pupil who keeps his years in Manchester always in pleased memory.

As you will have heard already, I have been in the field all the time of the war and I am grateful that I got safely through all the fights which many times have been severe enough. I came back middle of December, was at home at Christmas and began work again at the beginning of the year. Naturally, times at present are not very pleasant and the general disorder and the trouble about one's daily bread are rather depressing. It needs a lot of energy to do any work at all, but I hope my old interest

Newnham Cottage

in radium will soon be back again. To freshen up my memory I am at present re-reading your book once more and feel that this helps me more than anything to regain my old pleasure in scientific work. So far as I have read the progress on radioactivity during war-time, no very striking discoveries seem to have been made. A research made in Vienna on the number of alpha particles from radium does not look very satisfactory and I can hardly see an improvement on our old work. You will perhaps remember that I started on a redetermination of the number just before the war began and I think I will go on with it now.

It would please me very much indeed, if you should find time occasionally to write me a few lines and to let me know how things are in Manchester. I send this letter through Cook & Son, Post Box 706, Amsterdam, at a suggestion of Chadwick.

RUTHERFORD TO GEIGER

14 *June* 1919. I received your letter and am very glad to know that you have come through the struggle all right and are back at the old work again.

I have kept in touch as far as possible with my old researchers during the War and am glad to know most of them are safe and sound. The research men of the Laboratory here are very scattered, but Robinson and Florance have returned to their teaching duties. The latter spent most of his time in India and Mesopotamia and had several attacks of malaria. Our greatest loss was Moseley who was killed in 1915 in Gallipoli. We are intending to erect a memorial tablet to him in the Laboratory.

You will have heard that I go to take charge of the Cavendish Laboratory soon. J. J. T. is now Master of Trinity and is much occupied with general scientific work, but hopes to continue his researches in the Cavendish where he will have some rooms.

We are all feeling very rusty scientifically after the war, and it will be some years before we can get going properly, for apparatus is very dear and difficult to get. I think your experience is similar to everyone alive who has been in science. Robinson is gradually regaining his interest in pure science again, but it takes time to get into the old swing.

Marsden is now in New Zealand again. Russell is lecturer in Sheffield. I saw Bohr for a few days recently on his way to attend a conference in

CAVENDISH PROFESSOR [1919

Holland. His theories, as you know, are going strong. I hope you will take up the number of alpha particles again.

Well, I retain my old friendly feeling for my old researchers and hope that we may meet again when things have settled down to a more normal footing, but it will obviously be some time before this will be possible. I shall be glad to hear from you.

Early in June, Rutherford, who had received an honorary degree from Durham in the previous month, gave a Friday evening lecture at the Royal Institution on "Atomic Projectiles and their Collisions with Light Atoms". There he explained his discovery that alpha particles, bombarding nitrogen, drove out of the nitrogen atoms the quite different atoms of hydrogen, or rather the nuclei of hydrogen, known as protons. This was a startling result. True, the radioactive elements eject alpha particles, which are helium nuclei; beta particles, which are electrons; gamma rays which are electromagnetic radiations like X-rays, light and wireless waves. Nothing else comes from radio-elements. Now here was Rutherford declaring that he had knocked hydrogen out of nitrogen. It was a revolution!

One critic said that alpha particles could pick up an electron or two and so their range would be quadrupled, suggesting that Rutherford had discovered only alpha particles with longer ranges. But then Rutherford could not get a similar result with oxygen as for nitrogen. The critic justly added "the more epoch-making a discovery is, the more closely should all alternatives be examined".

It need hardly be added that further experiments proved Rutherford to be right. He had actually driven hydrogen (protons) out of the nuclei of nitrogen, and in doing so he had quite unwittingly transmuted or transformed nitrogen into oxygen. In any case this was the first successful attempt by man in the deliberate transmutation of matter.

Some newspapers missed the whole idea of Rutherford's discovery by the heading 'Nitrogen a Compound?' The point was that he had for the first time deliberately smashed up the *nucleus* of an atom.

Rutherford was fortunate in seeing a capable and distinguished

successor appointed to his Chair in Manchester. W. L. Bragg had shared
a Nobel Prize with his father for their discoveries relating to crystal
structure by the help of X-rays, and of the nature of such radiation
by the help of the crystals. His work on sound-ranging in France had
not only impressed his technical colleagues but had received praises
from the highest military authorities at the Front, who fully appreciated
its usefulness. W. L. Bragg turned rather longing eyes to research
work in Cambridge and with some diffidence toward teaching, of
which during the War there was but little experience. He faced and
shouldered the burden, and the anticipated success awaited him. He
wrote to Rutherford:

12 *May* 1919. I would just like to tell you how proud I feel that I am
succeeding you at Manchester. It is a tremendous honour for anyone
who has still his spurs to win, and if I hesitated at all about coming it was
because I naturally doubted my power to carry on your work. I do not
think a more magnificent opportunity for work was ever offered to
anyone in my position.

In due course he was destined to succeed Rutherford at Cambridge.

On 15 Aug. 1919, Mme Curie wrote to Rutherford discussing the
best name to be given to radium emanation, which Ramsay had suggested should be called 'niton'. Madame was rather in favour of
'radionéon'. Partly at the suggestion of Prof. Perrin, wiser counsels
prevailed and the emanations of the three radioactive families were
henceforward known as radon, thoron and actinon.

When Rutherford became Cavendish Professor he found that Sir
J. J. Thomson wished to retain his private laboratory and his invaluable
assistant Everett, together with some rooms on the ground floor for
a few research students, as well as some other advantages of the building
with reference to workshop and assistants. The two men met and
discussed the difficult situation; Rutherford then drafted a statement by
way of an agreement. It was full of alterations, erasures, additions, all un-
initialled; it was a document which would make a lawyer weep, but there
are the initials at the bottom, J. J. T., E. R. It was sufficient; it worked.

Prof. C. D. Ellis has clearly recalled Rutherford's early experiments at Cambridge:

Soon after his arrival at Cambridge Rutherford found time to follow up vigorously the experiments on the artificial disintegration of the light elements which he had started during the war period. In those days the experiments still involved only simple apparatus and were of the type Rutherford loved. A radio-active source provided the alpha particles, which fell on a foil of the material to be disintegrated, and the resulting protons, hitting a zinc sulphide screen, were detected by the scintillation method. The whole apparatus was contained in a small brass box and the scintillations were viewed with a microscope. I can well remember being surprised, in fact mildly shocked, that the apparatus was not more impressive, yet these experiments, so simple on the surface, required the highest experimental skill to make them yield dependable results.

Counting the scintillations was difficult and tiring, and Rutherford usually had one or two of his research students to help him. The experiments started about four in the afternoon and we went into his laboratory to spend a preliminary half an hour in the dark to get our eyes into the sensitive state necessary for counting. Sitting there drinking tea, in the dim light of a minute gas jet at the further end of the laboratory, we listened to Rutherford talking of all things under the sun. It was curiously intimate but yet impersonal and all of it coloured by that characteristic of his of considering statements independently of the person who put them forward.

In this simple apparatus contamination by radium emanation was avoided by blowing dry air through the tube T. The radioactive source R could be moved to or from the screen S, while the range of the ejected protons was measured by inserting thin mica plates before S. Nitrogen gave rise to protons of range 40 cm., aluminium 90 cm., while oxygen failed to produce any protons. The conclusion was inevitable. Alpha rays could disintegrate certain atoms and not others. Such disintegration meant the rupture of the atom and the inevitable result was transmutation.

The year 1920 was to Rutherford largely one of organisation and
reconstruction. He had taken with him to the Cavendish the 250 mg.
of radium lent to him before the War by the Vienna Institute of
Radium. Many of his great results at Manchester had been attained
with the help of this valuable material. He was fortunate in finding at
the Cavendish a keen helper in F. W. Aston who had developed and
improved the deflection of charged atoms by magnetic and electric
fields, first used effectively by Sir J. J. Thomson. If, for example, there

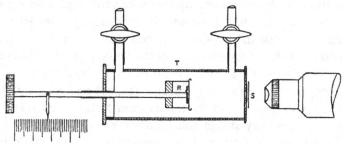

Transmutation apparatus

are two chlorine atoms of different masses, or weights, having the same
electric charge and the same speed, they would be deflected unequally
by a magnetic or electric field, and on arriving therefore at different
places on the photographic plate would make two separate and distinct
marks. In this way Aston found that there were two chlorine atoms
of mass 35 and 37. He measured also their proportionate amounts,
about two to one, whence the atomic weight, in this case 35·457, could
be partly explained. The subsequent developments of this great method
of investigation acted and reacted with the discoveries made by Rutherford and his co-workers.

RUTHERFORD TO BOYLE

1 *Jan.* 1920. We are in our home in Cambridge but still share it with
the workmen but we hope to have them out before long. It is New
Year's day and I have been exercising myself with C. Darwin in sawing
up a big tree in the garden and so feel quite virtuous, after a good day's
work. Eileen has gone to a dance and I go to Trinity to dinner so there
is a fair amount of activity in the household. My wife and I went the

other day to visit the Schusters—now Sir Arthur—see the morning papers. I gathered he was hoping to get you back again in Admiralty work. Possibly you may have heard something about it. I understand Shandon is partially closing down and Captain Gill is leaving. You may have heard F. E. Smith is to take Mac's place at the Admiralty. Well, I have thoroughly enjoyed my stay in Cambridge. The Lab. has been crowded both with students and researchers but all has gone pleasantly and I hope efficiently. We are very short of room and must extend before long. I have been buying a good deal of apparatus to fill up gaps and have been cleaning up the place generally. I have got my own work going but find it an almost impossible job to get enough scintillations to settle the mass of the recoil atoms from N and O. The use of necessary slits cuts down the effect so much. You will have seen in *Nature* that Aston by a development of the positive ray method has found isotopes in neon (20, 22), chlorine (35, 37) and mercury. I have seen all the photographs and the results seem certain. Aston is a good experimenter and very skilful and he deserves this success for he has slogged away for many years. I can quite appreciate how hard it is to keep your teaching going but be sure to get enough freedom for your researches.

Rutherford re-established relations with his old friend and admirer Stefan Meyer of the Radium Institute at Vienna, who drew a terrible picture of the state of Austria when the War ended.

RUTHERFORD TO MEYER

13 *Jan.* 1920. I would like to thank you for your kindness in sending me copies of the papers published in your Institute in war-time. I received them all safely and have read them with much interest. Your laboratory, I think, was the only one that kept work going steadily during the war. I was able to do a little work at intervals, but my time was much occupied in other matters. As you know, Lawson came to see me in Manchester and gave me detailed news of all of you, which I was very glad to hear. Lawson has now got a post in Sheffield and I hope is comfortably situated. He seems a very keen and prolific worker. I appreciate very much the kind way you looked after him in difficult times. You will have heard that I am now transferred to the Cavendish Laboratory and have now got into the working of it. I brought down your radium with

me and have been able to start my investigations again on the nitrogen problems. You will appreciate that it is very difficult work, but I am hopeful that I will be able to settle the question definitely before long. If the atom is not disintegrated by alpha particles I am of opinion it will not be done at all in our time. You will have seen in *Nature* about Aston's work on the isotopic nature of neon, chlorine and mercury. He has greatly developed the positive ray method and I have great confidence in his conclusions. He is a very skilful experimenter and has much experience with positive rays. You will appreciate what a large field of work this will open up and we may hope before long to decide which elements contain isotopes. Apparently so far no one has been able to obtain positive rays from the metals except mercury, but I am hopeful that this difficulty may be overcome by further experiments. I got during the war a copy of your book on Radioactivity and quite appreciate the very large amount of work you put into it. I hope, if I can find time, to bring my own book up to date before long. I gather from the papers that conditions as regards food are very difficult in Austria and particularly in Vienna, but I trust it is not as bad as has been reported. Give my remembrances to Madame and to my many old friends, including Hönigschmid, Paneth, Godlewski, Prof. Exner and others. I am hoping soon to write to von Hevesy whom I have heard from.

P.S. I am afraid some time must elapse before we can meet again to discuss radioactive problems but I am very hopeful conditions will rapidly improve from now onwards.

MEYER TO RUTHERFORD

Wien: 22 Jan. 1920. It was a very nice feeling to get your letter of 13 Jan. after so long a time and I thank you sincerely for the kind interest you show in it for us all. We had occasionally indirectly good news on your regard and were delighted to hear you got the professorship in Cambridge; in the last time we had some news by Lawson and by Ernest Goldschmidt, who met you in Cambridge and we felt always sure your feelings towards us, as ours towards you, could not be harmed by the psychoses of the surroundings. I was also pleased to hear you acknowledge that our laboratory kept working steadily during the hard times; that was not easy at all and some of the best men on both sides will never return to work again! But the so-called peace has aggravated the difficulties enormously and I fear, we will not be able to continue scientific work, if at all we may continue our life. Nobody who does not live here—with only Austrian coinage at his hands—can imagine the sad lot we were condemned to bear by a peace dictated only from nebulous hating and without knowledge of our country and our people. Of course there exist here persons, who made their business with war and with peace as everywhere, but he who has his fixed and reduced income is now a poor man. You think that conditions as regards food are not as bad as is told in your newspapers, but I can assure, that they are much worse than you can imagine it. Only a few numbers may illustrate this: It costs now 1 egg 10 Kronen, 1 kg. meal 50 Kr., 1 kg. butter 200 Kr., 1 liter milk 10–20 Kr., 1 simple cloth about 5000 Kr. etc.; and even at those prices it is nearly impossible to obtain the most urgent necessities. As long as our Krone has only about or less than 2% of its former value, there can't be much help. The construction of the new Austria without coal, without the necessary food, without the possibility to live by itself is a cruel nonsense. During the war we were at peace and friendship with Bohemia, Poland and all the other inventions; after the war they are constructed as new foes. My brother is professor of chemistry in the University of Prague, his poor wife died two months ago and I was not able to go to Prague, the politics did not allow me to be at his side.

If people here is ill by famine and by epidemics, which we can't suppress because we are not able to forward sufficiently soap, hot water, medicines etc., bolshevism and similar insanities will follow. Resolutions do not feed the people and the sentence "help yourself and we will help you" remains a simple word as long as every trial is suppressed by the victors and those who associated them after the victory. I am sure the gentlemen in England have no real idea of this kind of peace. As to my family and myself I have to tell you, that we got two children during the last years; my daughter Agathe is now four years old, my little boy, born November, 1918 and in view of the coming peace called Friedrich, begins to walk. The children are our sunshine, but it is hard to say how difficult it is to bring them up. This winter it would have been impossible to let them be in Vienna; babies older than one year do not get any more milk, we have no coals and much less wood to provide them a warm room and so we decided to let them stay in Ischl with the parents. My old mother is in Zürich with my sister Hertha, married to the vulcanologist Friedländer, who was expelled from Neapel, where he had founded and sustained with his own means the International Institute for Vulcanology. The old staff of the Radium Institute is now scattered. Hönigschmid is professor in München where also Fajans stays; Paneth is professor in Hamburg; Herzfeld has gone as Assistant to Sommerfeld and Fajans. Flamm is professor in Vienna, Technische Hochschule. Hevesy is still in Budapest, where we expect to see him soon. Godlewski, Smoluchowski, Patkowski and others were, as refugees during the Russian occupation from Galicia, our welcome guests. Poor Smoluchowski died; Godlewski and Patkowski remain our personal friends but are now in a hostile land, which wants, nobody knows what all from the restrained Austria. Dr Ulrich was expelled last summer from Joachimsthal—of course after the peace and on the beginning of the internal hostilities through the Czechs. He does not know the Czech language and as nobody in Joachimsthal speaks it, it was quite unnecessary, but this crime was sufficient to dismiss him. He is now in Vienna. I feel this letter is somewhat bitter, but I can assure you, your kind letter was like a draught of refreshing air and we will hope that your expectance that things will improve rapidly may be true. Your wonderful work, June number of Phil. Mag. has been received with enthusiasm; we also saw the note of Mr Aston in *Nature*. But I fear it will be soon impossible to get foreign literature. The annual publication of *Nature* costs now more than 1100 Kroner. Phil. Mag. is still more expensive and we can't afford the money. It would be very pleasant, if you would send us separates, as before the war and provide us the same from your scholars and colleagues. You will excuse that my English has made no progress, but I risk it to write in English knowing you prefer my bad English to my own language. Mrs Meyer begs to be remembered to Lady Rutherford and to yourself, as we do all.

RUTHERFORD TO MEYER

16 *Feb.* 1920. I was very pleased to receive your good letter but very distressed to hear of the bad conditions existing in Vienna. I hope that the funds collected in England will be of some little help in relieving distress. It seems to me, however, that things are on such a large scale that international help is required. I shall, of course, be very glad to send you any reprints from the laboratory but there is little to send at the moment, as the men have only just started work this year. I can

send you copies of *Nature*, if they would be of any service to you; please let me know. In your recent letter, you asked my opinion about a successor to Professor Exner. It is of course difficult for me to express any opinion, as, with the exception of Schweidler and Mache, I have little knowledge, either personal or scientific, of others you mention. Apart of his publications I know nothing of Ehrenhaft. I am personally of opinion that his work on the sub-electrons is not convincing, and I think you know Millikan is of the same opinion, but otherwise I know nothing of his scientific ability. I always supposed he was one of the band connected with the Radium Institute. The Laboratory here is in full swing and a number of interesting researches are now in progress. I am working hard at my own researches and hope before long to reach a definite decision on the nitrogen problem. We have had a very mild winter in this country and much sunshine.

In June 1920 Rutherford for a second time gave the Bakerian Lecture to the Royal Society. The first time he had given the complete foundations of the great subject known as Radioactivity. All subsequent work has been a development of the principles then enunciated. In this second lecture he laid the foundations of the transmutation of matter. He had once said that "unless it was possible to transmute matter by the action of swift alpha particles we are not likely to see it in our day". Now he had done it! He expounded step by step, with details and measurements, how he had used alpha particles to bombard the hydrogen nucleus out of the nitrogen nucleus so that what remained could no longer be nitrogen. He gave conclusive proof that the long range particles from nitrogen were indeed hydrogen nuclei, by measuring their deflection in a magnetic field. This was the first great step in the transformation of matter.

He also discussed the nature of the short-range particles that he obtained when he was bombarding either nitrogen or oxygen with alpha particles. He here made one of his rare mistakes in considering that they were isotopes of helium, a new type of particle of mass 3 and charge 2; or possibly of mass 3 and charge 1, an isotope of hydrogen. Ultimately these effects were shown to be due to an unsuspected type of alpha particles from radium C, few in relative number, but having an extra long range. Yet it is hardly to be regretted that this mistake was made, for it caused Rutherford to forecast the future in a remarkable manner which lent zest to his further investigations. Thus Chadwick wrote in the Royal Society obituary notice (Jan. 1938, p. 411):

The apparent occurrence of this new particle—part of the structure of the nitrogen and oxygen nuclei—led him to consider generally the constitution of atomic nuclei. Using as the ultimate constituents the hydrogen nucleus and the electron, he suggested that simple combinations of these might exist which had not yet been discovered. It may be of interest to quote his own prophetic words forecasting deuterium and neutron:

"It seems very likely that one electron can also bind two hydrogen nuclei and possibly also one hydrogen nucleus. In the one case, this entails the possible existence of an atom, of mass nearly two, carrying one charge, which is to be regarded as an isotope of hydrogen. In the other case, it involves the idea of the possible existence of an atom of mass 1 which has zero nuclear charge. Such an atomic structure seems by no means impossible....Such an atom would have very novel properties. Its external field would be practically zero, except very close to the nucleus, and in consequence it would be able to move freely through matter. Its presence would be difficult to detect by the spectroscope, and it may be impossible to contain it in a sealed vessel. On the other hand, it should enter readily the structure of atoms, and may either unite with the nucleus, or be disintegrated by its intense field, resulting in the escape of a charged hydrogen atom, or an electron or both."

A lecture in which the speaker clearly foresees four future discoveries, together with some of the actual properties as later found, is perhaps unique. He was forecasting the now known existence of neutrons, deuterons, triple hydrogen, and triple helium; or ^1_0n, ^2_1H, ^3_1H, ^3_2He, using a simple and useful method of nomenclature.

Rutherford and Einstein were appointed, both at the same time, foreign members of the Royal Academy of Sciences at Amsterdam. They stood at opposite poles of physical interests, on the one hand the concrete problems of the constitution of atoms, on the other side gravitational and cosmic theory. Between these poles was the careful and sympathetic scrutiny of critics like Lorentz. Both physicists had been recently encouraged by some striking discoveries; Rutherford by his disintegration of nitrogen by alpha particles with the ejection of protons; Einstein by some confirmation of his relativity theory by the observations at the total eclipse in Brazil. There was a suggestion from a leading mathematician and astronomer that these two men should work together at Cambridge, but the conjunction was not achieved.

At the 1920 Meeting of the British Association held at Cardiff, Rutherford suggested that the nucleus of the hydrogen atoms should be called the *proton* and, after some discussion, this name was supported by Sir Oliver Lodge and it is now universally adopted. So that we think of the hydrogen atom as possessed of a small charged nucleus around which moves at a relatively large distance a single electron which balances the charge and adds but a trifle to the massiveness of the whole atom.

Rutherford was constantly besieged with invitations for lectures most of which he had to refuse from lack of time. He was however glad, on the request of his young friend Bohr, to deliver three lectures in Denmark on the 'Counting of Atoms'; the 'Structure of Atoms'; 'Isotopes and their Meanings'. He received a great ovation and an honorary degree at Copenhagen.

In 1921 Rutherford was invited to become a Commissioner of the 1851 Exhibition, so that he might help to direct the valuable emoluments of that body towards the assistance of young men fitted to enjoy the opportunities which had already secured for himself a unique career. He valued this less than the joy of discovery and the hope of betterment of those burdened by poverty, insecurity and sickness. This view is illustrated by his address given on 'Radiology':

While the radiologists may justly view with some pride the great advances made in their subject in the last two decades, I think all are conscious that much more remains to be done if we are to reap the full possibilities of the application of X-rays, radium rays and other electrical agencies in medical therapy. We have arrived in a sense at a parting of the ways for the old haphazard organisation will hardly suffice if we are to keep a steady rate of advance in technique and knowledge. Under the present conditions, the radiologist is usually worked at full pressure and fully occupied with routine. It is almost impossible for him to spare either the time or energy to investigate the numerous important problems that arise in his work. Even if time were available, few institutions have the necessary laboratories and facilities for carrying out such researches. It is true that many hospitals have a fine X-ray equipment and a few possess comparatively large quantities of radium, which in most cases has been made available by the generous bene-factions of private individuals or by the ready response of the public at large. The management of many of these institutions is usually satisfied that everything necessary has been done when provision has been made for a good X-ray department and a supply of radium in charge of capable radiologists. A few are more advanced and recognise that if progress is to be made, it is of the utmost importance to give every encouragement to the prosecution of research in the problems of radiology. This not only entails the provision of a well-equipped laboratory, but even more important the provision of research workers. If the radiologist is to be in a position to carry out such researches, it is essential that all his time should not be taken up by routine duties, and also that the scale of payment be such that he can feel himself free to devote his spare time to research. Experience has shown that it is only in this way we can hope to obtain the services of a body of keen and progressive men, competent not only to carry out their work in the most efficient manner, but capable and desirous of still further advancing professional knowledge. It is only too by such a policy that we may hope to obtain the full return in service to the public for the numerous benefactions so generously made to many institutions.

RUTHERFORD TO BOYLE (AT OTTAWA)

23 *April* 1921. I was glad to hear from you again, and to know that you are getting your Department in such good shape. I was interested also to hear your items of Canadian news. I thought that it was a good idea of Mac's in Toronto to organise a Physics Conference. He would thoroughly enjoy the kudos of it all. I was very pleased to hear that they

had offered you the C.B.E. It is one of the misfortunes of living in such a democratic country that you are unable to sport the insignia of the Order.

I see F. E. Smith pretty often, and hear what is going on in the Admiralty. It is very interesting to find all the old devices appearing again as subjects for further research and study.

I had a fair holiday at the end of the Lent term, and got a week in Wales among a party of Cambridge climbers. I need hardly say that I was an on-looker, and took as little exercise as possible consistent with my self-respect. From there I went over to Brussels for the Solvay Conference, where we had a very pleasant week discussing the electron and its bearing in the structure of atoms, and magnetism. We had several representatives from this country, including Larmor, Jeans, Richardson, Barkla, W. L. Bragg; on the French side we had Madame Curie, Perrin, Langevin, and others; while Lorentz presided over our meetings. He is of course getting a little old now, but is a very good chairman and keeps us all working hard. Bohr was to have been there, but unfortunately he was suffering from overwork, and was not allowed to come. It is a great pity, for I was expecting him to lecture in Cambridge this term. This is now off, but we hope he will be well enough to come in October. You will have seen his recent letter in *Nature*. I shall be very interested to see how far he can deduce the arrangement of electrons on his theory. It sounds almost too good to be true in any detail. You will have seen a letter of mine and Chadwick's in *Nature* recently, in which we showed that we got very swift hydrogen atoms from a number of elements like aluminium, phosphorus and fluorine. There is a great deal of work still to be done in that direction, but it means a very large amount of scintillation counting. We have had recently a good deal of talk in the papers about the dangers of X and radium rays, which has had a natural effect on our assistants, so that it will mean we will have to take every precaution against undue exposure. I am arranging that all the men who do much work in this direction have regular blood-counts, so that they will not get unduly alarmed when they feel under the weather. I have been loaned a new quantity of radium by the Medical Research Board, and am arranging a good system of protection.

My wife and Eileen were in Italy recently. I was not able to join them after my stay in Brussels. You will have seen that the coal strike is still with us, and the power and light are to be cut off all the afternoons, so that experiments will be interfered with a good deal. I think it ought to be settled in a week, as we cannot afford the luxury of these perennial quarrels.

In 1921 Rutherford was elected Professor of Natural Philosophy at the Royal Institution in succession to Sir J. J. Thomson and every year he gave about four lectures, well illustrated with experiments, showing as far as possible his latest discoveries, skilfully leading up to them like a wise showman.

This year he was able to show photographs indicating the collision of an alpha particle with a proton, surely one of the most remarkable of human achievements. Moreover it was found that the laws discovered by Newton held good for these minute nuclei of atoms, as did the conservation of energy.

About this time Rutherford was endeavouring to find the neutron, and to establish artificial radioactivity, by experiments suggested to his research students. Thus Roberts was measuring exactly the heat generated in a hydrogen discharge tube, on the off-chance that protons and electrons combined to form neutrons with an energy change. As to artificial disintegration, the methods employed by Shenstone (*Phil. Mag.*, 1922, 43, p. 938) were ingenious and promising, but the search was for alpha particles rather than for negative or, still less, for positive electrons. The actual discoveries, in both cases, were made a full ten years later.

C. D. Ellis has recalled an amusing controversy when A. H. Compton was arguing in favour of a *large* electron, at a meeting of the Physical Society at Cambridge. This idea was hotly opposed by Rutherford, whose artistic views of Nature were outraged by such a suggestion. Finally he declared, "I will not have an electron as big as a balloon in my Laboratory."

It would have been interesting to learn Rutherford's views today on the new heavy electron, the mesotron—nicknamed the yukon.

Ellis has pointed out to me that Rutherford had a lively, innate sense of what might be termed the artistry of Nature, so that he almost knew what to expect. Yet he might partly conceal this delicate perception by a somewhat uncouth presentation. It is for this reason that the very same lecture might seem to some a perfect gem and to others a rough diamond.

The radium lent to Rutherford before the War became the annexed property of the British Government at the outbreak of hostilities. Although it was a personal loan, it was necessary by law for Rutherford to make a declaration of the fact that it was in his possession. On the other hand, Stefan Meyer pointed out that "every bit of the radium was a personal loan and that it never occurred to him that the British Government had anything to do with it", and he believed that Rutherford would find no difficulty in proving that the radium in question had nothing to do with the War and that it was placed at his disposal on friendly terms. This seems reasonable, but it was not in accordance with the law. However Rutherford replied as follows:

19 *Feb.* 1921. I note the suggestion in your letter about the small radium standard of yours in my holding. Can you tell me whether this was included in the list of 'enemy property' sent to the British Government? I omitted to make any definite mention of the same in my declaration. I regarded this as a personal loan from you to me to be sent back when I had done my measurements. Please let me know, as you will see it is difficult to make a move if it is officially registered through your Government. You will be interested to hear that owing to great improvements in counting, Chadwick and I have found that the long range atoms from nitrogen go 40 cms instead of the 28 which is the limit for ordinary hydrogen particles, using radium C. We also find long range atoms (probably protons) from fluorine, aluminium, phosphorus and boron, and possibly a trace from several others. No $4n$ 'simple' atom is affected, but only the $4n+2$ and $4n+3$. The surprising thing is the atoms from aluminium go over 80 cms and thus if they are 'H' atoms have greater energy than the incident alpha particle! Please regard this as private as I have not yet published it. You will also be interested to hear that G. P. Thomson and Aston here have by different methods found lithium has 6 and 7 for its istotopes.

MEYER TO RUTHERFORD

Vienna: 24 *Feb.* 1921. I was very much interested in your letter of Feb. 19 and I congratulate you to the astonishing results in your newest experiments. It is to be hoped now that in a not too distant future you may disclose the structure of all the simple atoms and it can't be foreseen what else will be found from that.

We are glad to have been of use to you in doing your splendid work through our radium. *Of course every bit of this was and is a personal loan to yourself, the big quantity as well as the small standard.* It never occurred to us to think it possible that your Government has anything to do with it. We never mentioned it to our Government and therefore it can't be officially registered. I don't quite understand what you meant with the 'declaration' you mentioned, but I suppose if there would be any difficulty you would be able to prove that our radium has nothing to do with the war and that it was at your disposal on friendly terms. I hope to hear soon from you again.

RUTHERFORD TO MEYER

14 *April* 1921. I was much disturbed by your statement of the financial side of the Radium Institute of Vienna and have been active in trying to raise some funds to buy at any rate a small quantity of the radium which the Vienna Academy so generously loaned me so long ago and which has been of so much aid in my researches. Through the kindness of the Royal Society I have been granted several hundred pounds to help you in these trying times by buying some of the radium from you. I am anxious that a reasonable part of the sum should go to defray the expenses of the Radium Institute. Please consult with the Vienna Academy about the matter and advise me where and how the money is to be paid for you to get advantages of the exchange. I shall of course require the written authority of the Vienna Academy to purchase 10 or more milligrams of their radium. I should first of all buy the small radium standard you loaned me before the war, amount, I think about 7 mgs., I speak from memory. Please give me your views on the question. I hope to hear from you soon.

This reply was received at the Vienna Radium Institute with 'vivid joy', and the question, always difficult, as to the price of radium was next discussed. The price in America for a gramme of radium at that

time was stated to be £24 a milligramme. (The present price is about
£4.) In Czechoslovakia the price was quoted at £30–40, "a price
that seems to be monstrous", stated Meyer. It must be remembered
that the pound sterling was then worth 3·83 dollars as against the proper
value of 4·86.

Rutherford obtained leave from the Administrator of Austrian
Property in England to release some of the radium loaned to him
before the War and he therefore bought, with the help of the Royal
Society, the said 10 mg. for £270.

Finally Rutherford raised funds to purchase all the radium and spread
the payment over a number of years and this money enabled the Radium
Institute of Vienna to tide over a period of depression and difficulty.

On 6 Dec. 1921, Ralph Howard Fowler married Eileen Mary
Rutherford. The wedding took place in the Chapel of Trinity College.

Rarely can a man have found a son-in-law so congenial in tastes
and outlook. Fowler was a Wykehamist and a Fellow of Trinity.
He was a man of great mathematical powers with a firm appreciation
of what might be termed the right and the wrong in physics, a master
in summing up all the experimental evidence and then proceeding
to a sound conclusion.

In due course Fowler became a Fellow of the Royal Society (1925);
he was awarded the Royal Medal (1936) and appointed Plummer
Professor of Mathematical Physics. Ultimately Rutherford had four
grandchildren named Peter Howard, Elizabeth Rutherford, Eliot
Patrick, and Ruth Eileen.

CHAPTER XI

THE ORDER OF MERIT

On 3 May 1922, Rutherford delivered the Eleventh Ray Lecture to the Institute of Metals on 'The Relation of the Elements'. He summarised the then recent advances in physics, beginning with the work of Sir J. J. Thomson and the fact that electrons were constituents of all atoms. He referred to his own discovery that every atom had a nucleus containing the bulk of the mass and always with a positive charge; to Bohr's theory of the planetary electrons round the nucleus; to Moseley's discovery of atomic number, giving the place from 1 to 92 of every type of atom, and equal to the number of unit charges on the nucleus; to the theory and chief properties of radioactivity; to Soddy's work on isotopes, inseparable elements with the same atomic number and nuclear charge, but with unequal nuclear masses. He next dwelt on an interesting point due to Harkins. There are in the outer surface of the earth nearly seven times as much by weight of the elements with even atomic numbers as there are of elements with odd atomic numbers. Here are the chief 'evens' in percentage of the whole constituents of the lithosphere: oxygen (8), 47·33; silicon (14), 27·74; iron (26), 4·30; calcium (20), 3·47; magnesium (12), 2·24. Total 85 per cent.

Odd atomic numbers, aluminium (13), 7·85; sodium (11), 2·46; potassium (19), 2·46. Total 12·8 per cent (Clarke).

It is clear that in the evolution of the elements in past ages, possibly in the sun before the earth broke away, there was a stability about the even-numbered elements to which the elements with odd atomic numbers could not aspire. Elements of low atomic weight predominate.

In the Thirteenth Kelvin Lecture given to the Institution of Electrical Engineers on 19 May 1922, and reported in *Nature* on 5 August, Rutherford spoke on 'Electricity and Matter' and the following passage is of much interest, now that the elusive positron has been foreshadowed by Dirac, discovered by Anderson and confirmed by Blackett.

It might *a priori* have been anticipated that the positive electron should be the counterpart of the negative electron and have the same small mass. There is not, however, the slightest evidence of the existence of such a counterpart.

Since it may be argued that a positive unit of electricity associated with a much smaller mass than the hydrogen nucleus may be discovered, it may be desirable not to prejudice the question by calling the hydrogen nucleus the positive electron. For this reason, and also for brevity, it has been proposed that the name proton shall be given to the unit of positive electron associated in the free state with a mass about that of the hydrogen nucleus, namely about 1·007 in terms of oxygen 16.

It is clear from these passages that Rutherford had the foresight to contemplate the subsequent discovery, quite a decade later, of the positive electron.

In 1922, Rutherford was awarded by the Royal Society the Copley Medal, the greatest honour in the gift of the Society.

He gave a lecture to the Institute of Metals on the 'Relation of the Elements', pointing out that in an ore of uranium the amount of radium present is but one part in three million as compared with the uranium, which conjures up a vision of that dauntless quest of Marie Curie among her tons of pitchblende. He stated that the Standard Chemical Company of Pittsburg had extracted 85 gm. of radium, and that the demand was large, especially during the War, for luminous paint containing radium for watches, compasses, dials and gun sights. He pointed out that uranium was a branching product, 96 per cent took the 'high road' through uranium, radium and so down to lead, while only 4 per cent turned into uranium Y, and thence into the actinium family, which in consequence was rare compared with the radium family.

He concluded by describing his latest discoveries of breaking up with alpha particles the nuclei of light elements, and of measuring the ranges of the protons ejected on disruption.

At the end of this year a Nobel Prize was awarded to his old comrade in the early days of radioactivity, Prof. F. Soddy, who wrote a letter of gratitude: "acknowledging the debt I owe you for the initiation into the subject of radioactivity in the old Montreal days. But for that, I suppose the chance of my ever getting the Nobel would have been exceedingly remote.... The paper has just arrived and I was delighted to see that Aston gets this year's Chemistry prize."

Rutherford was continuing his experiments on the disintegration of the light elements with alpha particles and was measuring the maximum ranges of the protons proceeding with speed in all directions, one from each disintegrated nucleus. For this purpose he inserted very thin sheets of mica equivalent in their stopping power to 5, 10, 20, 50 cm. of air. The apparatus employed is shown on page 275.

Mr G. R. Crowe, his assistant, gave me a spirited description of working with Rutherford:

He was shooting protons out of light atoms by means of alpha particles; carbon would not play. The next day was the turn of aluminium, and on some theoretical grounds Rutherford thought that the protons would have high velocity and a long range. "Now, Crowe, have some mica absorbers ready tomorrow with stopping power equivalent to 50 cm. of air." "Yessir."

On the next day: "Now, Crowe, put in a 50 cm. screen." "Yessir." "Why don't you do what I tell you—put in a 50 cm. screen." "I have, sir." "Put in 20 more." "Yessir." "Why the devil don't you put in what I tell you, I said 20 more." "I did, sir." "There's some damned contamination." "Put in *two* 50's." "Yessir." "Ah, it's all right, that's stopped 'em! Crowe, my boy, you're always wrong until I've proved you right! Now we'll find their exact range!"

In this way Rutherford found the length in cm. of the paths in the air of the protons from many elements—nitrogen 40, boron 58, sodium 58, fluorine 65, phosphorus 65, aluminium 90. This means that an alpha particle which will go about three inches in air will occasionally strike the nucleus of an aluminium atom so that a proton will leave the nucleus and travel a distance of nearly two feet through the air; thus the proton has more energy than that derived from the bombarding projectile! Rutherford and Chadwick worked together at all the problems which arise when various types of atoms are thus broken up.

RUTHERFORD TO R. W. BOYLE

9 Jan. 1923. I got your letter just before Xmas, at a time that I was hard at work preparing an address for the Science Masters' Association, which met in Cambridge about a week ago, 300 strong. We gave them a big programme of lectures including J. J. T., Pope, Priestley, Wood and others, and had all the labs. open. We are now just getting ready for the beginning of a new term. I will be giving my R.I. Lectures later in the term, on the subject "Atomic projectiles and their properties".

You may come across Henderson at Saskatchewan University. He is a quiet fellow but has got a good deal of research ability. You will see in the last Proc. Roy Soc. a very interesting paper by him on the "Capture of electrons by alpha particles". I have confirmed his results by the scintillation method and am generally occupied in seeing whether we can get definite information of the laws of capture and loss.

We have had a very good year in the laboratory, and made good progress. I have a number of good people as well as the usual uncertain material for preliminary training.

I am interested to hear that McIntosh is coming back to the fold. I imagine most of our scientific people in industry get bored from the lack of intellectual contact with their fellows. Rawlinson came up to see me a month or two ago. He seems to be making very good progress with his problem and hopes his theoretical investigations will be allowed for the D.Sc. in Manchester. F. E. Smith is doing very well as Head of the Scientific work for the Navy. He has got a well laid out laboratory in Teddington and, what is more, has the confidence of those in authority.

I hope you will come over to the Liverpool meeting of the B.A. I expect it will be a big meeting and I hope to see a number of my old friends there.

We spent a quiet Xmas at home with the Fowlers, and we are all very fit.

In February and March 1923, Rutherford gave six well illustrated lectures at the Royal Institution on 'Atomic Projectiles and their Properties' and the 'Life History of an Alpha Particle'.

Alpha particle strikes helium nucleus and they part at right angles (Blackett)

See p. 293

Alpha particle enters nitrogen which ejects proton and becomes oxygen (Blackett)

See p. 306

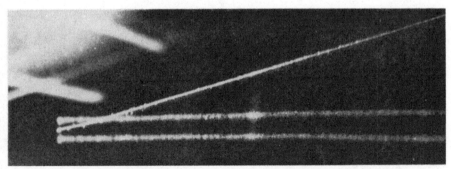

Alpha tracks from actinium emanation and actinium A atoms

See p. 324

He pointed out that an electron urged forward by one volt attains a velocity of 370 miles a second, and that exceedingly high speeds are attained with a hundred thousand volts—so light is the electron compared with its electric charge.

He was able to show some of the wonderful photographs obtained by Blackett and others, where a single alpha particle may be seen colliding with (1) a proton four times as light as itself, (2) a helium nucleus, of equal mass, (3) an oxygen nucleus four times as heavy. Strictly speaking none of these things is seen, but only their tracks along which the ions have been covered with mist when the moist air is suddenly expanded. In order to make this matter clear a photograph is shown, thanks to the kindness of Prof. Blackett, where an alpha particle strikes its equal—a helium nucleus—and they part company at right angles to one another, as would be expected from the laws of mechanics.

On 15 June 1923, Rutherford lectured at the Royal Institution on the 'Life History of an Alpha Particle from Radium', and a few extracts may be given:

It is now well established that the alpha particle expelled from radioactive bodies is in all cases a helium atom, or, to be more precise, the nucleus of a helium atom of mass 4 carrying two positive charges of electricity. It is only when the expelled nucleus is stopped by its passage through matter that it captures the two negative electrons required to convert it into the neutral helium atom.... There is a very wide range of stability exhibited by different radioactive elements. In a substance like radium A the average life of the radioactive atom before ejection of an alpha particle is about 4·2 minutes, for radium itself 2250 years, while in the case of a very slowly changing element like uranium the average life is of the order of 7000 million years.

The lecturer then referred to interesting experiments by G. H. Henderson (*Proc. Roy. Soc.* A, 1922, 102, p. 496), who fired in a high vacuum the alpha rays from a speck of radium through a small hole in a metal sheet and deflected them by a magnetic field. He obtained three bands on a photographic plate due to He^{++}, He^{+}, He; which is a shorthand way of stating that the main band was due to a helium atom less two electrons, the next to a helium atom less one electron, while at low velocities the third band indicated faintly a complete helium atom. Rutherford said that he had repeated and confirmed this result, using scintillations on a zinc sulphide screen. In effect, an alpha particle plays fast and loose with electrons when the opportunity occurs!

He referred also to Kapitza's wonderful photographs of the bent tracks of alpha particles in powerful fields of 80,000 gauss. The paths are not arcs of a circle, because the faster the particle is moving the harder it is to deflect it, so that the curving of the track is greater near the end where it stops.

The romance of Science certainly made a great appeal to Rutherford's mind, and in one of his lectures on the 'Life History of an Alpha Particle' he wrote:

We can follow in imagination the long life of the alpha particle—on an average many thousands of millions of years—as an integral and orderly part of the structure of the nucleus of uranium, or its descendants; the sudden cataclysm in the atom leading to its violent expulsion; its brief but exciting career of about a hundred-millionth of a second plunging through the atoms in its path, its long imprisonment in the mineral, and its release this evening to show its brilliant effects when an electric discharge passes through it. I hope you will agree with me that it is a fascinating story of a single atom of matter which, in its chequered career, had undergone so many vicissitudes.

He might have gone further and discussed its future escape into the atmosphere where it would diffuse, drifting higher and higher, until it would leave the earth altogether and wander through space for all time, or until some powerful star should draw it inwards and possibly reconstruct a heavier atom, so that the helium would again become a prisoner for vast periods of time as an alpha particle within a newly constructed nucleus. It is, however, typical of Rutherford that he kept his feet firmly on the ground and avoided the more speculative aspects of physics.

In the old-world preface to his modern work on *The Structure of the Atom* (G. Bell and Sons, 1923) Andrade paid a unique tribute to the master, who guided the young man's research at Manchester.

A Dedication to Sir Ernest Rutherford, Knight, Fellow of the Royal Society, Member of the Principal Learned Societies of the Old and of the New World and Nobel Laureate.

Sir,

I fear that what I offer you here is but an indifferent Plant, though 'tis grown from your own Seed. Since I laboured in your Elaboratory, working small things while those Choice Philosophers Moseley (of whom we may say, as the illustrious Newton said of Cotes, that had he lived we had known something) and Bohr were performing great ones, War and the Penalties of Publick Employment have long kept me from attempting to add to Knowledge by such Natural Experiments as my slender Wit can devise. In studying now at length to fit myself for a Venture I have gathered together much of what the Learned of our day have discussed in the matter of that Microcosm, the Atom, within whose exiguous Bounds is Space and to spare for Philosophical Speculations; and in the perhaps too Presumptuous Hope that what it has taught me much to collect may help others as Ignorant as Myself I have ordered my Conclusions into the Volume which I now present to You.

No Astronomer of these petty Suns and Planets has given us more and rarer News of them than Yourself, who first taught the Virtuosi to see in the Atom a Massy Nucleus controlling by electrick Laws his distant Servants the light-heeled Electrons. You will find here, then, much that is a picture of your own Thoughts; and as there is scarce a Man whose Portrait is being limned but feels a lively Curiosity to see what the Artist, be he never so unskillful, has made of his Task, so, I trust you will look not without Interest on much that I have endeavoured to portray of which you are the Originall....

Finally I bring you humbly this Book not so much in the Belief that it is worthy your Acceptance, as in the Hope that its very Imperfections and Errours may prove of Service, by suggesting to your Perspicacious Judgement Means by which they may be amended, and Experiments to resolve what I have set down doubtfully. The Principles which you have already establisht will remain to perpetuate your Name to Future Ages, and these I have endeavoured to elucidate to the Students in our great Science of Physicks. I have, however, small Doubt that you are so extending the Bounds of Natural Knowledge as to render what I have written Meagre and Incomplete even while I Subscribe Myself,

Sir,

Your most Humble Servant,

Edward Neville da Costa Andrade

At the Artillery College at Woolwich,
January, 1923

In 1923 Rutherford was President of the British Association, which met at Liverpool. In his Presidential address he pointed out that at the previous Meeting held in that city, in 1896, an account was given by a young Italian, Marconi, who had signalled for a few hundred yards by wireless waves, whereas at the present Meeting it was possible to broadcast a concert all over England. In those twenty-seven years complete order had been attained, and most of the gaps filled, in the chemical elements, while the first steps had been achieved in the disintegration of atoms by the deliberate purpose of man. He warned his audience that the laws which govern the massive world, in which we live and to which we are accustomed, may not hold good in the nuclei of atoms. In comparison with the past they were living in the heroic age of physics, but it had been suggested that we might appear to future generations as if still groping in the dark. This pessimistic suggestion found little echo in Rutherford's mind, for he knew full well how solid and lasting were the main achievements of physics.

He continued:

There is an error far too prevalent to-day that Science progresses by the demolition of former well-established theories. Such is very rarely the case. For example, it is often stated that Einstein's general theory of relativity has overthrown the work of Newton on gravitation. No statement can be further from the truth. Their works, in fact, are hardly comparable for they deal with different fields of thought. So far as the work of Einstein is relative to that of Newton, it is simply a generalisation and broadening of its basis, in fact a typical case of mathematical and physical development. In general, a great principle is not dis-

Liverpool, 1923

carded, but is so modified that it rests on a broader and more stable basis....

It is clear that the splendid period of scientific activity which we have reviewed to-night owes much of its success and intellectual appeal to the labour of those great men in the past, who wisely laid the sure foundations on which the scientific worker builds to-day; or, to quote from the words inscribed in the dome of the National Gallery: "The works of those who have stood the test of ages have a claim to that respect and veneration to which no modern can aspire."

Rutherford's speech at the British Association was for the first time broadcast throughout the British Isles, and while some found it unintelligible and asked the B.B.C. not to do it again, others enjoyed the experience of hearing an expert enthusiast speaking of his favourite topic, the Electrical Structure of Matter.

A select few of the members of the British Association belong to a Red Lions Club which holds an annual dinner as a relaxation after the more serious meetings and discussions of the week. The younger members of the club, known as 'jackals', presented to Rutherford a coat of arms suitable to the presidential dignity and to his scientific work. It is shown as a photograph, robbed of its brilliant colouring, with its motto *Atom virumque*.

The more orthodox banner of the President with abundant Kiwis is also given.

A very curious thing was discovered about 1924. The alpha particles from radium C generally travel about 7 cm. (about 3 in.) in air, but a very few out of a million go farther, some to 9 and some to 11 cm. Similar results hold for thorium C. Imagine a shaving brush with a million hairs about 3 in. long, with twenty-eight hairs sticking out 4 in., and five hairs about 5 in.—and there is a picture of the range of the alpha particles from radium C! It was puzzling at first to understand why Nature was so uniform and then occasionally diverged in this erratic manner. No doubt a long range particle has picked up some extra energy from the activated atom, which might otherwise escape as a gamma ray.

The amateur radio enthusiasts were allowed to use the short wave lengths because it was believed that the range of transmission would be small. The young experimenters soon discovered that there were wonderful possibilities with these short waves, which would carry great distances with incredibly small power at the source.

In 1924 a young McGill undergraduate, Jacobs, offered to send a wireless message to anyone in England that I liked to name. We drafted the following message (7 Feb. 1924): "Sir Ernest Rutherford, Cavendish Laboratory, Cambridge. McGill Radio Association sends greetings by transatlantic amateur wireless through their station. Please reply through station that delivers this. A. S. Eve."

The message was sent at night, picked up by Simmonds, Meadowlea, Gerrard's Cross, and forwarded by letter to Rutherford at Cambridge, who replied by the same route. The wave length was 100 metres and the power used 80 watts. It may seem easy now, but certainly it was no mean feat then.

On 7 Feb. 1924, there was a conference at the Royal Society on the 'Structure of the Atomic Nuclei'. Rutherford stated that he and Chadwick had found that when an alpha particle was projected at atoms ranging from copper (29) to uranium (92) it was under an inverse square law of force throughout its motion; and that the nearest approach to the very centre of the nucleus was 10^{-12} cm. for copper and 4×10^{-12} cm. for uranium. What an extraordinary thing that small nucleus of uranium must be, which contains among other things no less than eight alpha particles to be ejected before it can reach the stable product of lead! In the case of the elements lighter than copper the case is quite different and the alpha particle can approach much nearer to the centre of the nucleus.

Rutherford and Chadwick had by this time proved that the nuclei of all elements ranging up to potassium—except lithium, carbon and oxygen—could be disintegrated by the alpha particles of radium C which have energies equivalent to seven million volts.

In all cases there is a great repulsive force resisting the close approach of an alpha particle or proton to the nucleus. There seemed to be something in the nature of a barrier of very high potential which retained the nucleus intact, while checking the ingress of a foreigner.

The Franklin Medal was awarded to Rutherford by the Franklin Institute of Philadelphia and presented to him at the Foreign Office, London, on 14 July 1924, by Mr Charles Trevelyan, President of the Board of Education.

RUTHERFORD TO MEYER

19 *July* 1924. I am leaving for Canada in a few days and will not return to Cambridge until early in October. I am acting President of the British Association until the beginning of the meeting in Toronto but I will then be free from official duties. I hope to have a quiet holiday in the East and see some of the American Universities before I return. My wife and daughter and son-in-law [R. H. Fowler] are also going. Two days ago I received from Dr Pettersson an advance copy of a long letter he proposes to publish in *Nature*. Chadwick and I have definite evidence that many of the results in it are either wrong or wrongly interpreted. So I have written him a long letter giving him my reasons and suggesting further consideration before publication. I send a copy herewith to you to let you see the situation. You know me well enough to appreciate that I would not interfere unless I thought the situation was serious. I do not know Pettersson personally or his co-workers, but you do. He seems to me a man of originality and ingenious in his arrangements but I should judge he jumps to conclusions on insecure evidence. The subject of artificial disintegration is full of difficulties and pitfalls and wants investigators who are very careful in experiment and with good judgment. I am sorry to bother you over the matter but you will quite appreciate how important it is to Dr Pettersson and to your Laboratory and also to the subject not to go along wrong lines. I am feeling pretty fit but have much to get through before I leave.

There was a large gathering from Great Britain, Canada and the United States at the British Association Meeting in August at Toronto, where Sir William Bragg was President of the Mathematics and Physics Section. At one of the meetings Rutherford speaking on 'Atomic Disintegration' stated that physicists had accomplished much that, even as late as 1913, he had thought must be left for the next generation to deal with. Rutherford was the guest at a luncheon given by the meteorologists at which their distinguished president, Bjerknes, declared that had he been asked in his youth, which of two subjects would make the greater progress—the weather or the atom—he would have unhesitatingly replied, the weather! Exactly the opposite had been the case, and that was because (turning and bowing to Rutherford) abler men had been solving the problem of the atom.

After the Toronto Meeting Lady Rutherford and the Fowlers went on a riding trip in the Rockies, but Rutherford came eastwards and stayed with us at Glenmere, a cottage alongside a little lake in the country near Knowlton in the Province of Quebec. There he enjoyed bathing, deck tennis, and fly-fishing for the lively speckled trout in the lake, entering into all the interests of the party, and was specially beloved by all the children and young people.

After a tour of lectures at the American Universities he returned to Montreal and gave a talk to the McGill Physical Society, standing at the place where he had upheld, against all comers, the theory of the disintegration of the radio-elements. It is hard to describe the admirable effect of these lectures which Rutherford was ready to give to scientific societies—the inspiration to young students, the infectious vigour to the professors, the novelty of the experiments and ideas, the generalisation over a broad sweep of his subject—all these were put forth with volcanic enthusiasm, mixed with friendliness about human and personal affairs.

CHADWICK TO RUTHERFORD

Inveroran: Sept. 1924. I think I ought to tell you the results of the last experiments but I don't know when you will get this letter as I very seldom see a post office.

I corrected the number of disintegration particles from aluminium under as definite conditions as were feasible. As near as the experiments allow, the number agrees with that calculated on the assumption of an attractive inverse fourth and repulsive inverse square taking (1) zero force at 4×10^{-13} and (2) that the H particle appears when the alpha disappears. Of course the agreement cannot be very good on account of the counting errors and even in estimating the solid angles of the alphas used.

Blackett has got two more photographs which are somewhat clearer than the others. They show the track of the H particles and the track of the recoil atom, but no track for the alpha. If this is true it is a very fine addition to the evidence for the attractive field, and fits in very well with our expectations.

I tried the scattering apparatus using gold as radiator. The numbers were less than an inverse square field, about half I think, but there was no evidence of short range alphas. I made some attempt with aluminium and carbon but the results were not definite, owing to a slight contamination. What evidence I got seemed to show that the number of particles scattered through 135° by carbon is probably greater than the number scattered by aluminium. I feel sure it is true. I think it is exactly what must happen, but the evidence is very weak yet.

I think we shall have to make a real search for the neutron. I believe I have a scheme which may just work but I must consult Aston first.

I suppose you have received Pettersson's letter. He sent me a copy and also an additional short letter to which I made a very mild and conciliatory reply in the hope of inducing him to visit us. I don't think he will come, however.

We are having a most amusing time on these Highland roads. They are so narrow that one can pass another car only at occasional places. The other day we found a real hill in the way from Ullapool to Gairloch. I turned through a narrow bridge and found the road going up like the side of a house. I was so surprised I stopped altogether. It begins at 1 in 7 and the average gradient is 1 in 7·8 but fortunately it is only 1/3 of a mile long. We have now turned south again and I think I shall go back in a few days. I have to get my new rooms in order.

The Order of Merit was awarded to Rutherford at the beginning of 1925. In accepting the offer of it Rutherford had written:

20 *Dec.* 1924. I appreciate that this award is not so much in recognition of my own personal services to Science, for an individual can hope to accomplish but little, but rather a recognition by His Majesty of the importance of Science and scientific study to the welfare of the Empire.

For a lecture to the Royal Society of Arts Rutherford chose for his subject 'The Stability of Atoms' and described mainly the Bohr theory relating to the outer electrons connected with optical radiation, and with the inner groups which are related to Röntgen or X radiation. He touched lightly on the nucleus itself where he surmised that alpha particles might act as satellites to the rest of the nucleus, the size of which is at most 7×10^{-12} cm. even in the case of uranium, while the atom itself has a radius of about 2×10^{-8} cm., so that the atom is 3000 times as big as its nucleus in linear dimensions; while of course in volume it is 27 thousand million times as large.

The following extracts are typical:

The trend of physics during the past twenty-five years has been largely influenced by three fundamental discoveries that were made in the closing years of the nineteenth century. I refer to the discovery of X-rays in 1895, of radioactivity in 1896 and the proof of the independent existence of the negative electron of small mass in 1897.... The discovery of the electron and the proof that it is a constituent of all the atoms, gave us the first definite line of attack on the constitution of the atom.... It is convenient to consider that the structure of the atom consists of two parts, one the minute, charged nucleus which contains most of the mass of the atom, and whose electric charge controls the number and arrangement of the outer electrons, and the other the outer or planetary electrons which extend far from the nucleus and give the atom the dimension usually attributed to it. The motion of these electrons is responsible not only for the characteristic light seen in the spectra of the atom, but also for its high frequency X-ray spectra, while the position and motions of the more outlying electrons govern the ordinary chemical and physical properties attributed to the atom.

On the classical theory, it was impossible to devise a stable system of electrons in motion round a nucleus, since a moving electron radiated energy and would ultimately fall into the nucleus. In order to overcome this fundamental difficulty, it was necessary to invoke the aid of the quantum theory of radiation, first put forward by Planck in 1900, although its full importance and significance was not recognised until a decade later....

To explain the complex modes of vibration (as learnt from the observation of spectra) of the single electron attached to the *hydrogen* nucleus Bohr supposed that under appropriate conditions of excitation, the electron could be displaced from its normal orbit and occupy one of a series of stationary states, or levels of higher energy relative to the nucleus, which are defined by simple quantum conditions. The energy of the level relative to the nucleus is most conveniently expressed in terms of volts. For example, the energy required to move the electron from the normal state, or first level, to a great distance is equal to the energy acquired by an electron in falling freely between two points differing in potential by 13·54 volts. The corresponding energies of the successive levels (counting the normal level as the first) are connected by simple numerical relation, namely, 1/4, 1/9, 1/16 ... of the energy of the first normal state, i.e. 3·38, 1·50, 0·84, ... volts respectively. Then the difference of energy between the first (normal) and second states is 13·54 less 3·38, or 10·16 volts.

Rutherford went on to explain that if a collision or radiation raises the level and the energy of the normal electron into the second state, requiring 10·16 volts to do so, then that atom is said to be 'excited', and within a minute fraction of a second the disturbed electron will snap back to its normal level, orbit, or state, and in doing so it will radiate out the light of the exact frequency observed in spectroscopy. The connection is $E = hf$, where E is the change of energy, f is the frequency, and h is the known Planck's constant.

A similar scheme of things holds for Röntgen rays, where there are for heavy atoms several electrons in different levels, known from the centre outwards as K, L, M, N ... rings. Here again, as Moseley first showed, the numbers 1, 1/4, 1/9 ... hold. If high frequency radiation meets an atom it is apt to sweep away an electron from an inner ring. This constitutes absorption of radiant energy. An electron returning to that ring will give up a corresponding amount of energy; frequently the replacement is from the L to the K ring which will give up three-quarters of the total energy. This constitutes radiation. In the simpler cases of hydrogen, ionised helium and X-rays, Bohr's theory achieved a great success, and it still holds good.

About the time of this lecture Bohr's theory was shown not to be true in more complicated cases, and a thorough revision became necessary; so that physicists were plunged into the difficult atmosphere of matrices and wave-mechanics, thus securing the correct results without altogether comprehending the physical background involved in the mathematical procedure.

It was finally shown by L. de Broglie, Schrödinger, Dirac and others that "we have to abandon the description of atomic events as happenings in space and time, we have to retreat still further from the old mechanical view. Quantum physics formulates laws governing crowds and not individuals. Not properties but probabilities are described, not laws disclosing the future of systems are formulated, but laws governing the changes in time of the probabilities and relating to great congregations of individuals".[1]

Much of the new development left Rutherford unmoved, for it was outside his sphere. He was sympathetic enough and ready to avail himself of any conclusions which aided investigation. Yet, as he once said jokingly, they "play games with their symbols, but we, in the Cavendish, turn out the real solid facts of Nature".

Rutherford would have relished a recent remark of Kramers: "The quantum theory has been very like other victories; you smile for months and then weep for years."

RUTHERFORD TO R. W. BOYLE

16 *Feb.* 1925. I received your very kind telegrams of congratulations on the Order of Merit, but have delayed a reply until I had time to send you a reasonable letter. I have also received your letter of January 21st. I have of course been very busy answering the crowd of letters that have come in, and last week I had to go down for the formal investiture at Buckingham Palace. Frazer and I led the procession and we got through very rapidly.

You will be surprised to hear that my wife and I have decided to leave for Australasia probably late in July, and will not return until

[1] Einstein and Infeld, *The Evolution of Physics*, Cambridge University Press, 1938.

after the New Year. My primary intention is to see my people, but incidentally I shall give some lectures in the main cities of Australia and New Zealand. Under these conditions, I hope, if you come over to England, you will arrange to come here first so that we can see you.

I saw F. E. Smith the other day. He told me about the question of publication of your results. I quite agree with you that it would be a good thing to get some of this work published as rapidly as possible. Lehmann and Osgood his partner are very hard at work. They have had many trials and tribulations but, I have no doubt, will succeed in producing a really good paper at the end. This research has been a great experience for them; and Lehmann has developed into a very good glass-blower and a good experimenter. Boomer is hard at work on the 'helides' and thinks he has got some new and interesting results.

R. H. Fowler is away in Copenhagen where he is spending a couple of months with Bohr, who was very anxious for him to go over. He tells me that Bohr has now got the correspondence principle much more definite and is able to predict intensities accurately, even in the case of an optical multiplet. They are also hard at work thrashing out the old problem of the absorption of the alpha-particle. Ellis and others are hard at work on the gamma-ray spectra and the heating effect of the gamma-rays; while Chadwick and I are busy examining the field of force round atomic nuclei, particularly of the heavy elements. It is hard work but we hope to definitely settle how far the inverse square law applies. Kapitza, as you know, is preparing to do a big experiment for producing magnetic fields (by momentary currents) of the order of 1 million Gauss. This research is being financed by the Government, and involves the installation in one of the old Engineering Laboratories of a complete power station with a 1500 kilowatt machine as the source of the power. It is a very elaborate experiment and I hope it will come off all right. If so, there will of course be a great deal of work to do in the examining of magnetic properties in these very high fields. The whole scheme will cost a good deal and will be financed by the Department of Scientific Research.

We are all fairly well. Eileen and the grandson Peter are in the house. He is nearly two years old and is a lively vigorous youngster.

Four of the most interesting lectures that Rutherford ever gave are well reported in *The Electrician* for March and April 1925. They illustrate the thoroughness of Rutherford's work and the clearness of his exposition. He traced the history of the early efforts of men like Thomas Young, Kelvin, Maxwell, Rayleigh, and others, to find the size of a single atom. He gave an account of the successful modern experiments, by radioactive and other methods, to find the number of atoms in a given volume of gas, the size of a given atom, the mass (or weight) of a single atom of any element, and the unit charge (electronic) of electricity. He pointed out that all these quantities were known within an error of less than one per cent, better perhaps than we know the population of any stated country.

Rutherford gave another lecture, also at the Royal Institution, on 10 April, at which he first announced a result of the greatest importance. He said that P. M. S. Blackett at the Cavendish Laboratory had succeeded in taking photographs of an alpha particle which struck a nitrogen nucleus and *remained inside*, throwing out a proton, as a young cuckoo ejects a fledgeling from its nest. Clearly the nitrogen atom gains 4 and loses 1 in mass, and it gains 2 and loses 1 in charge. Hence the nitrogen atom has been transmuted into a heavier atom with a larger atomic number, one of oxygen! To Blackett thus belongs the great credit of having been the first to prove the synthesis, or building up, of a heavier atom from a lighter one. Of course, today, that is an everyday matter; but not then.

To the initiated, the whole affair is described by

$$^{14}_{7}N + ^{4}_{2}He \rightarrow ^{17}_{8}O + ^{1}_{1}H,$$

or nitrogen and helium give oxygen and hydrogen.

Oxygen of mass 17 is a rather rare isotope of the commoner oxygen of mass 16, and it has been detected by the spectroscope and by the mass-spectrograph.

In the year 1925 Rutherford was elected President of the Royal Society and held the office for five strenuous years. In that capacity he not only proved himself to be an excellent Chairman, but also on suitable occasions joined with zest in the discussions and drew out valuable contributions from others.

Rutherford received the news of his election with boyish elation. A student declares that it 'gingered up his lectures' to such an extent that the student resolved to follow where Rutherford led, nor has his resolve been without marked results.

In 1925 Rutherford paid testimony to the generous help that Prof. P. Langevin had given in the piezo-electric method of submarine detection during the War. Although the idea had occurred to many, it became a possibility only because of the development of amplifiers during the War period. Langevin was the first to make trials with this scheme in the harbour of Toulon, and he allowed representatives of the Admiralty (British) to be present and to assist in his experiments.

At an earlier date, about 1917, some of us advocated the view that Rutherford originated the scheme of a piezo-electric submarine detector, whereas Langevin had claimed priority. Rutherford's reply to me was short and final: "If Langevin says the idea was his, then the idea was Langevin's."

In July Rutherford sailed, with his wife, for Australia and New Zealand and has left the following account of his tour:

We left England on July 25th from Liverpool by the SS. *Ascanius* via the Cape. I was met by one of the Firm at the boat and they kindly allowed us to use the cabins de luxe on board the boat. This included a large cabin for each, a small sitting room, and a private bathroom. There were only about fifty passengers, about half of whom were bound for the Cape. We had a comfortable and rather cool journey to Capetown. There we stayed the night with Sir Carruthers Beattie, Principal of the University, and had a very pleasant dinner party in the evening. Next morning I visited the University and saw the work of both Schonland and Newbery. My wife spent the morning visiting special places and gardens.

We had a long and rather tedious journey to Adelaide but with only one or two days' bad weather. We arrived at Adelaide on September 2nd and there left the *Ascanius*. Half an hour after she arrived in port the crew and firemen joined the strike and the boat did not get away for nearly two months. We were met at the port by Professor Kerr Grant, Professor and Mrs Osborne and Professor and Mrs Robertson, all of whom I had met before. We had lunch at the Grants and my wife left that afternoon by rail for Melbourne and Sydney hoping to catch the boat for New Zealand at the latter place. Actually she failed to do so and spent the next four days in Sydney with the Gordon Craigs. Stayed at the South Australian Hotel, where rooms had been reserved for us, and found many letters awaiting me for plans in the West of Australia and New Zealand. Gave my first lecture in the evening on the 'Counting of Atoms'. A good audience and introduced by the Deputy Governor. After dinner went to the hotel with Dr Mitchell, Vice-Chancellor of the University and talked philosophy till 12.

On the Friday—the 4th—visited Robertson's new biochemical laboratory and the new physics and engineering buildings. I gave an informal talk to the people interested in physics—about 200—on work in the Cavendish laboratory. In the evening gave a second lecture when good speeches were made at the conclusion by Grant and Chapman (Eng.). Attended a reception by Miss Todd in the Adelaide Club.

Saturday the 5th. Answered letters and telegrams and visited Registrar. Went for a picnic with the Osbornes and Grants in a car and had a very pleasant day. In the evening dined with the Vice-Chancellor and a number of friends in the Club.

Sunday—motored by Osborne to see Miss Farr, and on return was taken out by the Vice-Chancellor to lunch at the Barr-Smiths, who have a fine house and garden a few miles from Adelaide. Left by train for Melbourne the same evening and was seen off by a number of friends.

MELBOURNE

A pleasant railway journey and was met at Melbourne by Orme Masson, Rivett, Barrett and a number of others. Photographs and cinema in operation. Lunched at the Melbourne Club where I was put up by Orme Masson. In the afternoon went to the University; had tea with members of the Department, including old friends—Glove, Rogers and Bells, and gave a talk to a large audience on the work going on in the Cavendish laboratory. Monday evening a reception by a number of representative Societies. The Governor and his wife—Lord and Lady Stradbroke, presided, and Lady S. made one of the main speeches—an excellent one. Unfortunately I had to receive with the President of the Victorian League, which I should have preferred to escape. Finished up by refreshments and general talk.

Tuesday—given a lunch in the Club by the Honourable G. Clarke, the President, and saw a good deal of him during my stay. Gave three lectures in Melbourne, two on the 'Atomic Constitution' and one on 'Radiology'. Attended a reception in the evening given by Lady Grice and in the afternoon by Mrs Allen.

On Wednesday lunched with my old friend J. A. Erskine, and on Saturday was taken by Bells for a motor excursion in the country. Returned in time to leave by the evening train for Sydney. My time in Melbourne was very much occupied with interviews with the Press and a number of visitors, amongst others several who claimed relationship but on no grounds that I could find out.

SYDNEY

Arrived Sydney on Sunday the 13th and stayed at the Sydney Club. The Club was comfortable but my host forgot to introduce me to any resident members and so had rather a dull time at ordinary meals. Gave one lecture in Sydney and a talk at the Physical laboratory, and then went by train to Brisbane.

BRISBANE

I was met at the station by a number of old friends, including Professors Steele and Priestley and put up at the Queensland Club, an exceedingly comfortable and pleasant Club and adapted for semi-tropical conditions. Next morning we were taken a drive by Professor Steele to the highest point in the neighbourhood and returned in time to give an informal lecture to the students. A good and appreciative audience. Visited the Physical and Chemical laboratories; in the afternoon there was a reception given by the Acting Vice-Chancellor in the open air. I was introduced to a number of students and had afternoon tea. Lecture that evening and the next, and next day lunched with a number of members of the Faculty. A pleasant but hurried visit and I did not have much time to see much of the city or country.

Returned by train to Sydney and was met some distance from the city by Professor Vonwiller with a motor car, and took up my duties there. Taken home for lunch and in the afternoon had a long run in the

country with a picnic tea. Gave in all three public lectures. The hall
was crowded and there was a good deal of discussion in the papers on
the fault of the University in not getting a bigger hall. Was given a
dinner on the first evening by Dr MacCallum, acting Vice-Chancellor
and had a most pleasant time. Had a formal reception by the Royal
Society and other bodies and made the usual speech. Was taken by
friends for a long motor journey through the Reserve Park and had a
very enjoyable day's excursion. Paid several visits to the new Physical
laboratory and the Engineering laboratory and was given a dinner by
the staff in the University Union. Saw a good deal of Vonwiller and
Bailey. Went to lunch with the Governor Admiral de Chair—a private
affair. Left Sydney earlier than was intended by the SS. *Niagara* and
rolled our way to Auckland. Had a comfortable berth but rather rough
weather most of the way. Met a number of New Zealand acquaintances
on the boat.

Rutherford left Sydney in S.S. *Niagara* on 24 September and four
days later arrived at Auckland, where he was met by his old Manchester
pupil, Prof. E. Marsden, who made all arrangements for his visit and
went with him on the tour. As Rutherford himself said it was a 'semi-
Royal' visit, with government free passes on the railways and a motor
car at their disposal. There were civic receptions to which Rutherford
raised no objection. "I am inured to them now", he said, but begged
them to "cut out all garden parties and reduce formalities to a mini-
mum".

As to his lectures they were given at Auckland, Wellington, Nelson
and Christchurch and they were well attended, for at Auckland "500
were standing and 500 could not get in at all".

Rutherford gave the Cawthron Lecture to the Cawthron Institute
at Nelson, the town of his old school. He first referred to "the gift of
64 acres of first-class land by Mr James Marsden, whom I had the
pleasure of meeting yesterday, to the Cawthron Institute, an example
which we hope will often be followed by other friends". The lecture
was on 'Electricity and Matter' and first dealt with the work of the
great pioneers, such as Faraday. He passed rapidly to modern dis-

coveries, such as that of the nucleus of the atom, with special reference to the work in 1911 by Dr Marsden in Manchester, who conducted, at Rutherford's request, an experiment which threw great light on this question. His words were:

If a thin sheet of gold leaf is placed in the path of a beam of alpha-rays some rays enter the atoms of gold and are scattered. That was only to be expected. But a remarkable and unexpected observation was that some of the swift alpha-particles actually came back again! That was a most surprising result considering the minuteness of atoms and the great swiftness and energy of the alpha-particles which were fired at them. It is just as surprising as if a gunner fired a shell at a single sheet of paper and for some reason or other the projectile bounded back again.

Previous to the civic reception at Christchurch, Rutherford watched a magnificent haka, or Maori dance, performed by Canterbury College students "who dragged the motor car of the world-famous scientist to the Municipal Chambers".

He visited his old school at Havelock and wrote:

Here I addressed the school children, planted a tree, was photographed and gave the youngsters a whole holiday. The school house had not been altered since my youth and the present master seemed an excellent fellow. Found the house we lived in had been pulled down, but Havelock very little changed in the interval. Left immediately for Nelson via the Rai Valley and the Wangamoa and had a most interesting drive over the mountains. All the bush had been cut down in the Rai Valley and burnt, but ultimately the grass had been overcome by a growth of fern so all the labour had been wasted, while a magnificent forest had been destroyed.

I found the house where I was born at Brightwater had been removed and was now occupied by a chicken run. Fortunately it had been photographed by Professor Easterfield before it was pulled down and my mother declares it is the correct house, except that a verandah has been added to it after they had left Nelson. My father declares that when he started up house in it, it was the best built house in the Nelson district; in any case it lasted nearly sixty years.

He visited his parents at New Plymouth and found them well, his father 86 and his mother 82, "both unusually well preserved for that age". On 9 October, a telegram came from his son-in-law R. H. Fowler announcing the birth of his granddaughter Elizabeth. The Nelson College Old Boys at Christchurch gave a complimentary luncheon to Sir Ernest and Lady Rutherford. Indeed he was so busy that he wrote "you will find it difficult to get any manuscript of my lectures out of me, for I have not time or energy to write it". He wrote however to Chadwick at Cambridge: "I was greatly amused to receive from Kapitza a magnificent photograph of your two selves taken on the day of your wedding. Both of you made magnificent portraits and I think that they should hang on the walls of the Cavendish Laboratory. The contrast between Kapitza in his everyday clothes and in his capacity as best man is highly diverting. I understand from Fowler that his top hat completed the toilet." So the whole month of October was passed in New Zealand, and, when he left, Prof. Coleridge Farr, F.R.S., of Canterbury College, aptly summarised the visit in a farewell message:

May I say with the utmost sincerity that it has been your inexhaustible good nature which has made your visit go so smoothly, and your unfailing readiness to fall in with any plans that have been made for you. You realise how intense has been the desire to meet and to see you, while the crowded audiences indicate the strength of the desire to hear you.

On the return journey S.S. *Maheno* was delayed by strikes so that the Rutherfords were late in reaching Sydney, but the P. and O. steamship company held up the ship for a few hours for them. The journey to Adelaide was a pleasant one, but Perth was like an oven (103° F.) and there Rutherford got through his lecture all right, but "lost a good deal of weight in the process".

There was a call at Colombo, and another at Aden, and four days' very pleasant stay at Cairo with a visit to Luxor. Rutherford declared:

If I go there another time I shall leave out the regular sights and go straight to Assouan, and stay there. One soon gets tired of inspecting tombs on a roasting day with a temperature of about 90° inside the tomb. My interest in the ancient Egyptians vanished very rapidly under these conditions. I am glad to say that the rest of the party were affected in a similar manner, so we struck out all the minor Kings and nobles.

At Cairo Rutherford received some news from the Cavendish:

KAPITZA TO RUTHERFORD

Cavendish Laboratory, Cambridge: 17 Dec. 1925. I am writing you this letter to Cairo to tell you that we already have the short circuit machine and the coil, and we managed to obtain fields over 270,000 in a cylindrical volume of a diameter of 1 cm, and $4\frac{1}{2}$ cms high. We could not go further as the coil bursted with a great bang, which, no doubt, would amuse you very much, if you could hear it. The power in the circuit was about $13\frac{1}{2}$ thousand kilowatts (13,500 amperes at a pressure of 1,000 volts) approximately three Cambridge Supply Stations connected together, but the result of the explosion was only the noise, as no apparatus has been damaged, except the coil. The coil had not been strengthened by an outside band, which we are going to do. Up to 200,000 gauss we have gone quite safely and we are not in very special difficulties.

At present all these experiments have been done with a higher speed machine, and I am very happy that everything went well, and now you may be quite sure 98 per cent of the money is not wasted, and everything is working.

The accident was the most interesting of all the experiments, and gives the final touch of certainty, as we now know exactly what has happened when the coil bursted. We know just what an arc of 13,000 amperes is like. Apparently it is not at all harmful for the apparatus and the machine, and even for the experimenter if he is sufficiently far away.

I am very impatient to see you again in the laboratory and to tell you all the little details, some of which are amusing, about the fight with the machines. Tizard and Sir Richard Threlfall are coming to the laboratory tomorrow.

CHAPTER XII

PRESIDENT OF THE ROYAL SOCIETY

"Before leaving Cambridge" said Rutherford in an after-dinner speech after his return from New Zealand, "I went in to see my old friend C. T. R. to say goodbye. He was slowly grinding by hand a large ground glass joint and I left him doing it."

Then, after describing his visits to the various colleges in New Zealand, he went on: "The voyage home was, I think, one of the most interesting sea passages I have ever had. We had a lot of young athletes on board and you know what a part deck tennis plays on a long voyage. Well, it was a real treat to watch these young men play. I have made a good many voyages in my time but I have never seen a better example of skill and power than these young men showed." A slight pause, and then, "I won the prize!!"

"But all good things come to an end and we arrived back in the old country and so to Cambridge. The first thing that I did after being away for some months was to look up my old friend C. T. R. I found him still grinding a large glass joint!"[1]

It must in fairness be added that Rutherford knew and said that the achievements of C. T. R. Wilson were of the very greatest importance in physics, and backed him for his Nobel Prize.

In January 1926 Bohr wrote to Rutherford his regret that Kramers was leaving Copenhagen and his delight in acquiring Heisenberg; owing to the work of the latter "prospects have with a stroke been realised which, though only vaguely grasped, have for a long time been the centre of our wishes. We now see the possibility of developing a quantitative theory of atomic structure." Bohr then referred to the important suggestion of Goudsmit and Uhlenbeck that electrons spin

[1] I am indebted to Professor A. M. Tyndall, of Bristol University, for the record of this speech.

on their axes and that this fact may account for the "fine structure of spectral lines".

On 4 March 1926, Rutherford was in the Chair as President of the Royal Society during a discussion on the 'Electrical State of the Upper Atmosphere'. Prof. S. Chapman dwelt upon exploration with balloons which had by that time achieved a height of 30 km. Thunderstorms were discussed by C. T. R. Wilson, while Sir Henry Jackson and Appleton dealt with exploration by wireless. C. T. R. Wilson said that, at any time, there were 1800 thunderstorms raging in the world, giving on an average 100 lightning flashes a second, thus balancing the steady leakage of negative electricity from the earth's surface in normal weather. It was pointed out how large a share this country had had in this work on the upper atmosphere, particularly referring to the work of Appleton in proving the existence of the Heaviside layer and measuring its height.

Prof. H. E. Armstrong wrote to the President: "Your attitude in the Chair is delightful; to have a President asking questions and prompting discussion is an astounding departure. You may restore a dead body to life, if you persevere and get a little human feeling into the show."

On 6 March 1926, Rutherford gave a lecture at the Royal Institution on the 'Rare Gases of the Atmosphere'. He pointed out that thirty years had then passed since Lord Rayleigh had lectured on argon to the Institution. Enormous labour was then necessary to extract even a moderate quantity from the atmosphere. At the time he spoke, argon was a by-product in the extraction of oxygen from the air and about a hundred million litres of argon were used annually in gas-filled lamps. Neon was in use in all cities for illuminating signs and in the United States hundreds of thousands of cubic feet were isolated and used for that purpose.

Lord Rayleigh discovered argon by noticing that nitrogen taken from ammonia was a half per cent less dense than nitrogen taken from the air. Why? Sir James Dewar pointed out to Rayleigh that Cavendish in 1775 had found a small bubble of gas remaining after removing other gases from some air. What was that bubble of gas? Rayleigh took some air and sparked it, thus forming nitric oxide which was absorbed by caustic soda. This was a tedious process but the remainder gas continued to increase in density, because argon is denser than nitrogen. Rutherford remarked with a smile: "Chemists generally assumed that Lord Rayleigh, being a physicist, did not know how to purify gases." Little they knew their Rayleigh and his immense caution.

Sir William Ramsay, having read Lord Rayleigh's papers, joined in the hunt. He collected the new gas more quickly because he used heated magnesium shavings to absorb the nitrogen. It was a better process. A joint account by Rayleigh and Ramsay of the discovery of argon was given at the British Association at Oxford (1894) and met with extreme scepticism. Ramsay later extracted from the air neon, krypton and xenon, three new rare gases. Rutherford proceeded in his lecture to tell the interesting history of helium, and touched on the three radioactive gases radon, thoron, actinon.

The year 1926 was the Tercentenary of the death of Francis Bacon, which was duly celebrated by his University of Cambridge. Rutherford was awarded an honorary Sc.D. and the words of the Public Orator began with Bacon:

That great man, the most illustrious of the Lord Chancellors of England, when caught at last by the cunning of his enemies, said he hoped there remained some quiet place for him in some Cambridge College where he might be at leisure for science....

Man, the servant and interpreter of nature, lives by the laws of nature and his own. The life of man, so long as he obeys the laws of nature, flourishes. If, on the other hand, we try to live without human laws, nature refuses us its benefits.

I present one whom you have long known, a high priest of natural science, censor of the atom, the flower of knighthood, our colleague and friend Sir Ernest Rutherford.

One of Rutherford's favourite quotations was taken from Bacon: "Human knowledge and human power are coextensive; for ignorance of causes prevents us from producing effects. Nature can be ruled only by being obeyed; for the causes which theory discovers give the rules which practice applies."

RUTHERFORD TO HEVESY

1 *June* 1926. I hear from Bohr that you have an addition to your family. Congratulations and best wishes to you all.

I do not think that I have written to you since you got your Chair in Freiburg. I wish you all success and happiness and will expect to hear of great works from time to time. As for Cambridge things go on much as usual but there is a good deal of activity in the Laboratory in various directions. Aston has got going with his new apparatus of great dispersion, and expects to fix the atomic weights of the light elements with an accuracy of 1 in 10,000 and incidentally to determine the mass of an electron. He had a great deal of trouble over the new method in tracing down the origin of a disturbance which affected the width and shape of his lines. The strange reason for this is now apparent and it can be allowed for if necessary. He now gets lines almost as sharp as spectrum lines.

Chadwick and I are working at the scattering of alpha particles and hope to publish some interesting results before long. I want to know a little bit more about nuclei before I retire from actual work.

I received, a few weeks ago, a copy of yours and Paneth's book on Radioactivity. I read it with much interest and think it reads quite well. With the help of Chadwick and Ellis I hope to produce a book before long on the radioactive radiations and their applications. My time is too much occupied to tackle big books again.

The grandchildren flourish and we are all very well. I hope to get a short holiday in the Tyrol after the International Research Council meeting in Brussels on June 29th.

RUTHERFORD TO SCHUSTER

21 *Nov.* 1926. I hope you are coming to the R.S. dinner on St Andrew's Day so that you may give me your moral support in the hour of trial. Would you feel inclined to make one of the speeches on behalf of the medallists? Hopkins will be the other. I know it is rather a bore, but you can be as short as your sense of decency allows. I hope you will do so, but if your mind and body revolt would you please send me a

wire with a Thundering 'No', as I have to make all arrangements the next few days. I am busy with the annual address—an awful task in a busy term.

RUTHERFORD TO MEYER

23 *Dec.* 1926. I was very glad to get your letter a day or so ago and to hear how you are progressing. We are fortunately all very well and will be spending the Christmas time in Cambridge. My two grandchildren are now growing rapidly and are at an amusing and interesting age. I have had a very busy term as the University is undergoing reorganization as a result of the Report of the Royal Commision on the Universities. This has involved a good deal of extra work in grafting on the new machine to the old. You will see in *Nature* that we celebrated in Cambridge on Dec. 18th J. J. Thomson's 70th birthday. The dinner was confined to the present and old members of the Cavendish Laboratory and we had a most interesting and successful evening with about 130 people at the dinner. *Nature* took the opportunity to publish a special supplement on the history of the Laboratory last week and you will see in this week's issue an account of the Celebration. As you know, these little digressions from routine involve a good deal of labour for everyone concerned. I was glad to hear from you about the prospects of an interchange of visits between this Laboratory and your Institute, to get at the bottom of the reasons of the differences in results obtained in the two Institutions. I thank you for suggesting that I should come over, but as you no doubt know, it is quite impossible for me to do so. My duties here as President of the Royal Society fully occupy my time. Chadwick has gone away for his holiday but he told me that he had received a letter from Pettersson. I have not had an opportunity to discuss the question with him but I am very doubtful whether he would be able to visit Vienna for the next few months. He is expecting an interesting domestic event in February and naturally cannot leave for some time. I will write you later when we have discussed the matter. I can quite understand that it might be best for Chadwick to visit your Institution and see for himself what is the cause of the divergences. I agree with you that it is highly important that this question should be amicably settled for I myself feel the whole subject of nuclear disintegration must remain in confusion pending a comparative investigation.

Experiments finally showed that, in the main, the Cavendish was right.

At the Cavendish dinner Rutherford referred to J. J. as "a star of the first magnitude, a central sun which does not shrink with age, but draws on some unknown source of energy".

Of Rutherford's own work at the Cavendish Bohr wrote:

Oftener than once it has been the unique power of Rutherford that has called forth a revolution in science by inducing him to throw himself with his unique energy into the study of a phenomenon, the importance of which would probably escape other investigators on account of the smallness and apparently spurious character of the effect. The confidence in his judgment and our admiration for his powerful personality was the basis of the inspiration felt by all in his laboratory and made us all try our best to deserve the kind and untiring interest he took in the work of everyone. However modest the result might be, an approving word from him was the greatest encouragement for which any of us could wish.

I remember on a visit to Manchester during the Armistice hearing Rutherford speak with great pleasure and emotion about the prospect of his going to Cambridge but expressing at the same time a fear that the many duties connected with this central position in the world of British physics would not leave him those opportunities for scientific research which he had understood so well how to utilise at Manchester.

As everybody knows, the sequel has shown that this fear was unfounded. The powers of Rutherford have never manifested themselves more strikingly than in his leadership of the Cavendish Laboratory, the glorious traditions of which he has upheld in every way. Surrounded by a crowd of enthusiastic young men, working under his guidance and inspiration, and followed by great expectation of scientists all over the world, he is in the middle of a vigorous campaign to deprive the atoms of their secrets by all the means at the disposal of modern science.[1]

In February 1927, Rutherford gave the twelfth Guthrie Lecture to the Physical Society in the Imperial College of Science. This lecture is of

[1] *Nature*, 18 Dec. 1926.

interest because he began to speculate on the constitution or structure of the nucleus itself, and referred, not very hopefully, to the services which the new quantum mechanics might possibly render in elucidating the many puzzling features attending the close collision of nuclei. He stated: "Great difficulties arise the moment we consider why the nucleus of an atom holds together and progress seems likely to be slow, because it seems clear that the ordinary laws of force between electrified particles break down at such minute distances."

C. D. Ellis has reviewed the scientific situation thus:

Perhaps the most characteristic example of Rutherford's genius during his Cambridge period is to be found in a paper, Phil. Mag. 4, 580 (1927), which at that time made little impression and is now almost forgotten. It puts forward a theory about the origin of the alpha particles, and the details of this type of disintegration, which was soon completely superseded by the wave-mechanical picture. Yet what is so interesting is that it shows how completely he has appreciated the difficulties that lay in the way of any explanation on classical lines. From his own experiments on the scattering of alpha particles and from those carried out in collaboration with Chadwick, he became convinced that the inverse-square electric field round a heavy nucleus must be valid to such small distances that the potential energy of a nucleus and alpha particle could reach values of ten million electron-volts. Even if an alpha particle, by some mechanism, just trickled away from the attraction of the main nucleus, it would find itself in this strong repulsive field, and could not possibly emerge from the atom with kinetic energy less than ten million volts. Yet the alpha particles from uranium have energies of less than six million volts. The hypothesis he suggested to overcome this difficulty is to-day unimportant, but the fact on which he based his argument is recognized to be one of the crucial points which show the need for some theory other than classical mechanics in dealing with these matters. Rutherford's ability to seize on one definite fact, to realize that, independently of all other arguments, it showed a fundamental error in the current physical picture, and then to follow this trail wherever it led, this ability was the measure of his genius. It is shown no less strikingly in this example than in his famous deduction from the fact that alpha particles could very occasionally be scattered through large angles. In the latter case he found the right answer, in the former he did not, but the real achievement was the realization of the experimental fact and insistence on its importance.

In 1926 the British Association met at Oxford under the Presidency of H.R.H. the Prince of Wales. Rutherford gave an account of his work, with Chadwick, on the collisions between alpha particles and light atoms, or rather nuclei of atoms, such as magnesium and aluminium.

There was a good course of lectures on Atomic Structure given by Rutherford at the Royal Institution when he referred to the work of their President, Sir William Bragg, on the ranges of the alpha rays from different radio-elements, and how they were stopped by the atoms in matter. Rutherford explained that matter was mainly *empty space*, so small were the nuclei and electrons, and so great, relatively, the distances between them.

In March 1927 Rutherford spoke to the British Wireless Dinner Club:

On one occasion, I was at the meeting of the British Association and said to Sir Arthur Keith, "a gathering like this must be of great interest to you, because you can study the skulls of the people round you!" Keith replied, "Yes, it is of great interest. You see that group over there", indicating Sir Oliver Lodge, Sir Arthur Schuster and Earl Balfour. "If you go and dig in this country you will find the kind of skull that Sir Oliver Lodge has. If you go to Nineveh you will find Schuster's kind of skull, and Balfour's is a typical Scottish skull."

In April 1927, there was an inaugural celebration at Buckingham Palace to mark the centenary of the birth of Lord Lister, pioneer of antiseptic surgery. Rutherford, as President of the Royal Society, presented an address to which His Majesty the King replied. There was a further celebration in honour of Lister at the Royal College of Medicine; "Sir Ernest Rutherford, wearing a scarlet cloak, was in the chair, a jolly and substantial high priest of physics, not of physic."

In August 1927, Rutherford learnt with deep regret that his old friend, B. B. Boltwood of Yale, had ended his life when in a temporary state of mental depression. In an obituary notice Rutherford dwelt upon

Boltwood's skill as a chemist and his success in the discovery and separation of ionium, the immediate parent of radium, as well as upon their first work at Manchester on the rate of growth of radon from radium.

The centenary of Volta was celebrated at Como (11–17 September) and a great International Congress of Electrical Science held. Rutherford recalled the communication made by Volta to the Royal Society in 1795, and his great discovery in 1800, for which the Society awarded him a medal. He pointed out that Volta made the first battery and that his name is perpetuated in the word 'volt'; he continued:

All scientific men, I assume, will gladly unite in doing homage to the memory of Alessandro Volta for his fundamental discoveries. To Englishmen in particular, the celebration of the Centenary is of especial interest and significance on account of the close connection of Volta with the scientific men of this country. His famous paper entitled "On the electricity excited by the mere contact of inducting substances of different kinds" was first communicated as a letter to the President of the Royal Society, Sir Joseph Banks, in 1800, and published by the Society. The importance of the discoveries of Volta was at once recognised and during his visit to England he was everywhere received as one of the greatest scientific men of his age.

The discovery of the generation of electricity by the combination of chemical elements was of great theoretical as well as of great practical importance. It first disclosed that intimate connection between electricity and matter which was so strikingly brought out later by the researches of Faraday in the laws of electrolysis. In this respect the discoveries of Volta provided the first step in the development of that 'Electrical Theory of Matter' which a century later was to dominate the progress of Physical Science.

In 1927, while Rutherford was President of the Royal Society, he was also Chairman of a Committee for the reorganisation of the National Physical Laboratory, the second Director of which, Sir Joseph Petavel, had been appointed at the end of the War.

The amount of time and thought that public men voluntarily give to these important problems is remarkable. As the Government finds the funds from the public purse, it must assume responsibility and retain control. On the other hand, the Government cannot pretend to have scientific knowledge sufficient to direct research. Hence the Royal Society assumed that responsibility from the outset, through a suitable Committee. Schemes were prepared by Sir Frank Heath, Henry Tizard and Sir Richard Glazebrook, which were interwoven and compromised in true British fashion. There was very little vinegar in the salad and Rutherford as Chairman was liberal with the oil, so that the resulting scheme suited all tastes and worked well.

At the Annual Meeting of the Royal Society, Rutherford presented the Medals and made his Presidential Address. He spoke of the passage of electrical waves over great distance round the earth; on the use of electrons in radio-pictures and in television; and on the then new wireless beam station at Rugby with short waves and high-powered electron-tubes—a skilful piece of electrical engineering by the Post Office. He added: "It is now possible, and will, it is expected, soon be practicable to connect any telephone subscriber in Western Europe to any telephone subscriber on the North American Continent."

Rutherford was awarded a prize of 1000 dollars by the Sigma Xi Scientific Society, meeting in connection with the Convention of the American Association for the Advancement of Science. The prize was awarded to him for his work as Director of the Cavendish Laboratory.

In October, Rutherford opened the new H. H. Wills Physics Building at the University of Bristol in the presence of their Chancellor, Lord Haldane; there was quite a strong group of physicists present—Bragg, Langevin, Eddington, A. Fowler, Max Born and Tyndall. Rutherford spoke on the advance of science and engineering due to the discovery of X-rays, electrons and ions. A lady afterwards praised Rutherford as being such a practical man in talking to them of something that all could understand—electric irons!

In 1928 Rutherford was elected an honorary Fellow of the Royal College of Physicians, on the occasion of the tercentenary of the publication of Harvey's great book "Exercitatio anatomica de motu cordis et sanguinis in animalibus".

In his annual course of lectures to the Royal Institution Rutherford drew attention to the two successive radioactive products, actinium and its emanation. Each ejects an alpha particle, but in such rapid succession that about a 500th part of a second elapses between the two events. This may be seen in the photographs of the trails of the alpha particles, shown by C. T. R. Wilson's moist-air expansion chamber. The positive and negative ions, due to the first shot, have been drawn apart by an electric field; the second shot occurs usually in a fresh direction, quite at random as compared with the first.

The *Sunday Dispatch* of 26 August gave a character sketch of the "Man who Chartered the Atom", from the pen of 'Rhadamanthus':

"See that man across the dining room?" asked my friend.
"You mean the hearty farmer who's enjoying his breakfast so much?" I replied.
"Farmer!" he cried witheringly, "Farmer—nothing! That's Sir Ernest Rutherford, who plays with atoms a deal more casually than you can play with billiard balls."
I looked at the man again unconvinced. His healthy colour, blunt features, shrewd eyes, heavy limbs, and even his easy tweeds, with their baggy pockets, all seemed aggressively agricultural.

The article went on with an attempt at describing his discoveries and continued:

Sir Ernest Rutherford wears his laurels lightly. His fellow-workers, from his senior assistant to the junior 'lab-boy', who is privileged to empty his waste-paper basket, adore him.
His smile is perennial, his kindly good-humour inexhaustible. Other men may lose their tempers, or their heads, or their courage. Sir Ernest is the same yesterday, to-day, and—apparently—for ever!
He is the best of good fellows, an admirable after-dinner speaker, a witty controversialist, a breezy lecturer, the confidant of his colleagues' ambitions, the arbiter of their jealousies, the composer of their quarrels.

The writer then expresses a belief that Rutherford would have been happier as a farmer. This is quite an illusion, for he delighted in his voyage of scientific discovery and enjoyed the sunshine of his success.

A research student, who worked for many years at the Cavendish, and had a great respect and affection for Rutherford, declares that the idea of an equable temperament and perfect control of temper is a libel. This student protests that "he never met anyone who could get quite so angry over trivial things as Rutherford—though he always apologised for it". This is contrary to my experience of earlier days and perhaps indicates the great strain of the work of his later years.

About the same time, an eminent surgeon, now dead, wrote to Rutherford at the end of a letter on Royal Society business: "You are a delightful dinner companion; and I promise, in return for the pleasure of last night, to deal with you with compassion if (supposing you do not succumb under the intellectual burden I am imposing on you) you should need a surgeon's help. Drink *lots* of water and escape us!"

RUTHERFORD TO MEYER

7 *Feb.* 1928. I am glad to inform you that the University has agreed to pay the sum of £3,000, for the Radium loaned to me by the Vienna Academy of Sciences and proposes to do so in six annual grants of £500 as arranged; the first payment to be made on March 31st, 1928 and at the same date in subsequent years. I am writing also to the Academy informing them of the matter and asking them to state to whom the annual payments should be made. If it is quite convenient, I would propose to send a cheque from the Department of Physics on March 31st of each year.

I am very glad that this matter has now been concluded. I was very distressed to hear of the death of Professor Lorentz last Saturday and I am going to Haarlem to attend the funeral on Thursday. I saw Lorentz last at Como, when he appeared to be in the best of health and full of intellectual activity.

The Council of the Royal Society of Arts awarded the Albert Medal to Rutherford for his "pioneer researches in the structure of matter".

The medal was presented on 19 July, by the President, H.R.H. the Duke of Connaught, at Clarence House, St James's.

In 1928 his father, James Rutherford, died at the age of eighty-nine. Only a short time before his death he was straight in the back and remarkably active for his years. Rutherford has stated that he felt that his father might have become a very good engineer, for he showed great skill in devising and controlling turbines and water-ways, in connection with his flax-mill.

In May, Rutherford lectured in Berlin to the German Chemical Society on "Atomic Nuclei and their Transformations". He stayed with his friend and old pupil Dr Otto Hahn of the Kaiser Wilhelm Institute.

At one time physicists looked askance at the periodic table as something as remote from physics as the Hebrew alphabet. It is now a vade-mecum, a Bradshaw, a Baedeker! Rutherford stated that at the time of his lecture thirteen elements were known to eject protons under bombardment. He pointed out that "it was probably not an accident that uranium and thorium are the sole survivors of the heavier nuclei in the earth to-day", meaning that it was due to their very long periods of transformation.

RUTHERFORD TO HAHN

8 *May* 1929. I got back home this morning at 10.15 after a very pleasant journey. We were very comfortable on the train and I slept peacefully through to Harwich. I feel very fit and have already done a good day's work but make haste to write to you on my safe arrival. I am getting this typewritten so as to save you the bother of deciphering my writing which is far from perfect. I had a most delightful time with you at your home and in Berlin. You and your wife could not possibly have looked after my comfort, both material and spiritual, better than you did. It was a great pleasure to meet you and so many old friends again under such pleasant circumstances. I thought it was very kind of you to prepare such original dinner cards. If you by any chance could send me copies of the old snapshots you used I would be very glad to have them.

A careful investigation was made by Rutherford and Aston as to the atomic weight of protoactinium—the immediate precursor of actinium—whose abbreviated chemical notation is quaintly Pa. It was found to be 231, and this figure fixes the atomic weight of the whole actinium family down to its ultimate lead (207).

In 1929, the Royal Society held a conference on Nuclear Structure opened by the President (Rutherford) who pointed out that fifteen years earlier the Society had held a discussion on the Constitution of the Atom. At that time he had talked on the nuclear theory of the atom and the evidence in support of it, while Moseley gave an account of his X-ray investigations. Soddy had spoken of isotopes and referred to the work of Sir J. J. Thomson and Aston using positive rays and getting two parabolas for neon, suggesting that neon consisted of two isotopes.

Since that time Aston had carried out a very careful measurement of the masses of many individual isotopes; alpha rays had ejected protons from many elements, perhaps twenty shots in a million being effective; and the theory of 'potential barriers' round the nucleus had been put forward to explain experimental facts.

Soon afterwards the President received a characteristic letter from the chemist, H. E. Armstrong:

My congratulations! You were very clear and most interesting on Thursday (at the R.S.) and Aston was good—but what muffs the others— as speakers. You should have a private rehearsal of these discussions so as to coach people at least to speak up, speak out and address the house, not the screen! You might, with advantage, add to your officials a teacher of elocution. Money spent in that way would give a far better return than much of that spent on research, so called!

This year (1929) the British Association met in South Africa under the Presidency of Sir Thomas Holland. Rutherford went to it with his daughter and son-in-law while Lady Rutherford went for a motor tour in the Tyrol with friends. The members of the Association had a strenuous time with visits to Johannesburg, Pretoria, Pietermaritzburg,

Grahamstown and back to Capetown. Rutherford was present and lectured at the centenary of the South African College, now the University of Capetown. He also opened the new Physics Laboratory of the Rhodes University College, Grahamstown, called the de Beers Institute of Mathematics and Physics, in honour of the generous donor.

Wherever he went Rutherford advocated research work both for its own sake and as a stimulus to sound teaching.

No university, he said, is worthy of the name, that does not do everything in its power to promote original research in its laboratories. It is the duty of the university to see that its professors and teachers are not overburdened with routine teaching, but are given time for investigation and provided with research laboratory facilities and the necessary funds for this purpose.

As Prof. Schonland once wrote: "Rutherford had no great liking for the attitude that scientific work could not be done because equipment or money was lacking; I remember him once replying to someone who made such a statement, that he himself could do research work at the North Pole!"

Prof. A. M. Tyndall also writes:

I was recorder for Section A and was therefore responsible for collecting papers for the programme. I approached Rutherford and mentioned my view that for South Africa where physicists were limited in number, the papers should not be too specialised. I asked him, therefore, whether he would care to choose a general title such as "The trend of Modern Physics". His reply was "The trend of Modern Physics? I can't give a paper on that. It would only take two minutes. All I could say would be that the theoretical physicists have got their tails up and it is time that we experimentalists pulled them down again!"

Some notes on the voyage are preserved in Eileen's letters to her mother:

S.S. *Nestor*....It has been very calm and hot for ten days and now is very chilly....Dad seems in great form and has enjoyed the trip enormously. He does look well. We land tomorrow and it will be nice to be on dry land again. We are expecting to get a week-end cable with some information about Peter—we get rather impatient to hear.

30 *July*. We are having a wonderful time in S. Africa. We go from lunch party to motor drives and then dinner parties, so although I am fatter than I have ever been, I feel a trifle worn out. Capetown was a lovely place and Johannesburg is fascinating too. It is all built on different terraces which makes a place attractive and feels rather like the hills outside Christchurch where the Seagers used to live. We are staying with very wealthy people here with a delightful, but not very big, house full of old furniture and lovely Persian rugs and a garden that would make you very happy. It drops away in terraces. The top part is full of English flowers and I have never seen or smelt such wonderful stocks in my life. Below are the cactuses and S. African rock-garden. It is supposed to be the most beautiful garden in Johannesburg, I was told. Daddy is staying at the main Club so will have a more restful time. He is very fit indeed and full of fun and cheerfulness; you would be pleased. All our arrangements are working out very well. So far we have had incredible weather, ceaseless sunshine and sharp mornings and evenings. We shall get a few days in Capetown at the end of our trip and are so looking forward to staying with the Beatties. . . .

Rutherford wrote a full account to his wife and from his letters the following extracts are taken:

Capetown: 23 July 1929. On Sunday Eileen, Ralph and a party climbed up Table Mountain and Eileen was as fresh as any of them, although Ralph and Schonland had had enough by the end of the day, so she is obviously pretty fit again.

I had a very pleasant morning, General Smuts called and took the Beatties and myself and other friends to the Botanic Gardens, a sort of natural park on Rhodes' property.

Thursday. We have had a very hectic week and there has been little time for rest or meditation. On Monday was the reception by the Mayor at which I spoke a few minutes—a pleasant function which passed off with considerable éclat. In the afternoon there was an opening Meeting at which the Governor General was present, when Hofmeyr, President of the South African British Association, gave an address quite eloquent but rather long....

On Wednesday the University gave a big afternoon reception—about 3000 guests at the new University Buildings at Groot Schuur, which are in a splendid scenic position up the mountain side with a view from the terrace of Table Mountain and in front of the Stellenbosch range of mountains—a magnificent panorama.

Rhodes has left his house Groot Schuur for this very purpose of new University buildings.

Today Ralph and I went to the Houses of Parliament and lunched with General Hertzog—Prime Minister—had a pleasant chat and met the Minister of Railways. This evening I have to give my Evening Lecture on the "Structure of the Atom" and so I did not go to the observatory this afternoon as I thought it better to have a quiet time after late hours yesterday.... The great event this morning was a discussion on Life by General Smuts, followed by many other speakers— three and a half hours in all and apparently it went off well....

20 miles S. of Kimberley. In train 8 a.m.: 30 July. We left Capetown by the second train at 9.30 a.m. and had a beautiful view of the mountains all day long, but were well on the Karroo by 3.30 and in the Veldt by 6.30.... We are now passing over the typical Veldt—everything fairly brown—with occasional kopjes in the distance. Nothing but a few cattle and a railway cottage at intervals....

There followed an interesting visit to the mines at Kimberley where the party was hospitably entertained, but unfortunately there was a high wind so that "the dust flew about in showers". On 4 August Johannesburg was reached:

The city has extended very much in recent years and now occupies an area of about 30 square miles. The suburbs are built on small hills or kopjes and many of the houses are very fine and have large gardens.... We went to the New Modder mine—one of the latest, best equipped and most paying propositions on the Reef. The manager, Mr Meyer, was a very fine man and he showed us the whole plan of the mine and its workings while we saw all the surface machinery for grinding the rock and separating the gold....

We got to Ladysmith after a slow and cold journey, about 11 a.m. and took a motor in the afternoon to see the sites of the battles in the Boer War; Waggon Hill, Spion Kop, etc. The hotel was comfortable with Indian waiters and Zulu houseboys. Started next morning about 11 for the Park, packed in the back of a Cadillac car, and made fast speed over rough roads and reached the National Park of Natal about 3. It was a good exhibition of fast but good driving. With Ralph and me behind, the springs often came together with a real bump. We must have averaged over 25 miles per hour. Found our rooms very comfortable. I have a rondavel to myself and Ralph and Eileen another close by. Meals are simple but good, and there are about 60 or 70 visitors who are able to dine at one time. The sanitary arrangements are of the old primitive type but well attended to, otherwise everything while quite simple is clean and comfortable. There is a fine view of the mountains on two sides and the higher points are covered with snow.

We have seen baboons who abused us from the hill tops, a wild buck and a climbing rabbit. Although midwinter many of the flowers are out, but the grass is still brown. Roses are blooming in the garden and the mimosa is in full bloom. Altogether this is a pleasant place for a holiday and the sun shines all day but the temperature falls rapidly at sunset and I find it difficult to keep warm between 5 and 6.30 when we have dinner....

Yesterday we made the main expedition to the Sentinel—near Mont au Source—three of us and a young fellow Debenham who came to the hotel with us and is taking a six months' tour round S. Africa. We took four Basuto ponies and two guides, one Charlie a Zulu, who is the best man here, and another a Basuto. Left at 8.30 a.m. on a fine day and rode wherever possible as there had been a fall of snow a week before, and it was believed the path near the cave might be impassable for horses and not easy for footsloggers. We first mounted several thousand feet by zigzags through a gorge, and after two hours reached the divide which separates Natal from the Free State—a fine prospect over a wide mountainous area. After hard slogging we reached the Gap, our lunch place and had plenty of tea. On the route Charlie had descended 700 ft. to a stream to fill the can with water and carried it up 1500 ft. to our lunch place. Previous to this we had found snow on our track away from the sun and it was expected to be much worse the last six miles on the journey. I parted with them at lunch and returned quietly with the Basuto and pony to the hotel arriving at 4.15 very tired as I had a bad cold coming on. The rest of the party returned at 8 very happy but stiff. They had reached the cave about 2.30 but found there was too much snow to climb the Sentinel, so they had tea and started home. They ate a prodigious dinner and went off early to bed and have rested today around the hotel. They had ridden and walked in all more than 25 miles on mountain paths and had done it in good time. They had a magnificent sunset and views over the whole range.

The next letter gives an account of the expedition to the Game Reserve in Zululand:

27 *Aug.* 1929. Eshowe is about 2000 feet high, a pretty and healthy spot, untroubled by malaria which affects the lower part of Zululand in summer. From Eshowe we started next morning for the Game Reserve between the rivers, the Black and White Umfalosi, about 20 miles. After 60 miles on the main road through fine country, we took a track for 30 miles to the Reserve. We had chains on the cars, for otherwise we could not have got through the worst parts, and we bumped our way slowly through tracks over the fields. We lost our way several times but ultimately got on the right road to the Tsetse Fly Camp near the banks of the Black Umfalosi. About 20 miles from the camp we passed through the 'Fly' belt where no cattle or horses can live but only man, goats and baboons! Over this region the country is undulating with low hills and covered with such deep grass waist high that it is difficult to see the track for any distance. The vegetation and flora is typical of Natal—varieties of cacti, euphorbia, prickly pear, thorn trees and the Kaffirbaum with deep red flowers and no leaves.

Arrangements had been made with the Administrator of Zululand to visit the station for research into tsetse fly under the scientific direction of Mr Harris who expected us, and had made arrangements for us to pass the night at his camp, which consisted of two buildings with cookhouse and galvanised iron roofs enclosed with wire netting to exclude the mosquito. Actually during our visit in winter there is no danger from malaria, and none of us saw or heard a single mosquito during our trip, and there was no need to take the quinine which I carried with me. I was given a bed in the diningroom of Harris's hut, while Ralph and Eileen and Mr Ellis had beds in a neighbouring hut, while

the 'girls' arranged for their own accommodation—two sleeping in
the car of which the front seat could be let down, and the other two
in a tent provided by Harris. After a walk and a chat, we had dinner
in relays of six. There were eight of us—three scientific staff—a visitor
Mr Jack—who is responsible for tsetse investigation in Rhodesia. The
menu consisted of stewed guinea fowl—provided by Harris, and fruit
tarts provided by the Ellises and plenty of tea. Harris had a Zulu cook
who was not very efficient so he acted as supervisor and distributor—
an amusing meal which reminded me of New Zealand days. We all
turned in about 8.30 and I slept well but got up at 6 next morning,
being provided with a cup of tea by Harris to fortify me till breakfast
at 7—porridge and kippers! provided by the establishment. They were
very anxious we should see a 'white' rhinoceros, a distinct species only
found in this game reserve and believed to be about thirty survivors
in all. These rare animals are easier to approach than the black rhino,
who is also much more pugnacious. There is little danger of attack from
the white rhino—who is not white, but a dirty grey, as I personally
verified. These animals were mostly on the other side of the reserve
about 12 miles away, so we arranged to go to another camp on the
White Umfalosi—12 miles away in a direct line but 20 miles by the
road by motor. The majority of the party including Ralph and Eileen
walked across with a native escort and Jack who carried a rifle, and saw
a great variety of game including hartebeest, zebras, warthog etc.
They had to wade across fords of the two branches of the Umfalosi
infested by crocodiles but did not see any. I went with Harris in his
motorcar—strongly springed, and he drove fast over rough tracks, so
we bumped our way along in unforgettable style. His cook was in
the back of the car and must have hit the hood many times for he
bumped about like a pea in a bucket.

We have arrived at the Marumba camp about 4 to find the walking
party arrived. This camp was not occupied and consisted of a couple
of small huts, and our party of twelve had to find accommodation.
The lorry has brought a couple of large ground sheets so a rope was
stretched between two trees and the canvas spread over and tied down.
With dry grass brought by the natives accommodation for five was
obtained. The girls camped in their motorcar and tent. Ralph and
Eileen in one of the huts, Ralph on grass and Eileen on a stretcher, while
I had a stretcher in the other hut. Harris slept in his motorcar. We slept

reasonably well but it is cold about 3 and Eileen and I found it cold underneath. We had a supper at 7 by the headlights of our motor—a stew of chicken, potato, peas, bacon and sausage with jam tarts to follow. We all turned in at 8.30. I got up to tea and a shave at 6 to find preparations for breakfast by the fire in order—bacon and eggs and sausage. It was a weird evening in this bare mud compound by firelight with Harris and his staff interviewing the native staff who control the game on the Reserve. About ten trackers went out before daylight to mark down a rhino, and they were to signal by smoke when contact had been made. We drove down by motor over a rough road to the river, and about 10 a.m. we saw smoke and found that a rhino and her calf had been located about a mile and a half away. We then removed our boots, crossed the river, and followed the trackers in single file without making any noise. They found traces of the rhino, and found her lying down among the trees. We managed to get within 20 yards without disturbing her, and saw what appeared to be a large greyish stone with a young one nearby. She rose up and faced us and looked nearly as big as an elephant and then the click of a camera startled her and she dashed off at high speed with the crash of the trees to mark her passage. Altogether everyone had a good view for a minute or so and it was a highly interesting experience.

Everyone was very excited and we returned after seeing zebra, hartebeests, and warthogs on our way. We returned to the camp loaded up and off back to Eshowe where we spent the night.

...I had a long talk with Harris on the tsetse fly problem and he told us what he was doing to reduce the number of the flies and study their habits. The main idea is to reduce the game in the reserves and so reduce the number of flies and their spread. The native hunters are systematically reducing the congestion in numbers and preventing the game passing out of the reserves. I saw lots of flies which seemed to like my stockings but had no serious bite. They say their direct bite is rather alarming but produces no more after effect than the bite from a mosquito. The control or reduction of the tsetse fly is the chief African problem, as no cattle or horses can live in their radius of action.

In a letter to her mother Eileen wrote:

You can't think how much Daddy enjoyed himself, he was the life and soul of the party.... The flowering trees in the country would fascinate you. All over Zululand, where there wasn't sugar, all the creases in the ground where there was moisture were full of bright green palms and Kaffir boom trees—a tree with a brilliant scarlet flower and no leaves. Most tropical trees and flowers seem to be red. Aloes with flowers growing like great candelabras are out everywhere. When we get to the Cape all the heather will be fully out....The *Euripides* is going to be three days late in leaving Capetown. We sail at dawn on September 15th, so we won't get to England till about October 2nd. It will be lovely to get home.

Rutherford about this time was faced with a curious puzzle. Radium C′ shoots out an alpha particle that will go three inches in air, uranium ejects one that will go but one inch. Why not then fire a radium C′ alpha particle right into the heart of a uranium nucleus? It was tried and it did not work! Some lame explanations were given and failed. Chadwick writes:

The difficulty was too deep-seated. Classical mechanics failed to describe these events. The explanation was given by Gurney and Condon, and a year or so later, by Gamow, who pointed out that wave-mechanics allowed the alpha-particle to escape from a nucleus through the energy barrier which retained it, and did not force it, as did the classical mechanics, to climb over the top.

Rutherford's old research student, Dr R. W. Boyle, had done valuable work during the War with the Anti-submarine Division in developing the piezo-electric submarine detector initiated by Prof. Langevin of Paris. Boyle had then returned to Alberta University as Professor of Physics and Dean of the Faculty of Applied Science, and he there continued research work on short sound-waves in water. His career suffered somewhat from the fact that for a time the Admiralty deemed it necessary to keep all his work secret and confidential, although as a matter of fact the main method had been described and published in Paris.

The Canadian Government built and equipped a large National Physical Laboratory for scientific and industrial research at Ottawa.

Dr H. M. Tory was appointed Director and Boyle joined him as head of the department of physics.

RUTHERFORD TO BOYLE

10 *Jan.* 1930. I was very interested to hear of your work at Ottawa and I saw only this week an account in *Nature* of the new buildings and staff. It must be very interesting work, but at the same time rather disturbing, as it prevents one getting on with a definite piece of investigation. The buildings seem to be planned on a fine scale and it is very interesting to see that Canada is developing so rapidly in these scientific directions.... As to the Cavendish Laboratory, the work goes on much as usual and we have a very good lot of men. Through Kapitza, we are erecting a liquefaction plant for hydrogen which has many novel features designed by him which we hope will make it very efficient. We are hoping to make the preliminary trials this term. Liquid hydrogen is primarily intended for Kapitza's own work on magnetic effects at low temperature. We shall not try and provide it at the moment for general research purposes. We are also developing a high tension laboratory in order to produce high speed electrons and positive rays. We have made a good deal of progress and hope soon to produce experimental results. We have spent a good deal of time in the Laboratory in developing methods of counting hydrogen particles in the presence of strong gamma rays. It is quite easy to count hydrogen particles by a number of methods, when gamma rays are absent, but it is a very difficult and unsolved problem to do the same in the presence of strong gamma rays from radium C and thorium C. We can now regularly count alpha particles and electrons automatically when required.

You will have seen that we gave Geiger the Hughes Medal of the Royal Society. He came over to receive it and spent a day in Cambridge. He is doing fine work with his new electron counter especially on the very penetrating rays in the atmosphere. He is now professor at Tübingen where he is very pleased with the laboratory.

On 29 Jan. 1930, the Council of the Institute of Electrical Engineers awarded Rutherford the Faraday Medal founded to commemorate the Fiftieth Anniversary of the Society of Telegraph Engineers (now the I.E.E.). The medal is awarded for conspicuous service rendered to the advancement of electrical science, without restrictions. It has been awarded to Heaviside, Parsons, de Ferranti, J. J. Thomson, R. E. Crompton, Elihu Thomson, A. Fleming, G. Semenza.

E. A. Milne wrote to Rutherford an interesting letter (1 Aug. 1930) discussing his friendly rivalry with Eddington in his efforts to arrive at a satisfactory mechanics of the universe. He concluded: "Your uranium argument combined with my dense hot core suggest that the sun was thoroughly disrupted to the centre when the earth was born."

On the death of Sir William McCormick, Rutherford was appointed by the Lord President of the Council to be Chairman of the Advisory Council of the Department of Scientific and Industrial Research. He put much thought, energy and enthusiasm into this work, for he regarded the prosperity of this country, of the empire, and of the world as dependent on the proper use and advancement of science and scientific method.

This position brought him into close touch with many industries and he was in frequent demand for speeches and ceremonies. Thus on 28 Feb. 1930, Rutherford opened the new High Tension Laboratory at the Metropolitan-Vickers Electrical Co. Ltd., Trafford Park, Manchester. The following are extracts from his address:

Since the experiments of Benjamin Franklin on drawing electricity from the clouds and his proof that lightning is an electrical manifestation, scientific men have always watched with interest, even with fascination, the beautiful and varied effects produced by high potential discharges in air and the marked changes which appear when the pressure of the air is lowered. In the Laboratory we have long had to content ourselves with the discharges produced by frictional machines and induction coils and the oscillatory discharges produced by a Tesla coil. With the advent of alternating currents high voltage transformers became more common. I remember seeing, in 1904, at the St Louis Exposition, an experiment in which a transformer giving about half a million volts was attached to an insulated bare wire stretched for several hundred feet. The experiment was carried out in darkness and I recall the striking corona discharge along the whole length of the wire, dissipating in the air the full power of the transformer. However, up to a few years ago potentials higher than 200,000 volts were seldom available in the Laboratory and it is only by much improved technique in producing high vacua that we have been able to maintain such a potential on the terminals of a highly exhausted tube....

Although the discharge in gases at different pressures has been the subject of close investigation for more than fifty years, we are still far from obtaining a complete understanding of its mechanism. In fact we are just beginning to recognise that many of the explanations, hitherto given, only contain a small part of the truth....

There is one aspect of this high potential problem that specially appeals to me. I refer to the application of very high potentials to highly exhausted tubes in order to obtain a copious supply of swiftly moving electrons and charged atoms. As far as I am aware, the highest potential so far applied to a single discharge tube is about one million volts and this is a difficult technical problem. In these respects, Nature can far surpass our puny experiments in the Laboratory. In order to obtain alpha particles as swift as those spontaneously emitted by radium, we should require vacuum tubes that will withstand a potential of more than 5 million volts while at least 3 million volts would be necessary to obtain as swift electrons and penetrating gamma rays as those emitted by radium. There is some evidence also that the very swift electrons and highly penetrating radiation that are found in our atmosphere would require for their production potentials of the order of a lightning discharge, namely about 1000 million volts....

What we require is an apparatus to give us a potential of the order of 10 million volts which can be safely accommodated in a reasonably sized room and operated by a few kilowatts of power. We require too an exhausted tube capable of withstanding this voltage. I recommend this interesting problem to the attention of my technical friends. I see no reason why such a requirement cannot be made practicable by the use of oil or air under high pressure, but these are problems for the future.

A small but distinguished company were then shown a twelve-foot discharge due to a million volts, for one of the objects of the new laboratory was the production and measurement of such high voltages.

Rutherford, within five years, saw his goal of safe, high-voltage apparatus reached and applied in his own laboratory to the disintegration of atoms.

The skill of Fleming and de Forest in their invention and perfection of electron tubes or valves brought wireless telephony into the field of everyday experience. In 1919 at a meeting of the Royal Society of Canada a woman singing in Montreal was just audible at Ottawa, 180 miles away. There was justifiable excitement when Rutherford, speaking in England in the year 1930, was easily heard by a large audience of the Royal Society of Canada in the Moyse Theatre of McGill University.

Speaking as President of the Royal Society, he said:

Mr President and Fellows of the Royal Society of Canada and Colleagues of McGill University.

Your President has given me the difficult task of making my voice audible to you over 3000 miles of land and sea—a truly formidable experiment even with the powerful help of modern Science. About three years ago I took part in the first radio conversation to America, between the Universities of Cambridge and Harvard. A friend of mine, who had perhaps suffered from the vigour of my voice, when he was told that I was heard clearly across the Atlantic, remarked 'why use radio?'...

It is more than 20 years since I was a youthful Professor in McGill University and an active Fellow of your Society and I am glad that you still regard me as one of your Fellows. While I have followed with great interest and sympathy the remarkable expansion of the Universities of Canada and the rapid development of science in Canada, I have always retained a specially warm feeling of affection for my old University of McGill and a lively remembrance of the happy days of work and play in the Macdonald Physics Building.

This address, quoted only in part, was to have been spoken at Cambridge at 1 a.m., Greenwich time, on 22 May, and to have been received on the previous day at 9 p.m. (summer-time) on 21 May, taking a fraction of a second to accomplish the journey. But a telegram was sent from Montreal to Rutherford, "Sunspot causing night static stop Please address Canadian Royal Society Tuesday afternoon 20th May 5.45 daylight saving 4.45 Greenwich Cable quick reply, Eve, Montreal."

And so it was actually sent and heard at 12.45 p.m. local summer-time, at Montreal. After his formal address there was a two-way conversation, audible to the whole gathering, between Rutherford and several of his old friends in Canada: Sir Arthur Currie, Principal of McGill University, Sir Robert Falconer, Principal of the University of Toronto, Dr H. M. Tory, then Director of the National Research Council of Canada, and A. S. Eve, the President of the Royal Society of Canada, who pointed out that ten years earlier they could just get wireless speech and song from Montreal to Ottawa, and that day (1930) quite easily from Montreal to Cambridge. "In ten years we hope to have not only your voice coming over, but also your picture on the screen."

Rutherford did not attend the Sixth Solvay Conference at Brussels, but he was not forgotten by his learned friends, who sent him their signatures on a postcard.

On 13 Nov. 1930, there was a dinner at the Guildhall of the Association of Scientific and Technical Institutions with Rutherford in the Chair, at which H.R.H. the Prince of Wales spoke. It was at this dinner that Rutherford told his lion and lamb story:

A farmer, in the far West of America, read in his local paper that in the Zoological Gardens of a neighbouring city there was on view, every day, the spectacle of a lion and a lamb lying down peacefully together. He was so curious about this interesting phenomenon, that he took the trouble to make a railway journey to the city and visit the gardens; and there, to his surprise, were the lion and the lamb, lying down peacefully together, almost in each other's arms. Curious and excited, he went to see the Keeper and enquired as to the genuineness of the show. "Yes," said the Keeper, "it is quite genuine; it has been going on for three months", and then, noticing the excitement and interest of his listener, he added, "but I don't mind telling you that there have been a good many replacements."

In 1930, Rutherford spoke very clearly and definitely on the "Transmutation of Matter" to the Royal Institution. When protons are packed

R. H. Fowler   F. W. Aston   Rutherford   G. I. Taylor

South Africa, 1929

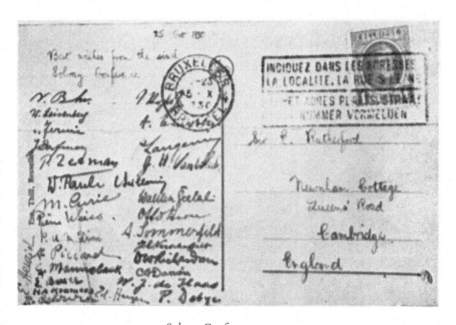

Solvay Conference, 1930

tightly in a nucleus there is a change of mass as compared with free protons. The mass-change can be expressed as an energy-change. "Most atomic masses are not integers and the fractional excess of the atomic weight over the whole number represented the mass-defect and energy-loss incurred in binding the nuclear constituents together."

It was now nineteen years since Rutherford had written his great book on *Radioactive Substances* and the subject had branched and grown like a healthy tree. It was desirable to obtain experienced help from younger men and in 1930 appeared *Radiations from Radioactive Substances* by Sir Ernest Rutherford, James Chadwick and C. D. Ellis (Cambridge University Press). It dealt, as the title suggests, with radiations and in an encyclopaedic manner. This is a book not for the armchair, but to be kept constantly by the physicist's side in his study or laboratory.

On 14 December 1930 Rutherford's daughter, Eileen Fowler, gave birth to a fourth child and all was going well when a clot of blood caused her sudden death on 23 December. Her mother was then on a visit to New Zealand. The loss of his only child, whom he loved and admired, aged Rutherford for a time; he looked older and he stooped more. He continued his life and work with a manful purpose, and one of the delights of his life was his group of four grandchildren. His face always lit up when he spoke of them.

CHAPTER XIII

LORD RUTHERFORD OF NELSON

In 1931, the New Year's Honours gave Rutherford to be created a Baron. He wisely decided not to change his name and became "Baron Rutherford of Nelson", thus linking himself with the town near which he was born and at which he went to school. It was characteristic of the man that the first thing that he did was to send a cable to his mother in New Zealand. "Now Lord Rutherford, more your honour than mine, Ernest."

In the description of his arms, as given by the College of Arms, it is easy to discern Rutherford's love of his old home in the Kiwi and the Maori; and the motto *Primordia quaerere rerum* fully expresses his search for the elements of matter.

OFFICIAL BLAZON OF THE ARMORIAL BEARINGS.

ARMS. Per saltire arched Gules and Or two inescutcheons voided of the first in fesse within each a Martlet Sable.

CREST. On a Wreath of the Colours upon a rock a Kiwi proper.

SUPPORTERS. On the dexter side a figure representing Hermes Trismegistus and on the sinister side a Maori holding in the exterior hand a Club all proper.

A Committee of the Royal Society, with Lord Crawford and Balcarres as Chairman, arranged for a portrait (see frontispiece[1]) of Lord Rutherford, retiring President, to be painted by Mr Oswald Birley. This was

[1] Published with the kind permission of the artist and the Royal Society.

done in the following year and a vigorous and pleasing three-quarter-length portrait of Rutherford has a prominent place among the portraits in the Royal Society rooms.

On 20 May 1931, Lord Rutherford spoke in the House of Lords on Oil from Coal, a vexed and difficult question.

There is the low temperature method, which might yield 7 per cent of oil, and the remaining material is a friable coke, which people will not burn so long as they can obtain a cheerful blaze from good coal in their home hearths.

There is the hydrogenation method which means high temperature and high pressure and a satisfactory yield, so that our home coals could thus be converted into oils.

Alas! the method is so expensive that so long as natural oil flows in abundance out of the ground, it is far cheaper to bring it to our shores by inexpensive sea transport. One plant takes long to build and is easily bombed; it is as cheap to build 32 tankers! Rutherford said: "Carbonisation would give us a useful smokeless fuel and some oil, but advance, apart from the improvement of technical methods, depended not on science, but how far the nation was prepared to pay for a smokeless atmosphere.

"The development of the hydrogenation method depended also on how far the nation was prepared to pay for its independence from oil supplies from other countries...."

This was Rutherford's maiden speech and it made a good impression. At that time we were importing, and had to pay for, £40,000,000 of oil every year, so that he was well justified in adding: "it was in the national interest that researches in the general utilisation of coal for the production of oil should be vigorously prosecuted."

The next day his old friend G. Elliot Smith wrote:

21 *May* 1931. Just a note of congratulations on the great success of your début yesterday. Your speech was splendid and your voice more clear and resonant than that of any other speaker. It was a pleasure to hear Lord Parmoor's appreciative remarks.

One result of this speech is clear. Rutherford was now not only the leading man in Science, but he was also a public man. For the rest of his life he could not shirk this double duty. He endeavoured to avoid entanglement in political strife and was regarded by all parties as a man to be trusted and respected.

To return to Rutherford's scientific activities, he selected for one of his Royal Institution lectures the subject of "Helium and its Properties". The use of this gas for filling balloons and airships, to safeguard them from fire and explosion, had brought it into prominent notice. Facsimiles are given of the notes that Rutherford used in the early part of his lectures and the following account of it is taken from *Nature* (25 July 1931):

The history of the discovery of helium presents some features of unusual dramatic interest. During an eclipse in 1868, Jansen and Lockyer noticed that the visual spectrum of the sun's chromosphere showed a bright yellow line of unknown origin. Later it was found that this line, and others that accompanied it, appeared not only in the sun, but also in many of the stars. Lockyer suggested that these lines were due to an undiscovered element, to which he gave the name helium.

Shortly after the discovery of argon, Sir Henry Miers sent a letter (1 Feb. 1895) to Ramsay pointing out that the American analyst Hillebrande had observed that a considerable quantity of gas, supposed to be nitrogen, was liberated by solution of certain minerals containing uranium. Miers suggested that the gas might prove to be argon and not nitrogen. Following his suggestions, Ramsay purchased about a gram of the mineral cleveite from a dealer for three shillings and sixpence and proceeded to purify the gases evolved and to examine the spectra. A number of new lines were observed and a spectrum tube containing the new gas, temporarily called crypton by Ramsay, was sent to Sir William Crookes for a detailed study of its spectrum. Crookes reported tersely, "Crypton is Helium; come and see it." Less than two months had elapsed, from the receipt of Miers' letter, to the announcement in the Paris Academy of Sciences, on March 26, 1895, of the discovery of helium on the earth—a discovery of profound significance to the development of physics. It was soon shown that

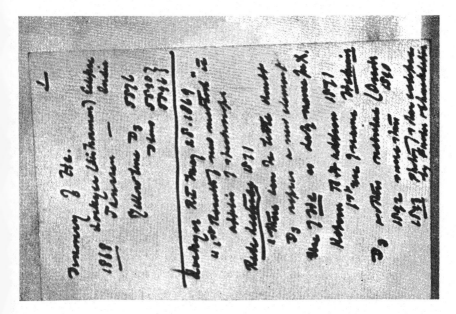

Lecture notes on helium

helium is a monatomic gas of density about twice that of hydrogen and of atomic weight 4. We now know that it is the first of that remarkable group of inert monatomic gases, namely helium, neon, argon, krypton, xenon, and the radioactive emanations which have played such an important part in helping to fix the arrangements of the electrons in the outer atom.

In 1903, Ramsay and Soddy found that helium was produced by the transformation of radium, and, as a result of a series of researches, Rutherford showed that the alpha particles which are ejected with great velocity from radioactive atoms are identical with helium nuclei. It is probable that the greater part, if not all, of the helium found in the earth and in the natural gases escaping from the earth owes its origin to the alpha particles expelled from the radioactive elements during their transformation in the earth's crust....

Helium is present in our atmosphere in small quantity, about one part in 185,000 by volume. In the early days, most of the helium used for experimental purposes was obtained by heating radioactive minerals and particularly the mineral thorianite from Ceylon. It was also found that helium is often present in considerable quantity in the gases escaping from hot springs and the natural gases from the earth's crust.

In 1914, Sir Richard Threlfall suggested to the Board of Inventions of the Admiralty that, on account of its lightness and non-inflammability, helium might prove of great service for balloons and airships. Prof. J. C. McLennan was asked to initiate experiments to see whether helium could be separated in quantity from the natural gases escaping from the earth in certain districts of Canada which were known to contain about 1 per cent of helium by volume. Arrangements were made on a semi-commercial scale to purify the helium by liquefying the methane and other gases present. The impure helium was concentrated in the non-liquefying portion. In this way, many thousands of cubic feet of helium were prepared and transported in cylinders at high pressure. About the same time, the Bureau of Mines of the U.S.A. began similar experiments on a large scale, using the natural gases of Texas, which are rich in helium. Large quantities of helium were separated by liquefaction methods, and the cost of helium was found to be sufficiently low to use it in airships in place of hydrogen. Apart from the cost of transport, the expense of separation of helium decreases with the concentration of the helium in the natural gases.

The commercial prospects of the use of helium in airships and other purposes have led to a search for rich concentrations of helium.

In a later lecture of the course, Rutherford announced that the wearisome watching and counting the effects of alphas or protons striking a fluorescent screen was now happily a thing of the past. Today every particle moves an indicator forward one point, and the particles can therefore be counted in the same kind of way that turnstiles count the number of visitors to the Zoo.

Some of the gas-filled valves (Langmuir's thyratrons) thus employed are sensitive to sound and visitors to the laboratory were requested to speak low, not an easy thing for Rutherford to do, so that there is a little humour in the photograph (Plate XIII).

Rutherford wrote an article to *The Times* on "Faraday as an Experimenter" as a prelude to the forthcoming centenary:

In the history of the physical sciences in this country, three great names at once occur to mind, Newton, Faraday and Maxwell. Of these, the first is one of the great mathematicians of all time, the latter an accomplished mathematician, and both were endowed with unusual powers of physical insight. The other, Michael Faraday, had no mathematical training and not a line of mathematics appears in his papers. Yet in physical and philosophical insight he possessed the essential qualities of the theoretical investigator without his technique, and was also an experimenter of unrivalled powers. The story of the life and work of Faraday has always exerted a strong appeal to the lay and scientific mind alike, not only as the record of a man of fine character and high ideals who rose from humble beginnings to the position of the greatest discoverer and natural philosopher of his age, but also as an example of the startling consequences that may result from the patient study of the workings of nature in the quiet of the laboratory, remote from the active affairs of life and without thought of material ends. Few if any men have accomplished so much for the advancement of knowledge or for the ultimate progress of the human race.

In September of this year, there will be a celebration of the centenary of the discovery by Faraday of electromagnetic induction. Although this is one of the most important and striking of his discoveries, which

"Talk softly, please"

has ultimately led to the development of the great industry of to-day, yet it represents only one of a great number of brilliant experimental investigations which added so much to our knowledge of Electricity and Magnetism and Chemistry....Faraday has often been called the prince of experimenters, and no one could deny his remarkable powers of penetration to the root of a subject by a number of well designed and well executed experiments....It is a pleasure to-day to read his *Collected Researches*, published in four volumes. There is everywhere such a freshness of thought and outlook, combined with a clarity of exposition, that it is difficult to recall that most of his work was written nearly a century ago at a time when our knowledge of Electricity and Magnetism was still in its infancy.

Faraday was more than a great experimenter. He was a great Natural Philosopher. He sought by his experiments not only to accumulate new facts of nature and to show their relations, but also to understand the physical processes which underlie the phenomena under investigation.

An opponent of the mathematical school of action at a distance, he continuously advocated the importance of the surrounding medium as the seat of the electrical and magnetic forces. This was visualised by him by the conception of curving lines of electric and magnetic force which are universally employed to-day.

The importance of these physical ideas was fully recognised by Maxwell who had the genius to formulate and generalise them in mathematical form and gave us for the first time a satisfactory theory of electromagnetic action.

The rest of this interesting article dealt with Faraday's experimental research and his written records. The reader cannot fail to be impressed by the fact that much that is said about Faraday and his work applies with no less force to Rutherford himself.

At his Centenary in Sept. 1931 there was a universal chorus of praise and admiration for this great man, Michael Faraday. It was then a hundred years since he had made the first transformer, winding a ring of iron with two coils, one connected to a galvanometer, the other to a battery. Change of current in the battery circuit created a transient current in the galvanometer circuit. Astounding!—but we have become familiar with it. Not so Einstein, who wrote: "I believe that future generations yet have as much to learn, as has been learnt in the past, from Faraday's conception of the electromagnetic field."

At the Meeting, Rutherford spoke of Faraday's powers of co-ordinating various branches of Science:

When we consider the life work of Faraday, it is clear that his researches were guided and inspired by the strong belief that the various forces of nature were inter-related and dependent on one another. It is not too much to say that this philosophic connection gave the impulse and driving power in most of his researches and is the key to his extraordinary success in adding to knowledge.

The more we study the work of Faraday with the perspective of time, the more we are impressed by his unrivalled genius as an experimenter and natural philosopher. When we consider the magnitude and extent of his discoveries and their influence on the progress of science and industry, there is no honour too great to pay to the memory of Michael Faraday—one of the greatest scientific discoverers of all time.

The celebration of the Centenary of the birth of James Clerk Maxwell was held at Cambridge on 30 Sept. 1931, soon after the meeting of the British Association in London and the Faraday Commemoration at the Royal Institution.

Rutherford wrote an account of Maxwell's life and work for *The Times*, and a few extracts may be quoted:

Maxwell not only adopted and developed the ideas put forward by Faraday, but had the genius to give them their true mathematical form. In his famous Electromagnetic Theory of Light, he was the first to foresee that electrical waves could be produced, and he identified their mode of propagation with the known properties of light.

It was not until ten years after his death that Hertz in a wonderful series of experiments showed that the existence of electrical waves could be demonstrated in the Laboratory, and determined their properties. At the beginning of his experiments, he had been unaware of Maxwell's theory, but soon found that all his results were in complete accord with the mathematical predictions of the theory....

Maxwell's main ideas and conclusions have stood the test of time, and illustrate his remarkable powers of scientific judgment and of physical insight into the processes of nature. When we recall the practical applications of electric waves as a means of world communication, and the still recent triumphs of broadcasting, so much of which we owe to the early experiments of Lodge and Marconi [he might have added 'and myself'], we should fittingly honour Maxwell as the great pioneer of radio-communication; for he had not only the genius to foresee that electric waves must be produced, but had given the complete theory of their generation and propagation long before their existence had been suspected by science.

To commemorate Rutherford's sixtieth birthday (30 Aug. 1931), Otto Hahn and Lise Meitner wrote an appreciation to *Die Naturwissenschaften* (1931, 35, p. 729):

He stands at the height of his creative power, highly esteemed by his colleagues, admired and loved by the great host of his pupils who work directly or indirectly in the sphere of radioactivity, and on this day will be paid to him in the highest measure the tribute of the recognition of the whole scientific world. We too, who have experienced the extensive growth and maturity of his great services, would like to take this opportunity of expressing our grateful respect.

What a wealth of brilliant discoveries and original creation does the career of Rutherford show! To present them in detail the whole development of physics and modern chemistry would have to be described....Professor Rutherford, to whom the highest scientific distinctions and honours have been assigned, has himself erected the fairest memorial in the minds and gratitude of his followers.

RUTHERFORD TO HAHN

2 *Sept.* 1931. On my return from a holiday by the seaside in Dorset with the grandchildren, I got your telegram of congratulations on my 60th birthday. It is pleasant to be remembered in this way by one's friends but I am glad to say I do not feel as ancient as my years suggest. In any case, we are told that time is an illusion! I have to report that I am still very fit and well and able at any rate to drive the other fellow to work!

You will have seen our papers in the P.R.S. on alpha rays and gamma rays. I am hopeful that the point of view of the paper by Ellis and myself on the origin of the gamma rays will prove helpful in throwing light on that subject.

We are all very well and the grandchildren are strong and growing apace. I thank you and Miss Meitner for your good wishes and hope you are all flourishing.

On 30 September, Lord Rutherford and Prof. Samuel Alexander were the chief guests at the Manchester University dinner held at Frascati's, with the Chancellor, Lord Crawford, presiding. In his speech Rutherford said that at Cambridge one might sit next to a man for forty years without knowing whether he had a wife, mother, or sister; whereas, before he had been at Manchester a week, he knew his colleagues and their wives and was calling their children by their Christian names. "I owe a great debt to Manchester University", he said, "for the opportunities it gave me for carrying on my studies. I do not know whether that University is really aware that during a few years from 1911 onwards the whole foundation of the modern physical movement came from the physical department of Manchester University."

In 1931 at a dinner of '1851 Scholars' held in London, when Rutherford presided, he said:

Only by chance, am I present. As a student of Canterbury College I was candidate for an 1851 Scholarship; there was another New Zealand candidate, MacLaurin. Both of us were strongly recommended by the examiners, and eventually it was decided that the candidate with the published thesis should have the scholarship. MacLaurin had published his in a little book; I had not. But MacLaurin had married a wife and therefore could not come, and that is why I am present as your Chairman.

At this time Rutherford was elected a Commissioner of the 1851 Exhibition and three years later he became a member of the Board of Management.

Years before, the Commissioners had promised three large contributions towards a Science Museum to be erected in successive stages. When the second payment was due it was realised that the Scholarships would have to be seriously curtailed and a case was presented to the Government asking for release from payment; this appeal fell on stony ground, and the case seemed hopeless, but at the next meeting of the Board, Rutherford, feeling strongly, again attempted to protest, but all he could blurt out was: "If it had not been for these scholarships, I would not have been." That did the trick. A new approach was made, the Chancellor of the Exchequer relented, and the Scholarships are still as numerous as they were before.

At an earlier date (1921) he reported to Lord Esher that he was in favour of the award of five Senior Scholarships (£400 for two years) to be given only to men of exceptional promise and not awarded at all unless very definite evidence were forthcoming.

In the last year of his life he was supporting the award of an 1851 Exhibition Scholarship every year to a suitable student from India, thus securing for that country a status similar to that of the Dominions. When he went to McGill he assured the Secretary that he would take a great interest in future Exhibition Scholars, and indeed he did so at Montreal, Manchester and Cambridge.

Rutherford never threw down the ladders by which he had climbed, but always acknowledged with frankness and enthusiasm the benefits he had derived from earlier days, whether it was at home, school or college—Cambridge, McGill or Manchester. For example, in 1932, he wrote an article for the *McGill News* and began with this kindly foreword:

I gladly accede to the request of the Editor of the *McGill News* to give a brief statement of recent work in Cambridge on the Transmutation of the Elements for two reasons, one personal and the other historical. With regard to the first, I have always retained a lively interest in the progress of McGill University since my happy days there long ago, and this interest has been maintained by the migration of many research students from McGill to work with me in Manchester and Cambridge. Now that nearly thirty years have passed since the first direct evidence was obtained of the transmutation of matter, I think we can view with some perspective the gradual development of our ideas on this subject. I should like to take this opportunity to emphasise that the credit of the first definite proof of atomic transformation belongs to McGill University. It was in the Macdonald Physics Building in the years 1902 to 1904 that Soddy and I accumulated the experimental evidence that the radioactive elements were undergoing *spontaneous* transformation. The disintegration theory, advanced in explanation of the observations, has stood the test of time and formed the basis of all subsequent developments.

The next stage in advance (1919) was a proof that certain elements could be artificially transmuted by bombardment with the alpha-rays of radium and it should be noted that it was in McGill, in 1903, that the true nature of this radiation, which has played such a great part in the development of physics, was first disclosed. There is a saying that 'it is the first step that counts', and it is clear that to McGill belongs whatever credit is due for the early ideas and experiments, which opened up the way into the unknown which all subsequent investigations have followed.

From the year 1930 onwards, Rutherford was in public demand to an ever-increasing extent. At the Royal Society on 20 April 1932, he gave a clear and sober account of the great results obtained by Cockcroft and Walton in the Cavendish Laboratory, and at the same meeting Chadwick described the steps leading to his discovery of the neutron. The proceedings were sedate compared with the lurid accounts in the press. It was explained also that lithium has mass 7 and a proton mass 1, so when a proton enters the lithium there is a nucleus of mass 8. This proves to be unstable and divides itself with violence into two nuclei, each of mass 4, our old friend the helium nucleus, alias alpha particle.

Thus
$$_3^7\text{Li} + {}_1^1\text{H} \rightarrow {}_4^8\text{He} \rightarrow 2{}_2^4\text{He}.$$
Lithium and hydrogen give 2 heliums.

On 30 April, Rutherford was called upon to respond to the toast of 'Science' proposed by the President of the Royal Academy of Arts.

Rutherford admitted that

Science has contributed little to that form of art which is so finely represented around us. After all Art flowered long before science or the scientific method had taken root.... Yet I think that a strong claim can be made that the process of scientific discovery may be regarded as a form of art. This is best seen in the theoretical aspects of Physical Science. The mathematical theorist builds up on certain assumptions and according to well understood logical rules, step by step, a stately edifice, while his imaginative power brings out clearly the hidden relation between its parts. A well constructed theory is in some respects undoubtedly an artistic production. A fine example is the famous Kinetic Theory of Maxwell, the centenary of whose birth we celebrated last year. The theory of relativity by Einstein, quite apart from any question of its validity, cannot but be regarded as a magnificent work of art....

I should like to mention one case where the artist—quite unknowingly—has been able to provide valuable data for science. The clay from which the Greek potter made his beautiful vases more than 2000 years ago always contained some magnetic oxide of iron. At a certain stage of the cooling, after firing, the iron particles are very susceptible to the action of magnetic forces, and orient themselves in the direction of the earth's magnetic field. The direction of this magnetisation was fixed permanently when the vase cooled and since we know the vase must have been always in a vertical position during the firing, the scientific man can find the direction of this magnetisation and thus fix the inclination or 'dip' of the earth's magnetic field at the time and for the place where the vase was made. By this curious observation we have been able to extend our knowledge of the secular variations in the earth's magnetic field to a remote epoch more than 2000 years before the importance of such measurements was recognised....

The importance of science to the modern state is increasingly recognised in these days, and there never was a time when science was in a more flourishing state, judged either by our extensions to our knowledge of the workings of Nature, or by its applications to industrial problems. It is a time of intense intellectual activity, rich in the development of new ideas and methods. For nearly three centuries this country has produced its full quota of great pioneers both in pure and applied science and it still continues to do so. Quite recently there has been much interest taken by the cultivated public in the metaphysical aspects of science, especially those of theoretical physics. Some of our publicists have boldly claimed that the old ideas which served science so well in the past must be abandoned for an ideal world where the law of causality fails, and the principle of uncertainty, so valuable in the proper domain of atomic physics, is pushed to extremes. The great army in its march into the unknown discusses with interest, and sometimes amusement, these fine spun disputations of what is reality and what is truth. But it still goes marching on, calling out to the metaphysicians "there are more things in heaven and earth than are dreamt of in your philosophy".

He concluded:

Even in these times of financial depression there is every reason to foster the development of Science and the Arts. "Man does not live by bread alone" and we may be sure that the position of our Empire in the world of to-morrow, if not of to-day, will be estimated not so much by its material wealth as by its contribution to the world in Arts and Science.

On 7 May 1932, the 46th Article on Scientific Worthies was contributed to *Nature*, on Rutherford, by Maurice, Duc de Broglie, a French savant who had himself made contributions to modern physics.

The article was a summary of Rutherford's scientific work and it contained an important statement:

Recent theoretical views suggest that a mechanistic view of Nature cannot be pushed beyond a certain point, and that the fundamental laws can only be expressed in abstract terms, defying all attempts at an intelligible description. *The philosophy of science has always swung between these two points of view.* The work of the great physicist (Rutherford) to whom these lines are dedicated shows, however, to what brilliant discoveries the method followed by Lord Rutherford can lead.

Marconi, Nobel Laureate for Physics (1909), was awarded the Kelvin Gold Medal by the Institution of Civil Engineers, a medal which had been previously awarded to Unwin, Elihu Thomson, Parsons and

Blondel. On 3 May 1932, this medal was presented to the Marchese Marconi by Rutherford, who first referred to the successive works of the pioneers, Clerk Maxwell, Hertz, Branly, Lodge, Popoff and Jackson, and continued:

It was at this stage, in 1895, that Marconi began his experiments in Italy. He was so firmly convinced of the practical possibilities of this new method for the transmission of signals that he came to England to develop his ideas on a practical scale. His initial experiments in England were carried out with the assistance of the late Sir William Preece and the Post Office.... Marconi gradually extended the range of signalling from two miles to twenty, and finally, in face of the predictions of some of the theories of that time, showed that signals could be transmitted over 2,000 miles across the Atlantic....

While it is natural in such a great technical advance that many scientific men from all parts of the world should have played their part, yet I think that all agree that the world owes much to that great pioneer, the Marchese Marconi, for his persistent faith in the possibilities of wireless, and for his inventive ability and power to overcome the many obstacles in his path, leading to complete realisation of his early dream within less than twenty years—truly a magnificent accomplishment!

I may recall that in my youthful days in Cambridge I was myself interested in electrical waves, and devised in 1896 a simple magnetic detector for these waves. I am glad to know that Marconi was able to develop and transform the germ of this simple device into a reliable and metrical detector which for ten years or so gave useful service to mankind....

I am sure that if Lord Kelvin were here to-day he would acclaim this award as to a kindred spirit, for Lord Kelvin combined to an extraordinary degree the quality of great theoretical insight with the power to realise his ideas in a practical form. It is interesting to recall on this occasion that Kelvin, in 1853, was the first to give the complete theoretical explanation of the oscillatory discharge of the Leyden jar, later verified experimentally by Feddersen, a theory which lies at the basis of all methods of generating waves.

In the Royal Society discussion (28 April 1932) on the 'Structure of Atomic Nuclei' Rutherford gave a useful summary of the situation as it then stood. Briefly, the points that he made were: actinium-lead has an atomic weight 207; the actinium family begins with an isotope of uranium; oxygen has three isotopes 16, 17, 18; carbon two, 12, 13; beryllium two, 8, 9; boron two, 10, 11; Urey and others have found that hydrogen has an isotope of mass 2. He said that wave-mechanics had given an explanation of the Geiger-Nuttall rule "connecting the velocity of escape of an alpha particle from a radioactive substance and the length of its average life". The shorter the life, the greater the velocity, and there is an exact mathematical relation. "Unfortunately it has not so far been possible on any theory to give any detailed picture of the structure of a nucleus." Gamma rays come from a change in the alpha-ray energy levels, probably. There are at least nine different groups of long-range particles from radium C', and five distinct groups from thorium C'. There is a mystery in the fact that some beta rays have a wide range of energies, though the next radioactive products seem to be identical.

This summary, not quite complete, indicates how Rutherford brought together facts for his audience, and how he faced difficulties which were naturally increasing with the extension of knowledge. There is at least as much complexity in a heavy nucleus as in a solar system.

BOHR TO RUTHERFORD

2 *May* 1932. By your kind letter with the information about the wonderful new results arrived in your laboratory you made me a very great pleasure indeed. Progress in the field of nuclear constitution is at the moment really so rapid, that one wonders what the next post will bring, and the enthusiasm of which every line in your letter tells will surely be common to all physicists. One sees a broad new avenue opened, and it should soon be possible to predict the behaviour of any nucleus under given circumstances. When one learns that protons and lithium nuclei simply combine into alpha particles, one feels that it could not have been different although nobody has ventured to think so. Perhaps more than ever I wish in these days, that I was not so far away from you and the Cavendish Laboratory, but the more thankful I am for your kind communication and the more eager to learn about any further progress.

It is indeed a very pleasant though modest task to speculate about the present situation in atomic physics. As you know, I have in these years been especially interested in the foundation of the new mechanics and its limitations. At the end of my Faraday lecture, of which I send a copy by the same post, I introduced a few remarks about the possible failure of energy conservation as regards electrons in nuclei, and in an account to the Rome congress, which I hope to send you soon, I have entered into more detail of the theoretical side of this problem. If it should be possible to excite electron emission from nuclei by means of the recently discovered powerful agencies, it would perhaps be possible to settle this fundamental problem. A few days ago I got a very interesting letter from Madame Joliot, who thought that she had got evidence of such electron emission under the influence of the radiation from beryllium, but I suspect that the beta ray tracks on her photographs were due to Compton effects in the atoms of the walls of the Wilson chamber rather than to nuclear excitation of the gas atoms within the chamber. In connection with the calculations discussed under our Easter conference, and of which I think Fowler has told you, I shall be very interested to learn whether any new evidence is obtained regarding the direct effect of neutrons and the electrons in the atoms through which they penetrate. I am preparing a small article about our discussions of the new aspects of the atomic collision problems of which the interactions between neutrons and electrons form so striking an example.

Our small conference was a very pleasant and instructive experience to us all, and we were especially glad to see a number of our English friends on that occasion and from Fowler to learn about things in Cambridge. As we spoke of at the unforgetful Maxwell celebration, my wife and I hope very much to see you and Lady Rutherford here in Copenhagen. We had reckoned on moving into our new home at Carlsberg this spring and hoped that you perhaps would visit us already this summer, but the restoration of the house has been more slow than expected, and we shall not get into it before July. We hope, however, very much that you and Lady Rutherford will come and stay with us there in the early autumn when the Danish scenery is still at its best. I need not say, how great an experience your visit to Copenhagen would be to all Danish physicists, and if you, as we hope, should be willing to give us an address about any subject you like, our scientific societies will consider it a great honour indeed if they may extend to you an official invitation. Above all, however, my wife and I myself would consider it a most lucky start of our new home if you and Lady Rutherford would be our first guests there.

At the opening of the newly erected and extended research laboratories of the British Cast-iron Research Association at Birmingham Lord Rutherford laid emphasis on the necessity of using

the comparative shelter of a protective tariff to develop particularly that research work which experience has shown is absolutely essential if any industry is to compete on equal terms in the world of to-day.... Even the hardest heads are aware that research is profitable and necessary, and fail to encourage it because they lack the means rather than the will....On one point I am quite clear, the public will expect that any industry that is protected by a tariff should be thoroughly aware of the need of organisation and of scientific research.

RUTHERFORD TO GEIGER

2 *Sept.* 1932. First let me thank you for your kind letter of congratulation on my 61st birthday and for your good wishes. They were happy days in Manchester and we wrought better than we knew. I also have received your article in *Metallwirtschaft* on my noble self and my virtues— but no mention of my sins—which I read as far as I was able with much interest and remembrance of old days.

I have just returned from a seaside holiday with the grandchildren in Dorset where we had a happy time and on the whole reasonable weather. Fowler was there and we all—children and grown ups— bathed in the sea together....I have heard about the Heidelberg offer and congratulate you on this honour. I can appreciate the difficulty of coming to a decision where so many factors are involved. You will have seen our recent papers on the beta and gamma rays in the P. R. S. The outlook seems promising....

A suitable method of obtaining protons in great numbers is to pass an electric discharge through gas at a low pressure. For example the passage of a current of one-thousandths part of an ampere through the tube corresponds to the bombardment of the cathode by about 6000 million million protons per second. To obtain an equal number of alpha particles we should require more than 100 kilogrammes of radium.

In 1932 Rutherford gave the Thomas Hawksley Lecture before the Institution of Mechanical Engineers on 'Atomic Projectiles'. It was a lecture full of interest and new material. It sounded a note of victory. In his own laboratory, the Cavendish, he and his students were using 600,000 volts to speed up bombardment with protons (hydrogen nuclei). In California, Lawrence and Livingston had invented the cyclotron which urged electrons forward in a spiral of semicircles in a magnetic field, and, by reversing at the right instant a modest field of 4000 volts, it gave the electrons a velocity due to many million volts, with further great possibilities. Mme Curie's daughter Irene had married Prof. Joliot and the Curie-Joliot discovery of the effect of alpha rays on beryllium had led to the identification by Chadwick of the long-sought neutron, a particle about as massive as a proton, but apparently without a charge. This weird particle had the very properties which Rutherford had foreseen and stated in his second Bakerian Lecture. The neutron would sometimes wander or be drawn into the nucleus of an atom and cause its transmutation. If it entered oxygen, the change was to carbon; if nitrogen, it transmuted into boron. Again if protons were hurled at the element lithium, some few would be captured and in that case the lithium atom split into two alpha particles which parted with great velocity, almost equalling that of the alpha particles from thorium C', some of the swiftest known.

In brief:

$$^{9}_{4}\text{Be} + ^{4}_{2}\text{He} \rightarrow ^{12}_{6}\text{C} + ^{1}_{0}n,$$

beryllium and alpha give carbon and neutron;

$$^{16}_{8}\text{O} + ^{1}_{0}n \rightarrow ^{13}_{6}\text{C} + ^{4}_{2}\text{He},$$

oxygen and neutron give carbon and alpha;

$$^{14}_{7}\text{N} + ^{1}_{0}n \rightarrow ^{11}_{5}\text{B} + ^{4}_{2}\text{He},$$

nitrogen and neutron give boron and alpha.

Prof. Joliot told me an interesting story of his work with his wife, Irene Curie, on the effect of alpha rays on beryllium. Although they both followed all publications with care and interest they had not read Rutherford's second Bakerian Lecture, because "in such lectures it is rare to find anything novel which has not been published elsewhere". Joliot stated that if he and his wife had read Rutherford's prophetic suggestion about the neutron in the Bakerian Lecture, it is possible or probable that they would have identified the neutron in place of Chadwick.

At last, in 1932, the public imagination was fully aroused; something had happened at Cambridge; Rutherford and his young research students J. D. Cockcroft and E. T. S. Walton had hurled protons, with a voltage exceeding 600,000, at lithium, and the lithium had split violently into two fragments, each helium. There were numerous interviews, articles in the papers, good pictures of apparatus in the *Illustrated London News* and the *Sphere* and best of all this telegram from the United States:

Lord Rutherford, Cavendish Laboratory Cambridge.

American Scientists inform us our story on work Doctors Walton and Cockcroft not sufficiently clear regarding nature experiments stop we have therefore sent this telegram each them and if you are familiar with our organisation we will deeply appreciate any assistance you may be willing give us stop Telegram follows quote the Associated Press of America serving about 1400 newspapers in North and South America would be deeply grateful if you will grant us an explanatory interview on your recent experiments in splitting the atom stop We have in mind nothing sensational and will be only too happy submit article your approval before sending New York stop We have a very good reputation among scientists for our handling of scientific news and frankly so far what we have sent New York has not been as clear as we would like otherwise we would not be bothering you with this request stop We would like to send a man to Cambridge to see you write his article there and show it to you stop please reply prepaid.

The answer has not been found; but some papers continued to talk nonsense about splitting the hydrogen atom.

The way in which the neutron was discovered is of great interest.

Prof. Bothe of Giessen found that alpha rays from polonium striking a layer of the light element beryllium instead of giving rise to proton gave a very penetrating radiation like gamma rays. The Curie-Joliots in Paris, husband and wife, repeated this, but put a screen of paraffin wax in the path of the apparent gamma rays. Protons were knocked out of the wax! Chadwick at the Cavendish found that the nuclei of many other elements besides hydrogen got shrewd knocks from the unknown radiation. On further investigation Chadwick, on 27 Feb. 1932, claimed that he had found the neutron, a chargeless body with about the mass of the proton, which will wander through most matter, but is specially checked by the hydrogen in water or paraffin.

The curious thing is that a neutron seems to ignore an electron—whereas it has hostility to a proton and deals it a blow to an extent depending upon their mutual speeds. As to other nuclei, it may knock them forward, or it may walk inside one and transmute it. Queer things, neutrons! They demand quite a revision of modern atomic physics.

Rutherford gave two lectures at Copenhagen on 14 and 15 Sept. 1932, and at the close it fell to his old pupil Niels Bohr to move a vote of thanks. He described Rutherford's successive triumphs as being

crowned by the fundamental discovery of the atomic nucleus which has caused so unexpected and immense development of physical and chemical science. He knew always how to use some new phenomena, which to others might appear insignificant, as a key to further developments. He was able to prove that the atomic nucleus may be broken up by alpha particles, by which a new epoch in science was begun, as it was shown, for the first time, that an element can be changed by external agencies. In his lectures he has given us a most vivid impression of the vigorous and fruitful attack on the problem of the inner structure of the nucleus, which he and his collaborators are making. When we think of the smallness of the nucleus, we remember that its size compared with the atom by Lord Rutherford was illustrated by a pin head compared with a well-sized lecture room, it is amazing how great an amount of knowledge has already been collected in this quite new domain of human experience, a domain which is perhaps the most important and most promising of all. ...

It is always a new and wonderful experience to hear him lecture, not only on account of the new and important progress which he every time, as in these lectures, has to tell about, but we always get a renewed enthusiasm for the beauty of the subject and a fresh courage to the best of our modest powers to contribute to the great common aim.

In the evening Bohr again warmed to his subject at the Banquet of Danmarks Naturwidenskabelige Samfund given in honour of Rutherford:

This afternoon I had the honourable but hopeless task in a few words to picture the unique position of Lord Rutherford in the world of science, a position which rests for all times on his many great discoveries, of which several may be said to be epoch making, if the word 'epoch' could be reasonably used in relation with the short time intervals he had allowed to pass between them. I shall not expand on this subject to-night, but Lord Rutherford's discoveries do not exhaust the picture of him, in which his pupils and his friends rejoice. When Lord Rutherford visited Copenhagen twelve years ago, I had opportunity in this society to reproduce a number of stories which circulate among the pupils in his laboratory, in the inspiring atmosphere of which I had the good fortune to work for a time. I shall not repeat these stories to-day, the more so because Lord Rutherford last time reproached me for telling stories against him. Still the jest of these stories is not a malicious one, but an emphasis of his purity of zeal and directness of methods in dealing with difficulties in his work and the deficiencies of his collaborators. If a single word could be used to describe so vigorous and manysided a personality, it would certainly be 'simplicity'. Indeed, all aspects of his life are characterised by a simplicity of a similar kind to that which he has claimed of nature, which he is able to discover, where others before were not able to see it. It is his simplicity and self-forgetfulness which allow him to use all his moods in the best way for the progress of his work. In his usual bright spirit he sees always new and hopeful openings in any kind of difficulties, and his courage and cheerfulness extend themselves to everyone who works with him; at the same time he knows how to use his temper, which he does not try to hide, to a sound criticism of his own efforts and those of his assistants. To the pupils in his laboratory it is often as if the sun suddenly began to shine when he arrives in the morning; but sometimes it is as if the sky was darkened by a thunder cloud....In such moments everyone is of course afraid of him, but not more so than we dare to confess. All know that, if he has time and leisure, as he surprisingly often has, he is just as willing to attend to the youngest student and if possible to learn from him, as he is to listen to any recognised scientific authority. Not least is this entire disregard of social standing, the background not only for the unlimited admiration of his pupils but for their complete devotion. These same characteristics which make him the truest of friends have not only been essential for his work and for the successful administration of the great institution he is leading, but have also proved most helpful in treating matters at large relating to scientific education and research in his country, from the government of which he has received the external recognition which corresponds to the unique position he has long ago acquired in the minds and hearts of all his fellow-scientists. We wish him all happiness which life can bring; that in many years to come he may retain his power and his youth for the progress of science and the benefit of mankind and to the pleasure of his friends.

A. SMITHELLS TO RUTHERFORD

12 *Nov.* 1932. Yes, I *have* enjoyed it very much—I mean your Hawksley Lecture. It was brought to my bed with the *Times* and I read the lecture right through first, though I didn't mean to.

The fact is I feel about your stories just as I used to feel about my favourite stories when I was a small boy—I want to hear it all again from the beginning, especially as I can always expect a little extra surprise bit. That is really the situation, my dear man, and it is not a case of second childhood in the bad sense.

How refreshing it is to see the great things for true science still being done by fine laboratory work and good honest mechanics (I mean the science!). It is really saying a good deal to say that I can quite well follow what you have done and said, but I believe I can.

As for this last version of your doings I think it is one of the best you have given. Having said all this I should naturally like to find

a few faults. I think I noticed a questionable piece of grammar somewhere but I cannot find it. I don't know why Coulomb's law should
break down when things get near any more than gravitation does—
but if you say it should—all right—I know my place!

After such reading my mind always fastens on the Proton—I wonder
what *you* have in your head—in your mind's eye—when you think
of it.

I don't know whether anyone has invented any guts for it.

The New York Times Magazine declared that a distinguished stranger
dined at the high table at Trinity College, Cambridge and fell into
conversation with his table companions. One of them he long remembered because of the unusual range of his conversation. He was
a large and heavily built man with a shaggy moustache—altogether
a most unscholastic figure to encounter at Trinity's high table—and
his conversation was in keeping with his appearance. It roamed far
and wide. Presently it was rambling about the Dominions, and by the
time the port had arrived it was alighting on the subject of farming, in
which it appeared to find itself most remarkably at home.

"Who was that Australian farmer that I sat next to?" the stranger
inquired after dinner.

"That was Lord Rutherford", he was told.

The journalist continues this imaginative and imaginary statement
with a truer picture:

To see him and to talk to him is to feel that nothing in the man's
appearance affords the faintest clue to his immense scientific prestige,
except his eyes. You feel that a man with eyes so alive, so shrewd, so
penetrating must have something out of the ordinary within him. And
so he has....

It has not fallen to many, even of the greatest scientists, to command
so enormous an influence in his own lifetime as Lord Rutherford....In
England his speeches are reported as fully as the Prime Minister's.
He is one of the best of speakers, with a robust wit and a gift of marvellous
clear exposition,...and he has a bluffness and heartiness in keeping
with his big physique.

Mary Anderson (Madame de Navarro)[1] has similarly recorded her impressions:

At the back of the Burkitts' garden is a small green door which leads to the house of *Lord Rutherford*, the eminent physicist. When going there, I was shown how I could save a walk by using it. Lady Rutherford has a lovely garden, in which she rightly has great pride. Their home is altogether fresh, gay and charming. Wherever he is, Lord Rutherford seems to radiate brightness. His zest for good stories, his recounting of boyhood experiences on the farm in New Zealand are refreshing. His power of concentration is remarkable; during their visit to us I have observed him at a game of billiards, or even bagatelle—while he is playing he is oblivious to all else; José[2] says he is the same at golf.

Lord Rutherford tells me that each night he extemporises a serial bed time story to his small grandson. The child (who is very like him) listens with intense concentration and if his grandfather is guilty of the slightest discrepancy, he pounces upon it at once. Truly " *The child is the father of the man.*"

There is something arresting about Lord Rutherford's appearance, whether it is the piercing expression of his eyes, or the natural energy of the man; his voice is resonant, his conversation animated and vivid. An unselfconscious, victorious personality. He takes his greatness in his stride and leaves it at that. That is how Lord Rutherford impresses me.

On one occasion the Rutherfords were guests at Broadway and Mme de Navarro invited them to go with her to Stratford-on-Avon to see *The Merry Wives of Windsor*. Rutherford was rather unwilling, because he declared that "I should only show my crass ignorance. You know, for forty years they've been trying to civilise me, and you see what a hopeless mess they've made of it!" However he went and enjoyed every minute, leading the house in his laughter and applause, even drumming with his feet. On the return journey to Broadway he held forth with enthusiasm and good judgment on all the points of the play, and at the end Madame asked, "Where is this crass ignorance you spoke about?"

[1] *A few more Memories* (Hutchinson).
[2] J. M. de Navarro, Mary Anderson's son.

It was at Broadway too that Rutherford met two distinguished generals, who had served in Gallipoli and Palestine, and the conversation turned on the campaigns of Alexander the Great. Rutherford joined in the conversation and eventually took the lead and gave an intimate account of Alexander's campaigns. The secret was that he had an excellent memory and visualised events so clearly that he could reproduce a faithful and vivid picture of what he had read long before.

A bare record of facts conveys but a poor idea of the spirit of enthusiasm and excitement pervading the Cavendish Laboratory at this period; it was all energised by Rutherford who would often telephone to some young research student after midnight to learn the results of an experiment and to discuss its bearings on the new ideas arising from Chadwick's discovery of the neutron, or Cockcroft and Walton's artificial transmutations by the bombardment of elements with protons. Only those who have lived through the experience and have themselves worked during such an extraordinary epoch of discovery can understand the happy exaltation of spirit due to high success in scientific research, to which they too have made some contribution. No one enjoyed all this more than Rutherford.

CHAPTER XIV

THE PEAK LOAD

1933 was a banner year for science. Anderson of Pasadena discovered that the cosmic rays striking heavy matter sometimes gave rise to positive electrons, having the same mass but opposite charge to the ordinary electron, which is negative. This new constituent of matter has been called, wisely or unwisely, the positron. Its existence was soon confirmed by photographs, taken by Blackett, of tracks in the Wilson expansion chamber. This is one of the most remarkable discoveries in the whole range of scientific experience. It seems to be true that a photon, that is a gamma ray, may pass near the nucleus of an atom and, disappearing, give birth at the same moment to both an electron and a positron. The gamma ray is then said to 'materialise'.

As Newton happily wrote: "The changing of Bodies into Light and of Light into Bodies is very conformable to the Course of Nature which seems delighted with Transmutations."

Early in this same year, Urey and his co-workers in the United States managed to separate out heavy water, built up of oxygen and hydrogen as usual, but with the hydrogen atom of double mass. The discoverers, surely entitled to christen their child, called the whole atom 'deuterium' and the nucleus thereof 'deuteron', or for a time 'deuton'. There was an agitation in England to use the terms 'diplogen' and 'diplon'. The symbol for this nucleus is $^2_1$H or $^2_1$D, as against $^1_1$H for the proton. The whole atom may be denoted by D, so that heavy water is D_2O.

The name matters little, the important point is that D can be separated from H during electrolysis, or rather D_2O from H_2O.

Some of this precious fluid, heavy water, was sent to Rutherford who used deuterons in place of protons wherewith to smash his atoms, and they were highly effective.

But apart from this, Rutherford and Oliphant, following a wise forecast of Gamow, built apparatus giving lower voltage and larger current and obtained disintegration with voltages as low as 40,000.

Another triumph of the same two workers was to drive protons into boron, and the total atom broke into three alpha particles, thus:

$$^{11}_{5}B + ^{1}_{1}H \rightarrow 3^{4}_{2}He,$$

boron and proton give three alphas.

So, at his Royal Institution lectures Rutherford had much to talk about that was new and fascinating. He also, greatly daring, took a mixture of hydrogen and helium gases, drove their ions along a partly exhausted tube with 20,000 volts, and deflected them with magnetic and electrostatic fields, producing thus *two* parabolas, or curves, of the *separated* gases; thus he demonstrated J. J. Thomson's famous positive ray experiments to a large audience.

There was also a discussion at the Royal Society on the nature and properties of heavy hydrogen and of heavy water. Soddy objected to the term 'isotope' being applied to heavy hydrogen because the term was first used to indicate that two elements were inseparable and this did not apply to deuterium and hydrogen. Bohr declared that they were both hydrogen, having the same nuclear charge.

Rutherford's concluding remarks dealt with this point:

I am sure you will all realise what an extraordinarily interesting subject this is. Though there are only a few people who have as much as a cubic centimetre of this heavy water, I hope some of our commercial chemists will see that we are able to get it by the litre, and then we can have a very interesting discussion in six months' time on the results that will follow from that. I commend that to any of our industrial friends who may be present. It requires a lot of work to get 100 c.c., and a fairly big bill for electricity at the price you have to pay for current in a rural district like Cambridge.

The general discussion has been most interesting, and I am sure you would like me to say a word about Professor Soddy's communication. Professor Soddy was the discoverer of isotopes, and quite properly he gave the definition first that they were more or less inseparable and occupied the same position in the periodic table; but a good deal of water has flowed under the bridges since he made that discovery, and now from the scientific point of view we do not regard an isotope as merely inseparable; we have come to bodies that have the same nuclear charge, and I think in that respect no one will doubt that heavy hydrogen has the same nuclear charge as hydrogen and has one electron attached to it; and therefore, unless we are going to change the whole nomenclature, I think the name isotope will have to be retained.

I can, of course, quite understand Professor Soddy's point, that he applies the name to things that are almost inseparable and with no differences except, possibly, almost second-order effects in the spectrum, and now the name is given to this heavy hydrogen, and apparently it can be separated, and has all kinds of different properties; but it is all due to the fact that one mass is twice the other; they are still isotopes. That is the essential point, and I should like to be able to convince Professor Soddy that in using the name isotopes we are not hurting his reputation but rather improving it.

On 3 February, the new Royal Society Mond Laboratory was opened by Mr Stanley Baldwin, then Prime Minister and Chancellor of the University of Cambridge. Rutherford furnished copious notes for the speaker's use and included the phrase 'try-out', which was adopted in the speech. There followed an amusing series of letters in The Times, some attacking Baldwin, and others defending him, for the use of such a word. This afforded some amusement to Rutherford, who was the real culprit, and he asked Baldwin how he liked the rôle of 'whipping-boy'.

The money for the laboratory was given by the Royal Society from the Ludwig Mond Bequest and it was devoted to building a laboratory largely for the development of the advanced type of physics due to Kapitza in the Cavendish Laboratory. Beyond the window on the right of the door was carved a crocodile by Eric Gill at the express wish of Kapitza. It is said that this animal never turns back, and was accordingly regarded as a symbol of Rutherford's scientific acumen and career. It is also said that the crocodile is regarded in Russia with mingled awe and admiration. Inside is a bas-relief bust of Rutherford, and a medallion of Mond, both done by Eric Gill.

In this laboratory Kapitza used a current in a coil which gave a magnetic field greatly exceeding all others contrived, but lasting only for a hundredth of a second—"a long time if you know how to use it", said Kapitza; "I want only a thousandth of a second for my experiment." The current during that brief period exceeds the total current output of the great Deptford power-station.

In his address Rutherford thanked the Chancellor for finding time in the midst of his heavy responsibilities to open the new laboratory and added:

The opening of this Laboratory is to me an important event, representing as it does, the culmination of my hopes during the last decade.... Kapitza thought that the very large momentary expenditure of power required could best be provided by short-circuiting a large electrical generator of special design and then breaking the current after an interval of about a fiftieth of a second by means of a special automatic break....I have taken an almost paternal interest in these new developments, but it is to the energy and enthusiasm of Professor Kapitza that we owe the forging of these new and powerful weapons to research which have added so materially to the possibilities of investigation in new fields at Cambridge.

RUTHERFORD TO HEVESY

3 *April* 1933. I find I have not answered your last letter. I am glad to hear of the success of your experiments on the alpha rays from samarium and wish you all good luck in your other researches along these lines. You will have seen in *Nature* the letter about the positive electron. I am very pleased that we have been able to get it in a laboratory process as we may be able to find out something of its properties. The results seem to me very definite and conclusive, but it is difficult to see at the moment how this new entity fits in with the theoretical scheme. Oliphant and I have been very busy on our experiments of transmutation with protons up to two hundred thousand volts. We are able to get about a thousand times as many particles as Cockcroft

The Royal Society Mond Laboratory

Chantry Cottage

See p. 397

and Walton at the same voltage. We have made a special examination of boron and hope to publish the results in a month or so.

We have of course all been very interested in following the progress of the new government in Germany and in particular the anti-semitic troubles. I hope you are in no way affected by this strange effervescence. I see that Einstein has resigned his Berlin post but I presume he is financially well fixed in the U.S.A. due to the special endowment there.

We are of course hoping to visit you in Freiburg in the middle of June if the situation in Germany has improved. Naturally we will not think of going if it should prove any inconvenience to you in these times. We have not yet made our plans but our general idea would be to spend a few days with you in Freiburg and then possibly go on to Switzerland for a short stay.

We have had one of our grandchildren Pat, aged 6, staying with us for a couple of months and we have all been very pleased to have such a jolly youngster round the house.

We are all very well. Fowler returns from California early in May.

Our experiments on the long range alpha particles from radium C have come out in fine style and we hope to throw new light on the origin of the gamma rays.

The Halley Stewart Laboratory at 30 Chesterford Gardens, Hampstead, was opened by Rutherford in April. This was a notable benefaction to King's College, University of London, by the Halley Stewart Trust, for research in the physical sciences and in particular for the prosecution of Appleton's researches on the electrical properties of the upper atmosphere. Rutherford took a great interest in this new departure, for Appleton opened up these new methods of attack on the constitution of the upper atmosphere when he was working in the Cavendish Laboratory in Cambridge in 1924. He had used wireless stations at Cambridge and at Oxford and proved the reflexion of radio waves from a height of about 60 miles, thus confirming the speculation of Heaviside and Kennelly, after whom this lower region is named.

Appleton also discovered another region, now bearing his name, at two or three times the height of the lower layer. Rutherford pointed out that these regions must have free electrons present, which cause reflexion and refraction; without such ionised regions it would be impossible for wireless waves to travel great distances round the earth as they undoubtedly do.

On 9 May 1933, Rutherford spoke in the House of Lords on the Rubber Industry Bill which was concerned with the question of compulsory contributions for five years to a Research Association. Rutherford's main point was one of general principle:

From some points of view the question before us is a small one. From another point of view it is one of real importance. It is a test of the attitude of the forces that govern this country towards scientific research in its application to industry. As a man of science who has had some opportunity of seeing the fruits of scientific inquiry into industrial problems I am convinced that research is a potent weapon for combating the evils of waste and inefficiency in industrial production. I hope that the House in passing this measure will give an unmistakable declaration of its faith in the application of scientific methods and knowledge as a means of keeping this country in the forefront of progress.

On 2 June, Rutherford gave the Boyle Lecture at Oxford dealing with the Transmutation of Elements, beginning:

The possibility of the transmutation of the elements has always exercised a profound fascination for the minds of men since the dawn of science. Following the rapid development of accurate knowledge of chemistry in the nineteenth century, the feasibility of the transmutation of the atoms faded into the background. Later Faraday wrote: "To decompose the metals, to reform them and to realise the once absurd notion of transmutation—these are the problems now given to the chemist for solution."

He then reviewed the progress of Physics, the discoveries of the electron and its ways, radioactivity, the nucleus, and the triumph of the transmutations of several elements.

After the lecture, questions were invited and a member of the audience pointed out that one of Rutherford's conclusions did not tally with the results of a certain physicist—naming a lady. Rutherford replied: "That was on account of her apparatus—it was like this—she had a very broad beam—." At this point the audience became hilarious, for there were many undergraduates present. The meeting broke up without further discussion.

On 11 May, Rutherford spoke on the radio from Broadcasting House, London to the Fifth Pacific International Science Congress meeting at Vancouver in British Columbia. This was a large undertaking requiring co-operation between the British Post Office, the B.B.C., Bell Telephone Co. of Canada and the Canadian Marconi Co. There was a seven-minute speech and three minutes of amusing impromptu conversation, between London and Vancouver, which was broadcast throughout Canada.

Rutherford dwelt on the importance of the Pacific Ocean and of its scientific exploration and development:

For science is international, and I trust will ever remain so; and it is meetings like yours to-day that help much to bind closely the ties between men of different countries and of different creeds.... Scientific research in the Pacific problems was inaugurated by the British Government and the Royal Society of London, more than a century and a half ago, when they together sent Captain Cook on his first voyage in the Pacific. With him went Sir Joseph Banks and his retinue of scientific observers. On his return from his famous voyage, Captain Cook was elected a Fellow of the Royal Society and awarded the Copley Medal for his scientific contributions. As a New Zealander I take pride in the initiative of the Royal Society in this famous voyage of discovery.

RUTHERFORD TO BOYLE

28 *July* 1933. I was very pleased to hear from you and glad that my radio address to Vancouver went off all right. I was pleased to have a chat with old friends under these unusual conditions.

I am sorry to hear that the hard times have affected the laboratory and I am sure Tory has a hard task in keeping things going. Undoubtedly things have improved a little in this country but the situation in U.S.A. at the moment does not look too promising and no doubt has a serious repercussion in Canada.

I daresay you have seen that we have been very active in the laboratory here the last two years in connection with transmutation. I find the position very similar to that in the old days in Montreal when every week brought a new development. Professor Lewis of California sent me a sample of his water rich in H^2 isotope and we have used it in our work. We find we can easily detect the transmutation effect of 60,000 or 70,000 volts. We have confirmed some of the results found by Lewis in California in connection with lithium, but have examined the matter in more detail with some interesting results. We find that lithium under proton bombardment in addition to the main range of 8·4 cm. gives two short additional groups of ranges about 7 and 12 millimetres; we do not yet know the interpretation thereof. Walton and Dee have been taking Wilson photographs of the effects and have confirmed the results which we hope to publish in the September Proc. Roy. Soc.

Work on the positive electron goes on rapidly and it looks as if the positive and negative electrons can be born in the strong nuclear field probably under varied conditions, e.g. gamma-rays, neutrons or electrons of sufficient energy.

At the Meeting of the British Association at Leicester, Rutherford described the recent great developments in physics and chemistry. He recalled that at Leicester in 1907 Lord Kelvin had declared the atom to be an indestructible unit, eternal in its nature. Since that time very many transmutations had been effected by deliberate experiment, but he added:

These transformations of the atom are of extraordinary interest to scientists but we cannot control atomic energy to an extent which would be of any value commercially, and I believe we are not likely ever to be able to do so. A lot of nonsense has been talked about transmutation. Our interest in the matter is purely scientific, and the experiments which are being carried out will help us to a better understanding of the structure of matter.

Primordia quaerere rerum!

On Tuesday evening, 3 October, there was a great meeting of 10,000 people at the Albert Hall when Rutherford made the opening speech from the chair:

This meeting has been called to consider a problem of great magnitude and of ever increasing urgency—the problem of the relief of German refugees.

The appeal is for a fund to be employed for the relief of students, University teachers and members of the professional classes for which co-operating associations have been formed....The Academic Assistance Council has a list of more than a thousand University teachers of all grades, men of science and learning, who have lost their posts and their opportunity to continue their specialised work....This problem must be approached in no petty spirit of national or sectional hostility, and with the complete absence of any spirit of political antagonism.... Each of us may have his own private political views, but in this work of relief all such political differences of opinion must give way before the vital necessity of effectively conserving this great body of learning and skilled experience which otherwise will be lost to the world.

Rutherford, with upraised hand, with a sweeping gesture towards Einstein, who was sitting by his side, said "Ladies and Gentlemen, my old friend and colleague, Professor Einstein."

The professor spoke in English, with much restraint, on "Science and Civilisation", pointing out how great had been the gifts of liberty to mankind—

liberty without which life to a self-respecting man is not worth living.... It cannot be my task to act as judge of the conduct of a nation which for many years has considered me as her own; perhaps it is an idle task to judge in times when action counts. To-day the questions which concern us are: How can we save mankind and its spiritual acquisitions, of which we are the heirs? How can we save Europe from a new disaster?...

It is only men who are free who create the inventions and intellectual works which to us moderns make life worth while....Men in their distress begin to think about the failure of economic practice and about the necessity of political combinations which are super-national. Only through perils and upheavals can nations be brought to further developments. May the present upheaval lead to a better world.

After Rutherford had thanked Einstein, there followed a great and moving appeal by Sir Austen Chamberlain:

Every now and again the world is shocked and moved by some great catastrophe of nature, but we are faced to-day with something worse— a catastrophe produced by the act of men—with destruction and suffering not less acute or widespread, not as a result of the uncontrollable forces of nature, but as a result of the selfish passions of man. We may protest, we cannot alter; but we can do something to alleviate the suffering which has been so needlessly caused, and secure for the human race a development of the gifts with which God has endowed His people, whether they be born in one country or another.

In 1933 Rutherford became President of the Academic Assistance Council, which aimed at raising a million pounds to alleviate distress. Every effort was made to avoid the incitement of political passions. In July 1935 the name was changed to the Society for the Protection of Science and Learning. Rutherford wrote two appeals for funds to *The Times*—'The Wandering Scholar' (3 May 1934) and an 'Appeal Letter' (16 Nov. 1934). He insisted on non-political action by the Society. He was ably supported by the two Honorary Secretaries, Sir William Beveridge and Prof. C. S. Gibson. When the funds were almost exhausted, the Imperial Chemical Industries presented £2500 to the Society.

Up to 1937 about 1600 professors and research students had been displaced and of these 507 were permanently placed and 308 temporarily; there were about 300 seeking positions. The Secretary, Mr Walter Adams, has stated how unsparing of time and effort, how efficient in Committees, how painstaking about individual cases, was their President up to the very end.

Sir William Beveridge, Master of University College, Oxford, has kindly contributed a note on this great undertaking:

The Academic Assistance Council was conceived at a week-end which I spent at Cambridge in May 1933 with George Trevelyan. Most of the week-end was spent on consultation with Trevelyan, Frederick

Hopkins and Rutherford in making the first draft of the Memorandum which was subsequently issued when the Council was launched.

It was Rutherford's attitude more than anything else which made it possible to launch this scheme with the hope of general scientific backing. I found him in a state of explosive indignation at the treatment that was being meted out in Germany to scientific colleagues whom and whose work he knew intimately and respected in the highest degree. It was his (and to a less extent Hopkins') support that obtained for the Council later the facilities of the Royal Society rooms for meetings, and it was also through him that Professor C. S. Gibson was enlisted to become the effective Honorary Secretary to the Council.

Rutherford himself was of course up to the eyes in other work and I remember that Lady Rutherford was rather vexed with me for urging him personally to take any part in the Council. The fact that he did finally agree to become a member of the Executive Committee of the Council, if established, and later its President, was decisive for its launching and its success.

It is hardly necessary for me to emphasise the immense support that the weight of his name and authority gave to all our work. Above all he stood for insisting that we should do something practical to help and that we should avoid politics and recriminations.

Lord Dawson of Penn, Lord Rutherford and Sir John McLennan were three moving spirits in starting the Radium Beam Therapy Research at the Radium Institute. The object was the treatment of cancer of the tongue and upper air passages by the gamma rays from five, or more, grammes of radium enclosed in a massive bomb, with a window through which the radiation could pass to the patient. The scheme was made effective by the loan of radium by the Radium Belge Co., and by financial support from the Medical Research Council, the Royal College of Surgeons, the Department of Scientific and Industrial Research, the Radium Institute, the Imperial Cancer Campaign, and other bodies.

Viscount Samuel has published in *Belief and Action* (p. 300) an interesting letter written to him by Rutherford in 1933 stating his

views on the Principle of Indeterminacy or Uncertainty. A few words
of explanation may be desirable. It had been shown by Heisenberg
(1927) that when an electron is under observation it is impossible to
determine with certainty both its position and its velocity. The product
of the two errors, of position and momentum, will always equal
Planck's constant h, a small known quantity. The smallness has nothing
to do with the principle at stake. This uncertainty is not due to the
observer's shortcomings, it is inherent in Nature. What an electron
may do when it is *not* under observation is a question which may readily
be asked, but can never be answered. We know nothing of the unobservable, and can only guess about that which is not observed. This
principle of uncertainty is believed by many eminent physicists to
upset the principle of causation—of cause and effect—which lies at the
root of science. It seems that Bohr, Eddington and Schrödinger take
this view, while Einstein and Planck take the opposite.

Rutherford wrote (1933):

While the theory of indeterminacy is of great theoretical interest as
showing the limitations of the present wave-theory of matter, its
importance in Physics seems to me to have been much exaggerated
by many writers. It seems to me unscientific and also dangerous to
draw far-flung deduction from a theoretical conception which is in-
capable of experimental verification, either directly or indirectly.

The subject is however still open to discussion and appendix IV in
Viscount Samuel's book gives an interesting account of his own views.

The beginning of the year 1934 afforded one of those startling and
almost unbelievable advances in physics, namely the discovery in Paris
by the Curie-Joliots of *artificial* radioactivity. The procedure was to
bombard the element boron, in any form such as boracic acid, with
alpha rays from any radioactive material such as polonium. Transmutation takes place, the boron turns into an unusual form of nitrogen
(13) and a neutron is ejected. So far there is nothing remarkable. But
examine the nitrogen! It is unstable and breaks down into carbon,
throwing out a *positron*. But that is not all. The nitrogen atoms do not

suddenly turn into carbon, but are quite deliberate about it. Half of these atoms have changed when a few minutes have passed. Indeed they are radioactive elements, made so artificially, but obeying the old Rutherford and Soddy law, with the period to half-value of 14 minutes. Rutherford was interested and surprised at the news, and at the slowness of the change, but he said "that shows how little we know about radioactivity".

In the same way aluminium exposed to alpha rays gave rise to radio-phosphorus, which broke up into an ejected positron and silicon; so too with magnesium, which gave rise to radio-silicon and that to aluminium. It is obvious that enormous possibilities were thus opened up, because other bombardments of other elements might be made with various nuclei as bombs, so that all kinds of artificial radio-elements might be made with widely varying periods to half-value.

In March 1934, Rutherford lectured at the Royal Institution on the 'Transmutation of Elements':

The first definite evidence that an ordinary element can be transmuted by artificial means was obtained in 1919, when it was found that some of the atoms of nitrogen could be transformed by bombardment with alpha particles. Occasionally an alpha particle may enter the nitrogen nucleus which then breaks up with the emission of a fast proton and the formation of a new element, an isotope of oxygen of mass 17. The resulting atom has a nuclear charge 1 unit higher and a mass 3 units heavier than nitrogen. A similar type of transmutation has been found to take place in a number of light elements.

A new and remarkable type of transformation has been found to occur when alpha particles bombard the light element beryllium. No proton appears but in its place a new type of unchanged particle of mass 1 called the neutron is liberated with high speed. Fast neutrons have re-markable powers of penetration of matter and in turn are able to produce a transformation in a number of elements through which they pass.

It is to be noted that *heavier* elements are thus produced in both these cases. The neutron can be used as a bomber:

$$^{16}_{8}O + ^{1}_{0}n \rightarrow ^{13}_{6}C + ^{4}_{2}He$$

or, oxygen and neutron give carbon and alpha particle, or helium nucleus.

Such equational statements of results are now very numerous. Every element has been transmuted up or down into heavier or lighter elements; and it rejoiced Rutherford to see the artificial transmutation of matter give rise to artificial radioactivity.

On 28 Feb. 1934, Prof. Dr J. Stark, an illustrious physicist, President of the Physikalisch-Technischen Reichsanstalt, wrote to Rutherford protesting against a letter published by *Nature* on 23 Dec. 1933, written by Prof. A. V. Hill. Rutherford collected opinions from eminent colleagues and wrote a friendly, firm reply, concluding as follows:

I myself would deprecate any criticism of a foreign Government in a publication like *Nature* except in so far as the matter in question concerns the interest of science as a whole. At the same time, to avoid misunderstanding, I think it desirable to put briefly before you the general attitude of scientific men in this country in connection with the migration from Germany of such a large number of scientific men, some of great distinction, whom we hold in high esteem for personal and scientific reasons. This country has always viewed with jealousy any interference with its intellectual freedom, whether with regard to science or learning in general. It believes that science should be international in its outlook and should have no regard to political opinion or creed or race. For this reason scientific men in this country were at first naturally inclined to be critical of a governmental attitude of mind that led so many men of science to leave their own country to seek suitable conditions for their work in foreign lands.

We all sincerely hope that this break with traditions of intellectual freedom in your country is only a passing phase and does not indicate any permanent change of attitude towards the freedom of science and learning.

Scientific men in this country have the most friendly feeling for their colleagues in Germany, and I see no reason why this should not long continue. We are all anxious to work in the most cordial way with our German colleagues, quite apart from any forms of government. At any time that *Nature*, for example, gave a mistaken impression of scientific matters in your country, I am sure that its Editor would be glad to give you an opportunity to state your reasoned view.

I am afraid this letter of mine is longer than I intended but I trust that you will understand that in stating my views, I am sympathetic with your desire for our two nations to understand each other and to work together in peace and unity.

In *The Times* (3 May 1934) Rutherford wrote:

In the conviction that the universities form a kingdom of their own, whose intellectual autonomy must be preserved, my distinguished colleagues formed the Academic Assistance Council one year ago and the Royal Society provided accommodation for the Council's offices. The occasion was the displacement of our fellow scientists and scholars from their university positions in Germany; but the problem with which the Council is faced is wider and deeper than that presented by the need for assisting these German teachers. Its ambition is to defend the principles of academic freedom and to help those scholars and scientists of any nationality who on grounds of religion, race or political opinion are prevented from continuing their work in their own country.

The series of political revolutions in Europe since the Great War has created a large body of wandering scholars; many, for instance, among the Russian or Italian *émigrés* have unfortunately through the absence of organised assistance by their university colleagues lost the means of continuing their scientific careers. But there are many whose talent and experience could still be effectively used, and their number has been tragically swollen during the past year by the expulsion from academic positions in Germany of persons possessing pacifist or inter-nationalist convictions or lacking that strangest of qualifications for the life of scholarship, 'Aryan' genealogies.

To incorporate the services of these wandering scholars in the other universities of the civilized world is more difficult to-day than in the Middle Ages when the 'communities of learners' were less hampered by administrative formalities, restrictive endowments, and incipient nationalist tendencies. Medieval scholars could migrate to other districts and the 'universitas' moved with them; the same catholicity of spirit has been fortified by the present crisis in both our ancient and our modern universities....

In 1916 the British Government had set aside a million pounds to promote industrial research and that fund became exhausted in 1934. A Conference was held in London, at which Rutherford presided, in order to put forward a statement as to past expenditure and future plans which could be presented to the President of the Board of Trade for the consideration of the Government. Rutherford stated the case:

While, no doubt, the concerns which promote and pay for these costly researches gain in many cases an ample reward, it should be remembered that the greater part of the benefit goes to the consumer in the provision of a better and cheaper article.... The improvements brought about by science are to the ultimate advantage of the consumer, and thus of the country as a whole. For this reason it seems to be proper and desirable that the State should give not only moral but financial support to promote the application of scientific knowledge to increase the efficiency of industrial processes.

The most notable incursion made by Rutherford into the field of Chemistry was when he received the Nobel Prize for Chemistry. He had no pretensions to be a chemist, but he had an uncanny habit of picking the right bottle from the shelves of the chemistry laboratory, whenever he wanted to extract a substance from a solution. When asked what he thought about being called a chemist, he replied that he had not the slightest objection provided his chemical friends did not mind. In another sense he might be styled as the greatest chemist the world has ever known. Was he not the first to transmute the elements at will?

One of his best lectures was on 'The Periodic Law and its Interpretation' delivered as the Mendeléeff Centenary Lecture before the Chemical Society at the Royal Institution, London (19 April 1934).

Rutherford began talking to the Society in a somewhat apologetic manner:

This year is the centenary of the birth of Dimitri Ivanovitch Mendeléeff (1834–1907) whose name is so indelibly associated with the great advance in our knowledge of the classification of the chemical elements....I feel less fitted than any of my audience to deal adequately with the chemical side of this subject, but I am encouraged by the reflection that I, either directly or indirectly, have been closely associated with several developments which have helped to give a much clearer understanding of the underlying meaning of this famous law of the elements.

He then referred to the early suggestions of Prout that all elements were built from hydrogen, but it was found that atomic weights were not whole numbers; Mendeléeff however recognised that there was

evidence of a definite kind of order in the variation of chemical properties when the elements were arranged in sequence of their atomic weights. The full paper, published in 1871, contained the classification of the elements into groups or periods essentially the same as those we are familiar with to-day.

The ideas of Mendeléeff at first attracted little attention,...it was not until the discoveries of the elements gallium in 1875 and scandium in 1879 and the proof that they had the properties of two of the missing elements predicted by Mendeléeff that the importance of his discovery was generally recognised.

The next great advance in our knowledge was the discovery of the inert gases, of argon by Rayleigh, and of helium, neon, krypton and xenon by Ramsay. It seemed clear that a new group of zero valency must be added to the periodic classification thus widening but not altering the scheme of Mendeléeff. I look back with some pride to the fact that Soddy and I were able to prove that the radium emanation must belong to the group of inert gases, although the amount of emanation available for our experiments was less than 10^{-13} c.c. This was made possible by the extraordinary sensitivity of the radioactive methods as a means of quantitative analysis. While the rapid discovery of this new group of gases was at that time of extraordinary interest, yet it was not until twenty years had passed that their significance was recognised in providing us with a clue of the utmost value in deciphering the intricate problem presented by Mendeléeff's law.

Rutherford then pointed out that the discoveries during his Manchester period—the nucleus, the Rutherford-Bohr atom, the atomic number—had thrown a completely fresh light upon the periodic table.

In an after-dinner address given in 1934 Rutherford spoke with unusual gravity of modern tendencies.

Everybody is aware that we are living in a world in which social and economic changes are occurring with startling rapidity and I am not quite sure that we all appreciate how fast these changes are taking place and the uncertain destiny to which they are leading us.

He used the analogy of a a chemical reaction being unduly accelerated by the carelessness of an experimenter, with the resulting explosion and oblivion! In the world of to-day there was a similar acceleration and "we could only watch with hope and some fear what would be the end of the experiment. Would it arrive at a stable end as everyone hoped?"

He touched on the rapid development of science and its relation to industry and then reverted to his favourite topic, the atom, and finally made some reflections on the newest physics, on wave-mechanics, causality, and indeterminacy. As this appears an isolated example of rash statement, it may be well to record his views. He pointed out that the old classical mechanics failed to explain the behaviour of atoms, so that it was necessary to revise our point of view:

The first point that arises is the atom. I was brought up to look at the atom as a nice hard fellow, red or grey in colour, according to taste. In order to explain the facts, however, the atom cannot be regarded as a sphere of material, but rather as a sort of wave motion of a peculiar kind. The theory of wave-mechanics, however bizarre it may appear—and it is so in some respects—has the astonishing virtue that it works, and works in detail, so that it is now possible to understand and explain things which looked almost impossible in earlier days. One of the problems encountered is the relation between an electron, an atom and the radiation produced by them jointly; the new mechanics states the type of radiation emitted with correct numerical relations. When applied to the periodic table, a competent and laborious mathematician can predict the periodic law from first principles.

Although in the new ideas there is accuracy of definition, it is necessary to be very careful in regarding the electron as a nice little piece of matter. It must be regarded at times as a sort of wave motion. While certain changes are taking place in connection with radiation it is impossible to explain the intermediary between the first and last stages. This is inevitable, because it is really all connected with the fact of regarding matter as having some qualities of wave motion. It is a question of indeterminacy. This idea has a reaction on the philosophic mind and has brought all sorts of queer ideas bearing on the problem. Some people suggest that the law of causation has failed; others want to see in it the origin of free will; others want to find in it the possibility of anything happening that they would like to happen! I confess, however, it may be that I am not properly attuned to these philosophic ideas—that these things do not worry me because I believe that whilst it is true that we cannot say that certain things will happen, there is the probability that they will. Our lives are all governed by probability.

He then spoke, for a time, of probability and continued:

Some writers appear to think that we know everything and that we can explain the universe on the knowledge of the moment. Actually, however, we know only a minute fraction of what there is to be known.[1] Perhaps that is as well, in order to leave something for the following generations to clean up.

He recalled radioactivity in the earlier days, which

was a great shock to scientific opinion and I appreciate it all the more because everyone wanted to jump on me in those days. The point is that the atoms, which scientists had spent a century in proving stable, proved to be unstable and broke up with explosive violence in going through a series of transformations. These were spontaneous changes. Since that time there has been artificial transmutation, and scientific men are now quite sure that, if they could get enough particles of the right kind, there is no atom in nature but would yield to that bombardment.

[1] "There was nothing wrong with the old inference that if I know all about the present I can forecast the future exactly; the trouble was the impossibility of knowing the present. Once this is seen, the whole argument becomes obvious, but nobody saw it until Heisenberg." C. G. Darwin (1938).

This forecast has already been verified.

In 1934 Rutherford gave one of his famous lectures at the Royal Institution where he lectured on heavy hydrogen (or deuterium) which has, like hydrogen, but one outer or satellite electron, but differs from it in having mass 2 instead of mass 1. To be more accurate, if an oxygen atom has mass 16 then ^1_1H has mass 1·0078, and ^2_1H has mass 2·0136. The one number is not quite twice the other. The difference 0·002 may not at first sight seem worth troubling about, but turned into energy and expressed in volts acting on an electron, it means a little less than a million volts. That is the kind of energy that holds together the proton and the neutron in the deuterium nucleus. Rutherford proceeded to illustrate the properties of this new gas discovered by Urey, Brickwedde and Murphy, and to describe and show some of the remarkable transmutations of matter carried out at the Cavendish leading to the discovery of a new form of hydrogen, tritium, ^3_1H and of a new helium, ^3_2He. To the uninitiated it may be explained that H stands for hydrogen and He for helium, that the upper number is the number of protons and neutrons in the nucleus, and the lower number the atomic number, or positive nuclear charge $(+e)$. Thus tritium has 2 neutrons and 1 proton in its nucleus, and consequently has the suitable charge $(+e)$ due to that 1 proton. Again, ^3_2He is not ordinary helium ^4_2He, but contains 1 less neutron in the nucleus. Rutherford showed that there was great advantage in using 'deuteron', the nucleus of deuterium, as bombing agent. If deuterons bombard deuterium the following changes occurred:

$$^2_1H + ^2_1H \rightarrow ^4_2He \rightarrow ^3_1H + ^1_1H,$$

which means that the deuteron gets mixed up with another deuteron and forms a new unstable helium nucleus and this explodes to give tritium and proton. But sometimes another change occurs,

$$^2_1H + ^2_1H \rightarrow ^4_2He \rightarrow ^3_2He + ^1_0n,$$

or 2 deuterons coalesce to form an unstable helium nucleus, which then breaks up into an isotope of helium and a neutron, which has mass nearly 1 and charge zero.

It will be noted that both protons and neutrons appear when deuterons bombard deuterons. These were actually shown at the lecture and were thus described in the press:

For the first time in a public lecture experiments were made to show the artificial transformation of the element lithium by protons and deuterons of energy corresponding to about 100,000 volts. The enormous emission of fast protons when ammonium sulphate containing heavy hydrogen was bombarded by deuterons was clearly shown by counting methods. The transformation apparatus was designed and operated by Dr Oliphant, while Messrs Watson & Son loaned a generator to provide a steady potential of 100,000 volts to accelerate the ions.

A new Central Research Laboratory of the United Steel Companies, Limited, was opened by Rutherford at Stocksbridge on 29 June 1934. It was a well-planned two-storey building with adequate equipment for investigating the physical and chemical properties of steel with its immense variety and possibilities.

In his address he pleaded for

generous treatment of the scientific men employed; because men clock in and clock out, it does not necessarily mean that they are doing what is most required. I know nothing more deadening to original ideas than keeping a man's nose firmly fixed to the grindstone. Even directors need a change, and young men should have opportunities of meeting other young men working in other parts of the country. Ideas are more likely to come from such meetings with colleagues than by holding men down to some work in which there might be no progress at all. No laboratory today is self-sufficing.

Rutherford always felt the greatest admiration for the work of Mme Curie, who died in 1934. The courage of that frail woman, tackling those tons of pitchblende single-handed, must always fill one with amazement. But Mme Curie was determined to get a pure source of radium and, in spite of the scepticism of many scientists as to the possibility, she persevered to a successful end. She had not only to struggle with the heavy manual labour, but there was also the great danger of injury to her health from the radium. In fact the writer tried a short time ago to get some radio-lead (radium D) from the Radium Belge firm, but was told they could no longer supply it. It was pointed out that they had previously provided some, but they said that it had been separated by Mme Curie, and that it was too dangerous a process for their men to undertake!

In the charming life of Mme Curie by her daughter Eve, it is stated that her mother died of pernicious anæmia, the result of prolonged exposure to radiations from the element with which her name will for all time be associated. Einstein has said, "Marie Curie is, of all celebrated beings, the only one whom fame has not corrupted." "She did not know how to be famous", writes Eve Curie.

Rutherford wrote an obituary notice in *Nature* (21 July 1934) and referred to her fundamental discoveries of polonium and radium, extracted from pitchblende, an ore which she found *had more radio-activity than that due to the uranium it contained. There must be a new element there!* Marie and Pierre Curie extracted first polonium and next radium, about three parts in ten million.

In 1904 a Nobel Prize had been divided between the Curies and Henri Becquerel, and in 1911 Mme Curie was again awarded a Nobel Prize for her purification of radium and the determination of its atomic weight.

Mme Curie (wrote Rutherford) retained her enthusiasm for science and scientific investigation throughout her life. She was an indefatigable worker and was never happier than in discussing scientific problems with her friends....Quiet, dignified and unassuming, she was held in high esteem and admiration by scientific men throughout the world, and was a welcome member of scientific conferences, in many of which she took an active part....The many friends of Mme Curie throughout the world, who admired her not only for her scientific talents but also for her fine character and personality, lament the untimely removal of one who had made such great contributions to knowledge, and, through her discoveries, to the welfare of mankind.

At a meeting of the Cambridge University branch of the Democratic Front Rutherford, who was in the chair, referred to the destruction of democratic forms of government in a disordered world.

It is a matter of great importance that those who believe in the present type of government should not stand idly by, but see if they can convince the waverers, or those who require convincing, of the great advantage of democratic government, at any rate in this country where we have experienced it so long.

At present there is a great feeling of tension, not only throughout Europe, but in a sense throughout the world. That feeling of tension and fear of war has been a striking mark of the last few years. This arises largely from the fear of the growing power of military aeroplanes with sudden and devastating attacks on defenceless cities, involving the destruction of combatants and non-combatants alike.

I am sure that the greatest possible relief from this fear of war would arise, if say, to-morrow we could ensure that aeroplane warfare could be abolished by consent of all the nations of the world. That would be a great epoch in history, but it may not be possible for a long period.... There is no question more important for the future than to see whether we can get some form of international agreement on the limitation of this air weapon which will undoubtedly grow in strength from year to year.

In Oct. 1934, a great scientific conference was held at London and Cambridge by invitation of the Royal Society. Members of the International Union of Pure and Applied Physics and of the Physical Society were invited, and many representatives came from Europe and America. There was much to discuss; indeed, so rapidly had the recent discoveries been made, that there was some indication of lack of theoretical assimilation. One notable change had gradually taken place. It used to be stated that the nuclei of atoms were composed solely of protons and electrons. There had been an alteration to the view that protons and neutrons were the main, or only, constituents of any nucleus. There was also some talk of the negative proton which either does not exist or else skilfully evades detection.

Rutherford presided at this Conference and gave a summary of the successive discoveries, concluding:

...We are as yet uncertain of the relation if any between the proton and the neutron. It has been suggested that a conversion of one into the other may occur by the capture or loss of a negative or positive electron, and even that under some conditions negative protons may be formed. It is probable that an accurate determination of the magnetic moment or spin of the neutron and proton will be helpful in throwing further light on these important questions....   .

When we consider the complexity of the structure of a heavy nucleus and the number of units that are condensed into its minute volume, it is obviously difficult to speak of the individual components as having a separate existence, for each constituent in a sense must occupy the whole nucleus. Bohr has pointed out that under such conditions the theory of wave-mechanics cannot be rigidly applied. Notwithstanding these difficulties and our uncertainty as to the detailed structure of nuclei, good progress has been made in several directions in interpreting in a general way some of the outstanding properties of nuclei.

It must be remembered that during these years Rutherford was the life and soul of the Cavendish, quickening every activity with fresh ideas and suggestions. One example of his scientific acumen may be given. In California various elements had been bombarded with high-speed deuterons and they gave a remarkable output of secondary particles. Although the experimental evidence was complete, he alone disbelieved the obvious explanation and declared that these particles came, not from the elements under bombardment, but from deuterons imbedded in them. A long and difficult investigation proved his surmise to be correct.

RUTHERFORD TO BOYLE

22 *Dec.* 1934. I am taking the opportunity before the laboratory breaks up to send you a note of good wishes for the New Year. Your last letter to me of November 12th gave an account of your adventures with blizzards in the early autumn, and reinforces the opinion that

I have long held that for six months of the year Canada is only suitable for polar bears!

You may have heard of the death of our old friend Lamb a few weeks ago. My wife and I had visited him a few days before his end and found him fairly cheerful but, of course, his increasing deafness and weakness on his legs made life a bit of a burden to him. He died suddenly and unexpectedly of a heart attack and the family and all of us feel that his end was a happy one. The funeral service, which was held in Trinity College, was a very imposing and dignified affair. As you know, I have known Lamb now for 27 years, and I always looked on him as one of the finest characters I have known, and one of the few men that grew old gracefully. So many are inclined to hang on and to grasp for power when they ought to be dandling their grandchildren!

I am glad to say we are all very well. We are staying for Christmas to have Christmas festivities with the grandchildren, and we then go down to our cottage in Wiltshire for a rest for a week or so. We are all very well and cheerful, and I managed to get a couple of days' golf with friends in Sussex at the close of the term, which has been rather a heavy one.

The research work has gone on very steadily, and we have made substantial if not sensational progress. The effect is very remarkable how the transformation by neutrons in many cases increases by slowing them down in their passage through hydrogen or hydrogen material like paraffin. The intensity of the radioactive bodies which are produced goes up in many cases about one hundred times, and this makes the neutron experiments very much easier.

We have been making an intensive study of the transformations of beryllium under bombardment with protons and deuterons. It has been very difficult to disentangle the complicated transformations which occur, and I think secondary bodies like $^3_1$H and $^3_2$He appear in some transformations.

I hope we shall have a chance of seeing you before long in this country. You ought to make a point that it is important for you to get in touch with the work going on in this country in the N.P.L. and the industrial research laboratories, and so wangle a trip over here. This is a recognised method of having a change of scene! Think of it—verb. sap.!

In December 1934, Rutherford gave the Ludwig Mond Lecture, at the University of Manchester, on 'The New Chemistry', and showed how the new field of nuclear chemistry was opening up with great rapidity. A whole new class of 'artificial' radioactive bodies had been discovered as a result of bombardment with various nuclei. He referred to the great success of Fermi and his collaborators using the neutron as disintegrator, for the neutron having no charge could wander into a nucleus, where a charged particle such as an alpha particle would be turned back.

Rutherford was at some pains to explain the difference between the old chemistry and the new:

The older chemistry with which you are familiar deals with the combination of atoms and molecules to make new compounds. The chemical attraction of one atom for another depends on the unbalanced electric or magnetic forces resident in the *outer* structure of the atom. The new chemistry has to deal with methods of changing one atom into another, that is, with the study of the modes of the transformation of the chemical atom itself when exposed to certain powerful disrupting forces. The most potent method so far employed depends on the bombardment of the atom by fast projectiles like alpha particles, or protons, or neutrons. Occasionally, one out of a great number of these particles by chance strikes the heart or nucleus of an atom. In some cases such a violent encounter may result in the breaking up of the atom into two or more lighter constituents—a veritable disintegration of the atom. In other cases a new atom of heavier mass may be formed by the incorporation of the bombarding particle into the nuclear structure.... During last year, definite evidence has been obtained that we can even transform atoms in the laboratory without the use of fast particles by using very penetrating radiations of röntgen or gamma rays....By these artificial methods we have liberated fast protons and alpha particles of higher energy than those liberated by radium and other radioactive elements. We have also brought to light the presence of unexpected and strange types of particles like the neutron and the positive electron, now called the positron. In addition, we have been able to detect the presence of new elements, or rather new isotopes of elements, previously not known upon earth, like a hydrogen of mass 3, and a helium isotope of mass 3. But even more surprising, we have been able to produce a new class of radioactive bodies which break up in novel ways—radioactive bodies which, if they ever existed in our earth, disappeared long ago. It may be that the radioactive elements uranium and thorium are the sole survivors of what may well have been a large group of unstable elements which are still being formed in the furnace of our sun.

The lecturer explained that, in his early experiment first proving the possibility of *artificial* transformation, the alpha particle striking a nitrogen atom caused the ejection of a proton while the alpha particle remained embedded in the nucleus. Yet the transformation probably occurred in two stages so that a fluorine nucleus existed for an exceedingly brief interval of time.

Most important of all are those transformations which give rise to *artificial radioactivity*, discovered by M. and Mme Curie-Joliot, the possibilities of which are now becoming vividly realised. Radio-nitrogen can be produced by bombing boron, in any chemical form, with alpha particles. That kind of nitrogen is unstable and changes into carbon, but the transformation takes place gradually. Radio-nitrogen continues atom by atom to shoot forth positrons so that the material decays to half-value, however much there be, always in 14 minutes.

It is possible that in time radio-sodium may be prepared in sufficient quantities to take its place in curative medicine in place of radium. It would have to be prepared frequently, for it decays to half-value in about 15 hours.

In January 1935 Rutherford gave the first series of John Joly Memorial Lectures at Trinity College, Dublin. The Provost, Dr E. J. Gwynn, was in the Chair and the subjects chosen were 'Radio-active Transformations: Old and New', and 'The New Hydrogen, and Artificial Transformations'.

He paid a tribute to his old friend, whom he regarded with great esteem and admiration.

Joly's alert and original mind was attracted to the problem of the effect of the generation of heat by the radioactive bodies present in the earth's crust on the geological history of our planet. To obtain reliable data,

he devised simple but ingenious methods for measuring the amount
of the primary radioactive bodies, uranium and thorium, in typical
rocks constituting the earth's crust. He was the first to point out the
far-reaching significance of this small but steady supply of heat on the
internal temperature gradient of the earth resulting in violent movements of the earth's crust. Indeed he was of opinion that the rise and
fall of continents and the elevation of mountain chains were intimately
connected with the heating effect of radioactive bodies over long
intervals of time. A fascinating account of these bold and original ideas
has been given by Joly in his books and papers.

At the annual dinner of the N.W. Centre of the Institution of
Electrical Engineers held at Manchester on 22 Jan. 1935, Rutherford
referred with pleasure to the time when he was one of a little group from
that city who were using electrical methods to foil submarine warfare
on our shipping. Turning to research he said that whereas in the old
days apparatus was simple and fairly inexpensive, today we must have
high voltage devices of a far more costly nature. He hoped to be
handling two million volts for this purpose in the near future. This was
necessary for bombardment—for firing very swift particles at certain
substances under investigation. He most gratefully acknowledged the
help and advice of Metropolitan-Vickers, for they had put their research
resources at his disposal and had been of enormous assistance. He then
drew attention to the great part played by electrons in electric currents,
and to the strange fact that the positive electron, or positron, was
insignificant. "Why is it", he asked, "that the positive electron takes
no prominent rôle in the transmission of electric energy? Always it
seems to be a case of negative electrons flowing through the metal,
never the positive ones. Somehow I always feel that the light positive
electron should be conspicuous in these electrical processes. But it
lies low and appears to do nothing."

He then referred to the discovery of the French physicists (Curie-
Joliots) in proving that some light elements under the action of alpha
particles gave rise to artificial radioactivity and the emission of positrons.
"May it not be that elsewhere in the Universe, or under circumstances

different from those familiar to us in electrical engineering, as we know it, the *rôles* of the positive and negative electrons may be reversed?"

It was at this dinner that Lord Crawford declared that he never disbelieved a scientist except when he spoke philosophy.

In his annual course of lectures to the Royal Institution Rutherford spoke on his old subject of 'Electromagnetic Waves'. He declared:

During the last fifty years there has been a great advance in our knowledge of electric vibrations, and we now have a detailed knowledge of the properties of electric waves over an enormous range of wave lengths extending from 1000 miles to a billionth of a centimetre. All these waves have the common characteristic that they travel at about 186,000 miles a second through space. They also possess certain other properties in common, being all scattered by matter and all absorbed, more or less, in passing through it.

In my youth it was supposed that the whole of space was filled with æther which was a perfectly elastic fluid and permitted the propagation of electromagnetic waves through it with the velocity of light.

In recent years, however, there has been a tendency to say that the æther is a pure invention, but I feel that for the purpose of my lecture, the æther is as necessary today as it has been formerly. The whole conception of waves loses its meaning unless there is a medium for them to pass through.

He gave another lecture to the Royal Institution on 29 March, speaking on 'Neutrons and Radioactive Transformations', and he showed remarkable experiments of neutrons passing readily through thick layers of lead, while they were absorbed by a comparatively thin layer of the metal cadmium. He showed that the hydrogen in paraffin wax also slowed down the neutrons. Most extraordinary of all was the breaking up of the nucleus of heavy hydrogen with gamma rays, producing a proton and a neutron.

Speaking of Chadwick's discovery of the neutron Rutherford stated:

These neutrons, which have remarkable powers of penetrating matter, are themselves very efficient agents for the transformation of atoms. Feather (of the Cavendish) has shown that both nitrogen and oxygen are transformed by the capture of neutrons, with the expulsion of a fast alpha particle. The types of transformation produced by the neutron are thus very different from those observed with the alpha particle. The capture of an alpha particle in general leads to the building up of a new nucleus three units *heavier* than before, while the capture of a neutron leads to the formation of a nucleus three units *lighter*, because an alpha particle is usually ejected.

In April 1935, Rutherford wrote an article for *Discovery* which summarised the work in modern physics up to that date. He mentioned that lithium and hydrogen nuclei can combine to give *two* alpha particles; boron and hydrogen give *three* alpha particles; fluorine and hydrogen give oxygen and helium; aluminium and hydrogen give magnesium and helium. He then referred to the showers of electrons sometimes produced by cosmic rays leading to the discovery by Anderson of the positively charged electron, or positron. Excellent photographs had been taken of the tracks in a C. T. R. Wilson expansion chamber of both negative and positive electrons curving in opposite directions owing to an applied magnetic field.

RUTHERFORD TO HAHN

25 *April* 1935. Thanks very much for your note and for sending me the copies of your recent articles on the neutron transformations of uranium. The examination of this point must have been very much in your line of territory, and I am sure you thoroughly enjoyed the opportunity of settling the nature of the transformation products. It is all very interesting. Things move so fast now that it is difficult to remember all the results obtained.

The account in *Nature* was merely a general statement, and there was no time to discuss the latest results.

I have just returned from a holiday in the country. You may have heard that we have built a little cottage in Wiltshire for our holidays, and my wife is very happy with this new holiday home.

Hevesy is coming over to Cambridge in a few weeks and giving the Scott course of lectures. As you know, he has left Freiburg and is now in Copenhagen.

I am glad to say we are all very well, including the grandchildren.

I am kept very busy not only with my personal research, but with many other scientific activities. You may have heard that Professor Kapitza has been forcibly detained in Russia by the Soviet authorities, so that I have to consider the carrying on of the work of his Laboratory during his absence.

The Rutherfords for five years spent their holidays in a cottage that they rented in Nant Gwynant, North Wales. This involved a long motor drive from Cambridge and it was desirable to have a country cottage within more reasonable reach. Lady Rutherford made a long search and found a delightful site near the village of Upper Chute, a few miles from Andover. It was on the slope of the downs on the borders of Wiltshire and Hampshire, not far from the Collingbourne Woods. Curiously enough, it was part of a farm called New Zealand. A few acres were bought which contained great elm trees, a rookery, and a pond, and had a fine view to the south. A fascinating country cottage was quickly built. Lady Rutherford with inexhaustible energy removed flints, made hedges and flower-beds, and filled them with all sorts of delectable flowering shrubs, and trees with autumn tints, besides flowers of all kinds in most natural settings.

Here Rutherford relaxed and regained energy, enjoyed reading and bridge, besides spending half the day in clearing bush, lopping dead limbs, felling trees and reducing them to firewood with axe, bill-hook and cross-cut saw.

A morning with him spent in this way was an invigorating delight; in the afternoon there might be a motor drive to Avebury, or Stonehenge, or one of the dykes or British camps with which Wiltshire abounds.

On 17 July Rutherford's mother died at the age of 92. She was a kindly, strong character, who had wisely brought up a large family, and given thought and care to their education. Her memory and good sense seem to have remained largely unimpaired to the end, and she had clear and strong views on modern education, small families and extravagance. Like many people of advanced age she had been a strong advocate of hard work as the "sovereign remedy for many evils of to-day".

When the British Association met at Norwich on 6 Sept. 1935, Rutherford opened a discussion on 'Nuclear Physics'. He referred to the work of Lauritsen in the United States who found that when lithium was bombarded with swift protons some gamma rays were produced equivalent to 16 million volt X-rays. He further called attention to the very important result obtained by Lawrence of California, who had shown that the bombardment of rock salt put its sodium atoms into a radioactive state. These give rise to gamma rays which may be of great value in biology and medicine, possibly in due course replacing radium.

There was a further discovery of importance when Fermi, also Otto Hahn and Lise Meitner, permitted slow neutrons to enter the nuclei of the two heaviest elements known—thorium (90) and uranium (92)—with the astounding result that they obtained new elements of higher atomic numbers (93, 94 and 95) than any elements hitherto known in Nature.[1] This was a crowning glory to that edifice the foundation of which Rutherford had so truly laid, and such discoveries gave him genuine pleasure.

In 1935 the case of Dr Kapitza became world-famous. He was a physicist, engineer, inventor and mechanic; loyal to his country. He left Russia in 1921 and after visiting some laboratories on the continent came to Cambridge and was awarded a Clerk Maxwell Scholarship for the period 1923–26. He began to devise methods of obtaining exceedingly high magnetic fields for a very short period. At first he charged accumulators (secondary lead batteries) of low resistance and suddenly discharged them in parallel, obtaining a current of 50,000 amperes for about a hundredth part of a second. He next used an alternating-current generator running at full speed with a current of 80,000 amperes shorted through his coils, whereby he got for a brief

[1] It has since been shown, at several leading laboratories, that when a neutron enters into a uranium atom there is an explosion and the atom really breaks with violence into lighter atoms, themselves radioactive.

period an intense magnetic field of 300,000 gauss. The switch gear for such work evidently needed careful design and manufacture. As he wanted to work at low temperatures, he designed a helium liquefier on an entirely new principle which is still working and giving two-and-a-half litres an hour. There was also a liquid hydrogen machine and another for liquid air yielding about twenty-five litres an hour of nitrogen. Rutherford declared that all his work was characterised by marked originality in conception with an unusual command of original technique.

In 1925 Kapitza had been elected a Fellow of Trinity College, Cambridge, and four years later, Fellow of the Royal Society. He married a Russian wife; they had children and he settled down to his work and family life in Cambridge. He had been to Russia several times from 1926 onwards, and in 1934 he again went to his native country, partly to attend the Conference held in honour of his great compatriot, the chemist Mendeléeff. Then came the bolt from the blue. A few days before his return to Cambridge, he was informed that he must stay and work in Russia. Now Kapitza without his apparatus or the apparatus without Kapitza was at a standstill. Rutherford wrote to Stanley Baldwin (29 April 1935): "Kapitza was commandeered as the Soviet authorities thought he was able to give important help to the electrical industry and they have not found out that they were misinformed." However that may have been, Rutherford wrote a letter appealing that Kapitza might be allowed, in the interests of science, to return and complete his work. To this the Soviet Government made the sagacious and fair retort that of course England would like to have Kapitza, and that they, for their part, would equally like to have Rutherford in Russia! Since Mahomet could not go to the mountain, the mountain had to go to Mahomet. Negotiations were begun and Professors Adrian and Dirac went to Russia to interview Kapitza and others. Finally the Russian Government bought the apparatus for £30,000—a fair and proper price—and Cockcroft had the apparatus packed and dispatched to Russia.

Not all this money came to the Cavendish Laboratory, because the British Government, through the Department of Scientific and Industrial Research, had contributed large sums for the generator and other apparatus. It was agreed that the Cavendish would not compete with Kapitza in his field of research, and the money received was devoted to other objects, in the Royal Society Mond Laboratory.

All these worries and troubles about his work and family fell heavily upon Peter Kapitza and Anna his wife, to whom Rutherford wrote most kind and consoling letters. Kapitza accepted the inevitable, settled down to work and determined to do his utmost for the development of Russian science. In 1936 he wrote to Rutherford, with something of Russian fatalism—"After all we are only small particles of floating matter in a stream which we call Fate. All that we can manage is to deflect slightly our track and keep afloat—the stream governs us!"

In December 1935 Sir Austen Chamberlain was installed as Chancellor of the University of Reading and an honorary degree was conferred on Lord Rutherford of Nelson, O.M., "the discoverer of new paths of knowledge".

In the same month, Rutherford opened the new Research Laboratory of the London, Midland and Scottish Railway Company at Derby. Many of the guests were also invited to attend at St Pancras the ceremony of unveiling the latest built engine of the Company which was taking their train to Derby. Lady Rutherford was requested to unveil the engine, which was named after her husband, but she thought that this would give great pleasure to her grandson, Pat Fowler, aged nine, a lover of engines.

So Pat went to St Pancras and, perched on a huge engine, drawing aside the veil, declared it to be named 'Lord Rutherford of Nelson'; after which Sir Josiah Stamp, President of the L.M.S., presented him with a finely executed model, complete in every detail.

Rutherford gave many public addresses and scientific lectures, but these formed the least important part of his activities. His real life was far more intimately bound up with his work in the laboratory, both with his own researches and with those which he inspired in the circle of able men who flocked to his guidance. As C. D. Ellis has pointed out[1] "the majority of the experiments at the Cavendish were really started by Rutherford's direct or indirect suggestion. Even in the later years, when he had so many outside responsibilities, he took a deep and personal interest in all the researches and stimulated all stages of the work to a remarkable degree."

In 1936, Pope Pius XI reconstituted the old pontifical Accademia dei Lincei (1603) and renamed it the Pontifical Academy of Sciences.[2] Seventy Academicians were named, three of them British—Lord Rutherford, Sir Charles Sherrington and Prof. E. T. Whittaker. The Pope stated that

Science, which is a true knowledge of things, never conflicts with the truths of the Christian Faith; thus—as those who have studied the history of science cannot but confess—it must be recognised that on the one hand the Roman pontiffs, together with the Church, have always favoured the researches of learned men, even in experimental matters, and that, on the other hand, these researches have paved the way for the defence of the treasure of supernatural truth, confided to that same Church. Wherefore, as the Vatican Council solemnly taught, "not only cannot Faith and Reason ever be in disagreement, but they bring one another mutual help"....

It is unfortunately true that in times very near our own, several learned men—certainly not through love of the truth—have left the Church as prodigal sons, and especially in the course of the last century, it was taught with false and bold reasoning that the teachings of human science are in open conflict with those of Divine Revelation. But it is a profound joy to Us to see that these prejudices are in our times so outgrown that hardly anyone can be found who, after having studied with any depth the positive sciences, continues to uphold this error.

It may well be asked what Rutherford really thought of this statement—did he agree with one half and smile at the other?

[1] London Physical Society, 1938, pp. 441–6.
[2] Nature, 1937, 140, p. 965.

I knew Rutherford rather well and under varied conditions from 1903 onwards, but never heard religion discussed; nor have I found in his papers one line of writing connected with it. His conduct was that of an honourable man, who was fair-dealing, kindly and helpful to others. There is the old story of the man who claimed that he belonged to that religion to which all wise men belonged, and on the question, Which? he replied, "That no wise man ever says."

Certainly Rutherford hoped that his work would live, but he had no wish that his personality should survive for all eternity. Yet there is one remark of his which deserves to be recorded:

Shortly before the death of W. B. Hardy, Rutherford was greatly concerned over his friend's health and said "Hardy is terribly ill". Then, turning to the Bishop of Derby, he said to him with earnestness, "You will pray for him, Pearce, won't you?"

CHAPTER XV

THE NEWER ALCHEMY

The 17 Jan. 1936 was the bicentenary of the birth of James Watt, and Rutherford gave the Watt Lecture in the Watt Hall in the great man's native town of Greenock. The subject of the lecture was 'The Transmutation of Energy' and Rutherford began by drawing attention to the good fortune of the young instrument-maker James Watt in making the acquaintance of Joseph Black of Glasgow University, the greatest authority on heat at that time, the man who had first introduced the idea of latent heat.

Rutherford pointed out that:

Watt began his experiments on the steam-engine with a breadth of scientific knowledge and outlook which markedly distinguished him from the engineers of that time. After exhaustive trials, he correctly concluded that a source of the wastefulness of the engine lay in the cooling of its walls by a jet of water used to condense the steam and thus create a vacuum necessary for the reintroduction of steam on the next stroke, so he arranged that the steam in the cylinder should be connected at the appropriate moment with an external separate condenser, kept at a low temperature.

Rutherford spoke of the work and ideas of Joule, Clausius and Kelvin and then passed to modern applications such as the use of mercury and its vapour (replacing water and steam) by the General Electrical Company in the United States. He touched on the conception of Einstein which linked energy and mass, so that matter is now regarded as a condensed form of energy. He called attention to a profound difficulty in the physics of the nuclei of atoms, where the laws of mechanics do not always hold, unless the balance sheet is made good by the introduction of an imagined particle called the *neutrino*, which is required to have no charge and little mass, but an accommodating spin. It is hardly necessary to add that the introduction of such an unobservable particle is far from pleasing to many physicists. Nevertheless Rutherford himself demanded the existence of radium B and radium D at a time when no physical properties could be attributed to them.

Finally Rutherford stated that an electron and a positron might sometimes combine and give rise to pure radiation, to two gamma rays. This was called de-materialisation. In a sense this is the reverse of the creation of matter, but the balance between mass and energy is still conserved, following Einstein's law.

This great statement links together energy and mass, so that, whenever one is converted into the other, strictly proportionate amounts are involved. The remarkable thing is that they are linked by the velocity of light, or more strictly by the square of that velocity, so that energy equals mass multiplied by the square of the velocity of light, or in shorthand, $E = mc^2$. This is, at the same time, one of the simplest and most profound relationships in the whole known realm of nature. A gramme of matter has 9×10^{20} ergs of energy; or one pound of any material has locked up within it about 15 million million foot-tons.

In his Faraday Lecture to the Chemical Society in February Rutherford stated that "no less than fifty radioactive elements have been artificially produced and many more will no doubt be found. The subject of radioactivity has indeed been born anew, and has entered on a new and vigorous phase of life."

Yet he had his regrets. He hoped that new 'families' of radioactive elements would be discovered, and wished he was young enough to help to unravel their complexities as he had previously helped to do with the natural radioactive elements.

In 1936 Rutherford was again prominent in an effort to maintain "the traditional loyalties of the freedom of science and learning" and the following letter appeared in *The Times* of 17 March:

Without international goodwill there can be no stable peace in Europe. It is clear that public opinion in this and other countries needs to be convinced that Herr Hitler is sincere in his professions of peace.

Something must therefore be done to restore the confidence upon which goodwill depends. We cannot at this juncture refrain from suggesting that one of the greatest obstacles to a good understanding between this country and Germany, one of the major causes of diffidence concerning her proposed re-entry into the League, is her harsh treatment of the Jews and increasing interference with religious liberty. Not only is it intolerable that other countries, and especially those bordering on Germany, should, despite their economic difficulties, be obliged to cope with a flood of refugees, but the knowledge that those personal and spiritual rights which Englishmen hold most dear are being daily violated by the German Government is a formidable obstacle to cordial co-operation. Is it too much to hope that Herr Hitler will have the greatness to remove that obstacle?

We are, etc.,

Allen of Hurtwood.
G. P. Gooch.
Gilbert Murray.
Rutherford.

In his 1936 report on 'The Scholar Exiles', as President of the Academic Assistance Council, Rutherford stated that

Of the approximate 1300 scholars displaced in Germany—that is, nearly one-sixth of the university staffs—700 have emigrated. Of these 363 are now permanently re-established elsewhere, and 324 are in temporary positions or supported by special maintenance grants....

Conditions in Germany for the displaced scholars have become much worse recently. 'Non-Aryans' have been forbidden to accept invitations to lecture to learned societies, or to universities outside, and have been made to send excuses on grounds of ill-health. Scholars have been denied access to public libraries, so that some cannot even continue their research privately....It would seem that our displaced colleagues within Germany must either suffer extinction, or flee as destitute refugees, or be assisted to emigrate and re-establish themselves with the aid of an organisation like this Council.

In his Royal Institution Lectures for 1936 Rutherford, as usual, had plenty of new things to show. There were photographs of the tracks of positrons ejected from radio-phosphorus. He explained how Chadwick and Goldhaber had broken up the deuteron, or nucleus of heavy hydrogen, by the use of gamma rays—the first example of transmutation of an element effected by radiation alone, and leading to a good method of estimating the mass of the neutron. He showed photographs of the trails of proton and triton, that is the nuclei of hydrogen and hydrogen of mass three, going out in opposite directions for unequal distances, as obtained by Dee when he bombed deuterons with deuterons.

There was another photograph of helium and hydrogen of mass three moving asunder when lithium (6) absorbed a neutron; two alpha particles when lithium (7) combined with a fast proton; while lithium (7) and a deuteron gave two alphas and a neutron, thus

$$\mathrm{^7_3Li + ^2_1D \rightarrow ^8_4Be \rightarrow ^4_2He + ^4_2He + ^1_0n,}$$

where beryllium is the intermediate unstable product.

He specially directed attention to the result obtained by Lawrence of California, who, bombarding the element bismuth with deuterons with the aid of his cyclotron, obtained radium E, which turns into polonium—the very same radioactive element that Mme Curie had first extracted from pitchblende. Here was a natural radio-element prepared by artificial radioactivity. A bridge was built across a gap.

When Rutherford spoke or lectured he was, at the start, nervous, largely because he was anxious about a troublesome throat, but when he warmed to his work, particularly on his favourite topics, he was carried away with his subject and he used simple and direct English, clear and forceful. His writing was devoid of frills and combined clarity with power. He found fault with those willing to bestow months on an investigation and yet grudging hours for the publication. In a recent speech Lord Tweedsmuir has recalled a remark made to him by Rutherford that he did not consider a discovery complete until it had been described in simple and correct English.

At the official banquet of the Chemical Engineering Congress of the World Power Conference (May 1936), Rutherford proposed the toast of the Congress, saying:

I cannot claim to be a chemist, but, on the other hand, I have often raided the borders of the chemist like a moss-trooper and returned with some loot. We all recognise that the borders between chemistry and physics are getting very shadowy and it is very difficult to label a man as a physicist, a physical chemist or a mathematician....

Bellingham is one of the great triumphs of the chemical engineer.... The production of artificial silk and the development of plastics—still in the initial stages—are of great importance today and will be in the future. These things do not happen by accident, but are the results of long research in the universities and other places. The Department of Scientific and Industrial Research, of which I am Chairman, has financed a special laboratory which has been engaged, during the last ten years or so, in high pressure chemistry. Indeed without pure science the possibilities of industry would wither.

On another occasion he stated:

In this age of severe industrial competition, other things being equal, success will go to the Nation which is so organised that it can utilise rapidly the application of science for the improvement of existing industries and the rapid initiation of new ones.

In April 1936 came the good news that Sir Herbert (now Lord) Austin had presented to Cambridge University for the Cavendish Laboratory the sum of £250,000. This gift followed from a recommendation made by Mr Baldwin, Chancellor of the University. This generous donation came as a complete surprise to Lord Rutherford and gave him great satisfaction, as did also this letter from the donor to the Chancellor,

I have for several years been watching the valuable work done by Lord Rutherford and his colleagues at Cambridge in the realm of scientific research. Knowing that as Chancellor you are keenly interested in obtaining sufficient funds with which to build, equip and endow a very much needed addition to the present resources, I shall be very pleased indeed to present securities to the value of approximately £250,000 for this purpose.

After the initial pleasurable excitement, came thought and planning; some money was to be spent in building and some reserved for endowment.

RUTHERFORD TO HAHN

5 *Nov.* 1936. I was very interested to get your letter, and thank you for sending me your book *Applied Radiochemistry* containing the gist of your Cornell Lectures. I have only had time to glance through it and read the first few chapters, but I can appreciate you have got a lot of good stuff in it, with much of which I have only a slight acquaintance. I was particularly interested in your estimate of the quantities of substances like mercaptan which can be detected by their smell. I thought that the olfactory nerves were more sensitive than they apparently are, but I suppose they have about the same relative sensibility for molecules as the eye has for 'visible' quanta.

I am sorry that I have unwittingly upset you by my reference to the thorium-neutron transformations. Actually I wrote my paper when I was in my cottage in Wiltshire on holiday, where I had no papers to refer to. I had glanced through the Joliots' paper and thought that their evidence was rather vague, and I gave expression to that view in my address. I had for the moment forgotten that you had actually published your letter in *Naturwissenschaften* before their paper appeared, and I quite agree that they ought to have specifically mentioned your definite statement about the $4n+1$ series. I can assure you that I had read your papers, but you can well imagine in the terrific rush of papers on neutrons that my memory is occasionally defective! I shall make a point of putting the matter right when I have the opportunity.

This brings me to another point on which I was going to write to you. I shall be giving a Royal Institution lecture, probably in April, in which I shall draw attention to the effect of neutrons, etc. on uranium and thorium, and I would be very much obliged if you could give me briefly (preferably in English) your considered opinion of the present state of that subject, on which I know you and Miss Meitner have spent so much time. This will be very helpful to me, and I need not then discuss the immature work of others.

I am glad to say that I am very fit and well, but I am kept very busy in a multitude of ways. You will have heard of the substantial gift of Lord Austin to the Cavendish Laboratory, and we are now in the midst of considering plans for a new research block and for possible reconstruction of parts of the old Cavendish. Our new High Tension Laboratory is nearly completed, and we hope to begin to install our new apparatus to obtain 2 million volts D.C. early in January. Oliphant is busy preparing this new installation, which is experimental, but which I hope will work all right. As 2 million volts will probably give a spark of more than 30 ft. long, you will appreciate that plenty of headroom is required!

With kind regards to Miss Meitner and yourself, and best wishes for the success of your work.

P.S. You probably know that Professor and Mrs Eve are now living in London, and I see them as much as I can. He is remarkably fit for his years—about 74—and plays a game of golf that would not disgrace a youngster!

Rutherford first played golf, in his student days, at Royston with J. J. He could hit the ball, but not in the right direction, and came to the conclusion that he was not then old enough for the game. According to Sir Harold Carpenter, he found the game a valuable relaxation at Manchester and played on the Withington course until an increasing trouble in one knee kept him for a time from the game. When he returned to Cambridge he was a regular player on the Gogmagog Hills with R. H. Fowler, F. W. Aston, G. I. Taylor, R. V. Southwell and other friends. They generally played a game with three balls—not a threesome —because the players might number five or six.

On the course, they were known as 'The Talking Foursome', but behind the conversation was a desire to win and considerable skill in the effort. Rutherford had an uncanny way of arriving near the pin when least expected. Those who were prepared to face an early breakfast on Sunday morning—and some rebelled—were amply rewarded by an entertainment amid blue flax and skylarks, in which golf played no inconsiderable part.

After 1928, it became a custom to organise a party to enjoy two or three days' golf, immediately after the Oxford and Cambridge Rugger Match. The first visit was to Frilford Heath, and then for two or three years the games were on the courses near Brighton, until some cold and windy weather drove the players to prefer the very beautiful course at Ashdown Forest.

It is agreed by all who were lucky enough to be there that Rutherford was on these occasions absolutely at his best, "reminiscencing about everything under the sun" both discreetly and most indiscreetly. It need hardly be added that the party dominated the hotel in the evenings and that other guests were curious to learn the origin of these irresponsible and exuberant beings.

Sometimes a golfer would drop out of the party and Rutherford would exclaim, "Ah! he couldn't stand the rough and tumble of a Trinity foursome."

At the hotel he would begin a story, at the top of his voice, unconsciously regardless of others present: "Now this is just between you and me—this mustn't go any further—" and proceed with his yarn, with a very attentive and large audience, who were deeply disappointed when he lowered his voice to a whisper.

Rutherford was never shy with strangers. Invited once to a dinner at Pembroke, he arrived early and found the Master's drawing-room empty except for an ecclesiastical personage. The two men introduced themselves and were soon in friendly conversation. Later, in the college hall, Rutherford was called on for a speech and described the meeting: '"I'm Lord Rutherford" I said and the other man said "I'm the Archbishop of York"—and I don't suppose either of us believed the other!'

It was of another archbishop that Rutherford asked what books he had lately been reading. The archbishop explained that his multifarious duties prevented him from reading as much as he would wish. "Ah, yes" said Rutherford "you must lead a dog's life".

On 12 November, a large audience heard Rutherford give the twelfth

With J. J.   After breakfast

On the sands

Norman Lockyer Lecture at the Goldsmiths' Hall on 'Science in Development'.

He contrasted his Grove battery of the days of his youth, which he had to prepare afresh daily to get a current, with the recent developments at the Cavendish then preparing to obtain many million volts to drive their bombarding nuclei; he compared the old hand air-pump with the modern types of high diffusion pumps. He spoke of the use of electrons, those faithful little slaves of ours, in oscillators and rectifiers which made possible those miracles of modern science, radio-telephony and broadcasting; while the cathode-ray tube played an indispensable part in television....He added:

It is sometimes suggested that scientific men should be more active in controlling the wrong use of their discoveries. I am doubtful, however, whether the most imaginative scientific man, except in rare cases, is able to foresee the result of any discovery....Taking a broad view, I think that it cannot be denied that the progress of scientific knowledge has so far been overwhelmingly beneficial to the welfare of mankind.

He further suggested the possibility of

forming a Prevision Committee of an advisory character, composed of representatives of business, industry and science, who could form an estimate of the trend of industry as a whole, and the probable effect on our main industries of new ideas and inventions as they arose, and to advise whether any form of control was likely to prove necessary in the public interest.

A competent Committee of this kind could, no doubt, have foreseen the coming competition between motor and railway transports, which had such serious effects on the latter, and have advised the Government on the need of adjustment of competing claims before the difficulties became acute.

About this time Andrade tried to persuade Rutherford to write an account of his life and scientific work, but unfortunately it was postponed, for Rutherford wrote to him:

17 *Nov.* 1936. I note what you have to say about a possible autobiography of my scientific career. I am much too busy and possibly too

youthful to think of it now, but I have had it in my mind for some time as a possibility, especially in connection with some historical lectures I gave at short notice recently.

In this same letter Rutherford added a note on some then recent advances in modern physics:

Within a month of Chadwick's proof of the neutron, Feather in the Cavendish showed by expansion chamber work that neutrons were very effective in disintegrating both oxygen and nitrogen, and this was followed up by Harkins in the U.S.A. The main merit of Fermi was his rapid trial whether neutrons would produce radioactive bodies, immediately after the Curie-Joliot discovery.

A letter to Stefan Meyer throws some light on Rutherford's activities:

11 *Dec.* 1936. I was very pleased to see that Hess had been given a half quantum of the Nobel Prize, and I have received a very pleasant letter from him.

I have had a very busy time but for the last few days have managed to get a day's golf in the country. I was present in the House of Lords yesterday when the letters of abdication of King Edward were read and short speeches were made by various members. It was a very crowded House, both with members and visitors, and was an unforgettable experience.

We are holding the Cavendish Dinner on Saturday and making it a special occasion as J. J. T.'s 80th birthday falls on December 18th. You may have seen that he has published a very interesting book called *Reminiscences and Recollections*. He gives an interesting account of British men of science whom he has known in his long life.

In his early days, Rutherford did not smoke, but when he first went to Cambridge he decided (11 Aug. 1896) that tobacco soothed his nerves when under the strain and excitement of his eager experimentation. He finally became a heavy smoker, generally preferring a pipe.

The tobacco tin when received was opened and the tobacco spread out on a newspaper to dry. So far as is known, this heavy smoking did him no harm, but perhaps his hand was not now so steady, when doing an experiment, as in his younger days. He always said that he only smoked half of every pipeful.

The story goes that one day Crowe was busy fitting a new gold leaf to an electroscope, always a tricky operation, when Rutherford, impatient at the time the process was taking, exclaimed "Here, give it to me". He sat down and made several attempts, but each time his shaky hands prevented it. At last he gave in and said, "Here, Crowe, you had better do it, my nerves are not good today". A few days later he was attempting the same operation and failing for the same reason, when Crowe rashly exclaimed "Nerves not too good today, sir?" Whereupon there was an irritable reply, "Nerves be damned, you're shaking the table!"

Rutherford continued his lively interest in the treatment of cancer with a beam of gamma rays proceeding from five grammes of radium. The results obtained were satisfactory enough to justify further financial assistance and reorganisation. He advocated that the investigation should proceed and develop under the direction of the Medical Research Council—a step which has been achieved. Thus he wrote:

RUTHERFORD TO EVE

20 *April* 1937. I have just returned from the country after a good holiday. Chantry Cottage will be at its best in about a month's time when we hope to make a flying visit during Whit week.

I am intending to be present at the next meeting of the Beam Committee on Friday, April 30th, when I understand Smith and Mellanby will be there. I believe that our difficulties with regard to finance next year and for future developments would be greatly reduced if the whole Radium Beam Research was governed as a Committee of the Medical Research Council under Mellanby. You can appreciate the difficulty of a Government Department being responsible for considerable financial contributions to an outside body on which they are only slightly represented. If things are to go forward, I feel we must have the full support of the Medical Research Council and also the D.S.I.R. to a secondary degree.

The next letter refers to a matter of the greatest importance. It is now possible to introduce into the human body by various methods some of the new artificial radioactive substances which may be perfectly innocuous in themselves and yet irradiate for a known period of time the tissues or other parts of the body. There is an immense field of research involved. Furthermore such substances, in the minutest quantities, can be traced as they spread through the system and may be profitably employed in this manner as 'indicators'.

RUTHERFORD TO HEVESY

10 *May* 1937. I have of course read your papers that dealt with the circulation of phosphorus and deuterium in the body, and I thought your results of very great interest for throwing light on matters that appeared in the old days to be outside the region of possible investigation. Of course I have not sufficient biochemical knowledge to understand the possible interpretation that can be placed on the data.

Our High Tension Laboratory is completed and we are hard at work getting the apparatus in working order. We can get at present 1,300,000 volts D.C., and hope to begin work with it shortly; we also have in view the production later of 2 million volts D.C. We also expect to install a cyclotron in some months' time.

We spent our last vacation at the Cottage, and go down, as you suppose, at Whitsun for a few days when the garden ought to be looking at its best. We have New Zealand visitors with us at present, and my wife and Fowler are busy arranging to take the grandchildren to the Coronation. I myself have decided to have a quiet time at home and hear it on the wireless!

Rutherford was interested in the prevention of tropical diseases and asked the Secretary of the Royal Society for particulars about Dr Saul Adler, who hunted germ-bearing insects in Sicily, Malta, Greece, Crete and Palestine. This skilful doctor was told that a certain dangerous insect did not exist in Athens. Adler, knowing its habits, found it the first night at dusk. How then had the other savants failed? It is said that the insects appear at the hour of cocktails!

During the summer at Cambridge, Rutherford began his Presidential address for the joint Meeting in India of the British Association for the Advancement of Science and the Indian Science Congress; hence this letter to Prof. J. L. Simonsen:

18 *June* 1937. I understand that you are coming on the Indian Tour at the end of the year, when I have rashly promised to be President of the combined British and Indian Associations.

Before long I shall have to consider, I suppose, the preparation of an address, and of course I shall want to make some references to the Indian Association of Science. I have been given by Howarth your Presidential Address in 1928, which gives the history of this body, in which you took such a valiant part. I presume that there have been no substantial changes in its organisation since that time.

I should be very greatly obliged if you could make any suggestions of matters to which you think I ought to make reference in my address. While I have not finally decided on the arrangement of it, I think it probable that I shall, after references to the Indian Association, deal in a general way with the organisation of research in this country, following this up of course by remarks about the scientific activities of the Indian Government. My references to the latter will necessarily be rather brief, but it may be important for me to put in some perspective the early efforts of the Indian Government in this direction. Possibly you could give me some pointers as to what it would be desirable to mention in appreciative terms. It is not likely that I shall, except indirectly, refer to what might be done in the future in India, but I shall leave it to be inferred from my remarks on the organisation of research in this country and on the activities of our Dominions along similar lines. In conclusion, I shall probably comment on the modern work on transmutation, and the great changes that are taking place in the technical developments required for many of the researches in this field.

I know that a visit of the B.A. to India was one of your pet projects long ago, and that you are very interested in its success. If, by any chance, you could come to Cambridge, I should be very glad of the opportunity of having a good talk with you about the whole subject. I am going for a short holiday to my country cottage in a day or two, and shall be returning on July 5th. I should be very glad to see you again, and to put you up, if you can come to Cambridge. Naturally I would value your advice on questions of travel and dress.

This is already a long letter, but I am sure that I can rely on your good offices in giving me your opinions.

In the summer of 1937 the garden at Chantry Cottage saw a novel type of horticulture. Just under the surface there are numerous heavy and close-lying flints. In order to dig a hole for trees or bushes, or to prepare a flower bed, long hours of toil with pick and crowbar are necessary to dislodge the flints. After two years of this a new scheme was adopted. This was to make a hole with a crowbar about three feet deep and to insert two or three sticks of high explosive—gelignite—with a detonator and about two feet of fuse. When this was lighted all took shelter behind a tree, and in about half-a-minute there was a dull thud, and stones and earth went skywards. The method was very efficient and contributed some excitement and amusement to all. Of course Rutherford enjoyed it as much as a schoolboy.

Outdoor life always charmed Rutherford, whether playing golf, bathing, motoring, or tree-felling. Once, as he turned homeward after clearing bush, he quoted the lines:

> Sleep after toil, port after stormy seas,
> Ease after war, death after life, doth greatly please.

H. Wyndham Boyle, in an article on "Lord Rutherford's Restless Quest", wrote in 1936:

Lord Rutherford looks like a peer—large, big-boned, his heavy grey moustache, thinning hair, twinkling wide-open eyes, ruddy out-door complexion and rather old-fashioned clothes put you in mind of a county grandee, up from the West. He has an air of natural authority and dignity which are best shown off by the scarlet benches of the House of Lords. When he goes there to make one of his rare speeches he commands instant attentive respect—due to an aristocrat of learning and a great experimental scientist.

He speaks, too, in a rich, easy, country sort of voice. When you hear him, you are puzzled to place the very individual burr which affects all he says. It seems a peculiar Rutherford manner of speaking; you cannot assign it to any particular part of Britain....He wears his years very lightly and vigorously. He has no time to worry about such little things as growing old. He is too much absorbed with the present. . . .

He is proud of his position. He is innocently delighted by the Order of Merit to which he brings such distinction. In fact he is rather like a nice schoolboy, who has won a scholarship, concerning the recognition which his enormous learning and monumental research have brought to him.

Best of all is to see him in his home, a very comfortable home looking over the 'backs' at the colleges across the River Cam. . . . It is a triumph of all a home ought to be.

Lord Rutherford is untidy and forgetful. A nuisance in the house, I suspect. He litters papers about, and behaves just as a great man ought. And Lady Rutherford tidies up after him and wonders why so much untidiness is really necessary.

Which is all as it should be.

In July 1937, Lord Rutherford took his oath of allegiance to the new King. He had soon afterwards the pleasure of acting as one of the supporters to two friends when they were introduced to the House of Lords—Lord Cadman and Lord Chatfield. Rutherford wrote on 10 June:

I note in The Times this morning an account of Baldwin's introduction. You will be thoroughly amused at the procedure, and the way that the Garter King of Arms treats both you and your two supporters as schoolboys, and orders them in a loud voice to don and doff their hats. As a matter of fact, I think that the Garter King of Arms puts both the new peer and his supporters through a preliminary canter before the official time, so that each one shall know his part.

In 1937, Rutherford's last book, entitled The Newer Alchemy, was published by the Cambridge University Press. It was based on the Henry Sidgwick Memorial Lecture delivered at Newnham College, Cambridge, in the November of the previous year. The book begins thus:

The belief in the possibility of the transmutation of matter arose early in the Christian Era. The search for the Philosopher's Stone, to transmute an element into another, and particularly to produce gold and silver from the common metals, was unremittingly pursued in the Middle Ages. The existence of this idea through the centuries was in no small part due to a philosophical conception of the nature of matter which was based on the authority of Aristotle. On this view, all bodies were supposed to be formed of the same primordial substance, and the four elements, earth, air, fire and water, differed from one another only in possessing to different degrees the qualities of cold, wet, warm and dry. By adding or subtracting the degree of one or more of these qualities, the properties of the matter should be changed. To the alchemist imbued with these conceptions, it appeared obvious that one substance could be transmuted into another if only the right method could be found.

In the early days of Chemistry, when the nature of chemical combination was little understood, the marked alteration of the appearance and properties of substances by chemical action gave support to these views. From time to time there arose a succession of men who claimed to have discovered the great secret, but we have the best reason for believing that not a scintilla of gold was ever produced....Transmutation was a hopeless quest with the very limited facilities then at the disposal of the experimenters.

The book contains, amongst other things, some of the most superb photographs of nuclear collisions that were ever taken, the work of Blackett, Dee and Gilbert. There are the trails of an alpha particle striking an oxygen nucleus four times as massive as itself; the disintegration of nitrogen by an alpha particle with the resulting proton and oxygen nucleus; a neutron and helium (3) produced when deuteron encounters deuteron; boron and hydrogen combining to form three helium nuclei, all in the same plane, with their collective energies of motion corresponding to the energy of disruption.

Not less amazing is the photograph of a six-million-volt beam of deuterons issuing from a cyclotron, due to Prof. E. Lawrence. Rutherford pointed out that

the atomic nucleus is a world of its own in which a number of particles like protons and neutrons are confined in a minute volume and are held together by very powerful unknown force....

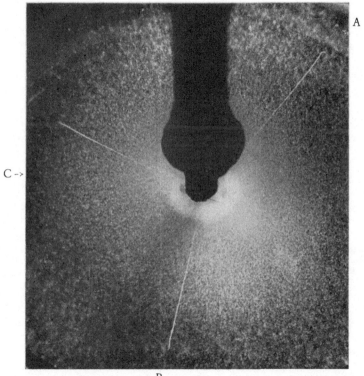

A

C ->

B

(*Photographed by* P. I. Dee *and* C. W Gilbert)

Boron and hydrogen → three alphas

Bohr has suggested that the nucleus should be regarded as an aggregate of indistinguishable particles which is capable of vibration as a whole and has well-marked energy levels....

In the interior of a hot star like our sun, where the temperature is very high, it is clear that the protons, neutrons and other light particles present must have thermal velocities sufficiently high to produce transformation in the material of the sun. Under this unceasing bombardment, there must be a continual process of building up new atoms and of disintegrating others, and a stage at any rate of temporary equilibrium would be reached. From a knowledge of the abundance of the elements in our earth we are able to form a good idea of the average constitution of the sun at the time, 3000 million years ago, when the earth separated from the sun. When our knowledge of transformations is more advanced we may be able to understand the reason of the relative abundance of different elements in our earth and why, on the average, even-numbered elements are far more abundant than odd-numbered elements. We thus see how the progress of modern alchemy will add greatly to our knowledge not only of the elements, but also of their relative abundance in our universe.

On 21 Aug. 1937, Rutherford published his last contribution to *Nature* on "The Search for the Isotopes of Hydrogen and Helium of Mass 3". These rare isotopes had been detected by Oliphant and Harteck when bombarding heavy hydrogen with heavy hydrogen, or rather deuterons with deuterons, for only the nuclei are concerned with these transformations. The new gases had not been collected however, so the search began for hydrogen of mass 3, tritium, by heroic measures.

In Norway 13,000 tons of water were treated by electrolysis until less than 100 pounds of water remained, and this would contain much deuterium, as D_2O. It was hoped that tritium, T_2O, would also be there. It was therefore a great disappointment when Aston could not find tritium after careful examination with his mass-spectrograph. Tritium probably exists, but it has not been collected.

RUTHERFORD TO GEIGER

5 *Sept.* 1937. It was very good of you to write on my 66th birthday which I spent at our country cottage in Wiltshire. My wife and I are

very happy there and I shall spend a month there to have a good rest before we leave for a tour in India at the end of November. I am President of a joint meeting of the British Association and the Indian Association and about 50 or so scientific men will go from England and take part in the meetings at Calcutta. I hope we shall enjoy this first visit to India in the good weather.

I am glad to say we are all very well and physically I feel in very good form and occupy myself very pleasantly in the country—tree-cutting, motoring, etc. I am very glad to hear you are enjoying your work in Berlin and hope you will have a successful time. It does not seem long ago when I first met you in Manchester, but I have to encounter hints that I am now one of the ancients of my profession. I thank you for your kind congratulations on my birthday. Give my kind regards to your wife and I hope you and your family are all well.

P.S. We are very busy in the Cavendish installing high voltage apparatus and also a cyclotron for work on Transmutation.

During the summer of 1937 Rutherford was collecting material for his Presidential Address to the joint meeting of the Indian Science Congress (on the occasion of its Silver Jubilee) and of a representative gathering of the British Association. This address was ultimately read in January by Sir James Jeans who was nominated President after Rutherford's death. It is fitting to consider his address at this place because it was one of the principal undertakings of the last few months of his life.

Rutherford's Presidential Address was his swan song in which there is a subdued note of triumph. He had indeed built far better than even he had expected.

The first part of his lecture dealt with Indian affairs—science and development:

While science has no politics, I am sure that it is of good omen that our visit happens to fall at a time when India is entering upon an era of responsible co-operative government in the success of which both our countries are deeply concerned.

In earlier days in India, research was largely confined to the great

Cyclotron magnet

Million-volt generator

scientific services, initiated and maintained on a generous scale by the Indian Government, for example, the Survey of India, the Geological Survey, the Botanical Survey, the Departments of Agricultural and Meteorology and many others.

He named several pioneers such as Everest, Pratt, Burrard, Lenox Conyngham of the Survey; Holland in geology; Walker, Simpson in meteorology; Roxburgh, Wallich, Prain, Hooker in botany; Troup, Pearson, Robertson in forestry; Ross on malaria; Rogers on cholera and leprosy—researches which gave new hope to the people of India; and, at the Universities men like Sir Alexander Podler, Sir Alfred Bourne, Sir Jagadis Bose and Sir Prafulla Ray.

The Indian student has shown his capacity as an original investigator in many fields of science, and in consequence India is now taking an honourable part and an ever increasing share in the advance of knowledge in pure science.

He made special mention of those Indians who by their merit had become Fellows of the Royal Society—Raman, Saha and Sahni—and particularly of that mathematical genius Ramanujan, also a Fellow of Trinity College, Cambridge: "but for his premature death, it may be said of him, as Newton said of Cotes, that we had known something."

Rutherford spoke with some just pride on the Commissioners of the Exhibition of 1851 assigning a scholarship to India, for he himself was an instigator of the award. He continued:

It is imperative that the universities of India should be in a position not only to give sound theoretical and practical instruction in the various branches of science but, what is more difficult, to select from the main body of scientific students those who are to be trained in the methods of research. It is from this relatively small group that we may expect to obtain the future leaders of research both for the universities and for the general research organisations....This is a case where quality is more important than quantity, for experience has shown that the progress of science depends in no small degree on the emergence of men of outstanding capacity for scientific investigation and for stimulating and directing the work of others along fruitful lines. Leaders of this type are rare, but are essential for the success of research organisation. With inefficient leadership, it is as easy to waste money in research as in other branches of human activity. . . .

It is clear that any system of organised research must have regard to the economic structure of the country. One essential feature at once stands out: India is mainly an agricultural country for more than three quarters of her people gain their living from the land, while not more than three per cent are supported by any single industry. . . . Research on food stuffs has a primary claim on India's attention. . . .

Rutherford passed next to the need for research on radio–communication, an important matter in a large country like India, and he pointed out that fundamental work on this subject had already been done by Saha and Mitra.

He considered industrial research in Great Britain such as that of Bessemer on steel and Parsons on steam turbines, adding that in a number of cases, British Science gave ideas to the world, but it was left to other nations to develop them by intensive research and to reap the industrial benefit. He made special reference to the investigations on fuel and the electrical, chemical and building industries and on the organisation of co-operative research—"a bold experiment which has undoubtedly proved a great success".

He spoke with authority for he had been for eight years Chairman of the Advisory Council of the Department of Scientific and Industrial Research in Great Britain. Continuing, he said:

In this address, I have to a large extent confined my attention to research in pure science, agriculture and industry. I am, however, not unmindful of the pressing needs of India to alleviate the suffering of the people from attacks of malaria and other tropical diseases. I know that India herself is giving much thought to these vital problems in which science can give her vital help.

In the second half of this address, Rutherford dealt with the triumphant progress from the discovery of radioactivity in 1896, down through the main stages described in this book, ending with the marvellous successes of the Newer Alchemy—the mastery of that "age-old problem of the transmutation of matter"—a domain in which he moved as king.

He stated that nearly a hundred of the artificial radioactive bodies are known and that, after a short period of stability, the atoms break up with the expulsion of either an electron or a positron.

As Fermi and his colleagues have shown, neutrons and particularly slow neutrons are extraordinarily effective in the formation of such radio-active bodies. On account of the absence of charge, the neutron enters freely into the nuclear structure of even the heaviest element and in many cases causes its transmutation. For example, a number of these radio-elements are produced when the heaviest two elements, uranium and thorium, are bombarded by slow neutrons. In the case of uranium, as Hahn and Meitner have shown, the radioactive bodies so formed break up in a succession of stages like the natural radioactive bodies, and give rise to a number of transuranic elements of higher atomic number than uranium (92). These radioactive elements have the chemical properties to be expected from the higher homologues of rhenium, osmium and iridium of atomic numbers 93, 94 and 95.

Rutherford then contrasted the simplicity of the apparatus which he used in Manchester in 1919 with the vast torture-chambers of the atom contrived today.

At Cambridge a great hall contains massive and elaborate machinery, rising tier on tier, to give a steady potential of about two million volts. Near by is the tall accelerating column with a power station on top, protected by great corona shields—reminding one of a photograph in the film of Wells' *The Shape of Things to Come.*

CHAPTER XVI

THE END

RUTHERFORD TO EVE

Chantry Cottage, Upper Chute, Andover, Hants.: 1 *Oct.* 1937. Thanks for your note re Gask and also for your kind thought in presenting me 3 of your special collars. They fit me very well and I shall probably get a number of them for my travels. Many thanks.

I shall probably be in London next Wednesday and will lunch at the Athenæum. I shall write to Gask to see if he can meet me on that day. I have seen a paper to be given by Cockcroft at Paris this week. He estimates that the 'maximum' radioactive source to be produced by 1 million volts and a *strong* beam of ions is under one curie but that Lawrence, with his 150 ton cyclotron, may produce a source up to 10 curie. This seems the likely limit of possibilities by present methods.

I have made a still further clearance of the blackberry patch and the view is now quite attractive. Leave for Cambridge in an hour—fine day.

This was the last letter that I received from him. It indicates that the artificial production of radioactivity, using radio-sodium, may soon give the equivalent of 10 grammes of radium, but this material (radio-sodium) decays to half-value in fourteen-and-a-half hours and must be renewed almost daily at an expense probably equivalent, at least, to the interest on the prime cost of radium, which decays to half-value in 1600 years.

In his last year, in spite of his good spirits, his hard work may have been drawing too heavily on his nervous energy and left him almost a tired man. As I said good-bye to him at Chantry Cottage in September, I noted, with a start, a certain lack of tone in his hand.

Yet three weeks later, at the Athenæum, Rutherford, Gask and I had lunch together to discuss the rival prospects of radium and artificial radioactivity for the treatment of cancer. Afterwards Sir James Jeans, Sir Richard Gregory, Sir Frank Smith and a few others, gathered together in the Smoking Room and we all agreed that never had Rutherford been in better form: it was but three weeks before his death.

On Thursday, 14 October, Rutherford was unwell and summoned his doctor next day. Prof. Ryle was called in consultation and the patient was taken to the Evelyn Nursing Home in the afternoon. On the way he told his wife that his business and financial affairs were all in order. Lady Rutherford told him not to worry, for his illness was not serious. To which he replied that life was uncertain. An operation was performed on Friday evening by Sir Thomas Dunhill and there seemed every prospect of good recovery. From Saturday afternoon onwards there was steady deterioration.

Within an hour or so of his death he said to his wife: "I want to leave a hundred pounds to Nelson College. You can see to it", and again loudly: "Remember, a hundred to Nelson College."

He hardly spoke after that and on Tuesday evening, 19 October, he died peacefully.

The following authoritative medical account was written for this book:

Lord Rutherford had enjoyed a very healthy life and his large frame, his genial manner and great capacity for work gave outward expression, to the last, of his physical and mental well-being. He had reached an age when certain inevitable accidents or degenerative changes become common and a stroke or failure of the heart or other organs may bring a swift cessation to active days. None of these, however, obtained in his case and the tragedy, if it can be called a tragedy to pass quickly from this world without forebodings and with powers undimmed, was heightened by the fact that his fatal illness was occasioned by a small mechanical, although notably dangerous, accident of an anatomical kind.

Lord Rutherford had a slight hernia or rupture which had previously given no cause for concern. On October 14th he was troubled by flatulence and nausea, but his bodily functions were otherwise normal. That night he vomited. He sent for his doctor, Dr Nourse, on the next morning and by mid-day had also been seen in consultation by

Professor Ryle. Although the signs were not then conclusive, a strong suspicion of commencing intestinal obstruction in association with the small umbilical rupture was entertained. He was moved to the Evelyn Nursing Home that afternoon and at 6.45 p.m. was also seen by Sir Thomas Dunhill, who had been summoned from London. He was operated on without delay. It was found that a portion of the wall of the small bowel had become imprisoned in the rupture without obstruction to its whole lumen, a so-called Richter's hernia. The circulation quickly returned to the injured tissues as soon as they were freed, and it seemed to those present that the outlook should be good and that there was no special reason, apart from the patient's years, to fear secondary complications. He was very well the next morning, but by the afternoon of October 17th vomiting had recurred together with a quickening of the pulse and signs of paralytic distension of the bowel. With the introduction of a small tube in the stomach by the mouth, siphonage of regurgitated contents was established and from that moment over-distension of the stomach and further vomiting were entirely obviated. There was increased anxiety leading to another consultation between Dr Nourse, Professor Ryle and Sir Thomas Dunhill, on October 18th, but it was decided that no further operation was justifiable. In the earlier part of October 19th, there was a slight improvement, but later, heart and circulation failed and the end came peacefully, a consequence of intestinal paralysis and intoxication, which would have taxed a younger man and against which the various measures employed were unavailing, excepting in so far as they were successful in conferring a very real measure of comfort and freedom from distressing symptoms.

Men of his generous build are always viewed anxiously as subjects for abdominal surgery, but the nature of the trouble, and its apparent prompt relief by operation, had raised hopes at first that in this case the issue would be a happy one. It had even been considered possible that Lord Rutherford might become well enough to join the British Association's Congress in India. This had been his strongest wish when the need for the operation had been made known to him. At the time of his premonitory symptoms, that is to say, on the day before his doctor was summoned, Lord Rutherford had had treatment from his masseur, but there is no reason to suppose that any measures were employed which aggravated the condition.

The care of Lord Rutherford during this brief illness was felt as a great privilege alike by his doctors and nurses. He was patient, courageous and uncomplaining, still capable of a joke, still alert enough in his mind to take an interest in the mechanics of the relieving siphonage employed.

He humorously reminded his doctors at one stage that they were treating a professional colleague, for it will be remembered that in 1928, on the occasion of the Harveian Tercentenary Celebrations, Lord Rutherford, in concert with Lord Balfour, and Professors Pavlov and Wenckebach, had been accorded the very rare distinction of being made an Honorary Fellow of the Royal College of Physicians. It may, indeed, be said that the veneration due to so great a scientist did actually engender in the thoughts of those who attended him a sense of that rare relationship which is experienced when a younger colleague is called upon to advise and help a senior whom he has long included in the category of Master. Inevitably, too, there was a wistful sense of regret that science should fail at the last to save one whose devotion and contribution to scientific method and knowledge were so enduring and outstanding.

The news of Lord Rutherford's death was received with consternation. There was to the many who knew him a deep sense of personal loss, there was to all who knew about him a realisation that they had lost a leader who could never be replaced.

Sir Frank Smith called on the Sub-Dean of Westminster to propose burial in the Abbey and this suggestion was immediately granted. The body was cremated, and it was announced that the ashes of Lord Rutherford of Nelson would be buried in the nave of Westminster Abbey, just west of Sir Isaac Newton's tomb and in a line with that of Lord Kelvin.

In the Abbey there were many personal friends and leading men. The simplicity and greatness of the man were reflected in the service.

The ten pall-bearers were chosen to represent the chief spheres of Lord Rutherford's main activities; the High Commissioner for New Zealand, the Vice-Chancellor of the University of Cambridge, the Presidents of the Royal Society, the Royal College of Physicians and the British Association, representatives of McGill University, Trinity College, Cambridge, the Department of Scientific and Industrial Research, the University of Manchester and of the Institution of Electrical Engineers.

At Trinity College, Cambridge, a memorial service was held in the Chapel at which the Master of Trinity, Sir J. J. Thomson, old friend and guide, read the lessons.

In Moscow there was a gathering of scientific men to do honour to the memory of Lord Rutherford. The chief address was given by Prof. Peter E. L. Kapitza, now Director of the Institute of Physical Problems, formerly Director of the Royal Society Mond Laboratory at Cambridge.

There is unusual interest in the obituary notice of Rutherford written for the Institution of Electrical Engineers by his son-in-law, Prof. R. H. Fowler:

His untimely death at the early age of 66 removes from us the outstanding experimental scientist of the age. For nearly forty years he had directed a physical laboratory; first at McGill University, Montreal, then in Manchester; and finally for eighteen years in Cambridge. For almost the whole of that long period, coterminous with what Rutherford himself described as the heroic age in physics, his laboratory, whichever it might be, was the resort of physicists from all over the world, and played the leading part in the development of our knowledge of atoms and the atomic structure of matter.

Seldom can anyone have started his career at a moment more auspicious, for Rutherford entered Cambridge as an advanced student in 1895, within a year or two of the discoveries of X-rays by Röntgen, of radio-activity by Becquerel, and of the electron by J. J. Thomson. Seldom can anyone have been better endowed to grasp, or more gloriously successful in exploiting, the opportunities that crowded fast upon him.

Ideally equipped for directing a physical laboratory, he was capable at once of intense sustained individual research, of suggesting and inspiring with his own fire cognate researches of others over a very wide field, and, particularly in later years, of organising the team work required for elaborate attacks on many modern problems. His genial but dominant personality, his exacting demands for the best, the inspiration of his personal research, and the generosity with which he suggested and directed the work of his staff and students, created an atmosphere in any laboratory he directed which no one who experienced it will forget, or, alas, ever hope to meet again.

The three periods of his professorships, in Montreal, Manchester, and Cambridge, correspond roughly with the three major phases of the development of atomic theory which will always be associated with his name. The first, at Montreal, was concerned with unravelling the intricate phenomena of radio-active change and the chemistry of the natural radio-active elements. For this work Rutherford received the Nobel Prize for chemistry in 1908, and it remained to the end a good joke against him, which he thoroughly appreciated, that he was thereby branded for all time as a chemist and no true physicist. The second, at Manchester, is associated mainly with the discovery of the nucleus and the development of the nuclear model of the atom— once called the Rutherford–Bohr atom, and now so universally accepted that it is just 'the atom', the shortened title being perhaps the greatest compliment that a scientist can ever pay to a scientific theory. The third, at Cambridge, was devoted to the study of the structure of the nucleus itself, developing from the disintegrations of nitrogen by alpha particles which were first observed by Rutherford himself at the end of his Manchester period. The third period culminated perhaps in 1932, that *annus mirabilis* of nuclear physics, which saw the discovery of artificial disintegration by protons, of the positron and of the neutron, the first and third being Cavendish contributions, while other members of the laboratory took an important share in the work of placing that strange entity the positron securely on the map. During this period, Rutherford was as much engaged in the direction of the large-scale operations of the laboratory as in personal research of his own, and was highly excited and delighted by the laboratory's triumphs. The last few years were rather years of reconstruction and preparation for a new attack with still more powerful weapons on nuclear problems. Rutherford's own fire seemed unabated to the end.

On 3 Jan. 1938, the Jubilee Session of the Indian Science Congress, held jointly with the British Association, was opened by the Viceroy at Calcutta. Sir James Jeans as Chairman read the Presidential Address prepared by Rutherford as already in large measure given, but he first of all paid him this tribute:

Those of us who were honoured by his friendship know that his greatness as a scientist was matched by his greatness as a man. We remember, and always shall remember, with affection his big, energetic, exuberant personality, the simplicity, sincerity and transparent honesty of his character, and, perhaps most of all, his genius for friendship and good comradeship. Honours of every conceivable kind had been showered upon him, so that he could not but know of the esteem in which he was held by the whole world, and yet he was always simple, unassuming and ready to listen patiently to even the youngest and most inexperienced of his pupils or fellow-workers, if only he were honestly seeking for scientific truth.

He recalled the statement of Niels Bohr that Rutherford's achievements are so great that they provide the background of almost every word that is spoken at a gathering of physicists.

Rutherford lived in a period which he himself used to describe as 'the heroic age of physics'...which opened up a new road which led no one knew where—but obviously into a very different territory from that which nineteenth-century physics had so industriously and thoroughly explored.

Rutherford directed his colossal energy and tireless enthusiasm to all the best new problems in turn....

In his flair for the right line of approach to a problem, as well as in the simple directness of his methods of attack, he often reminds us of Faraday, but he had two great advantages that Faraday did not possess—first, exuberant bodily health and energy, and second, the opportunity and capacity to direct a band of enthusiastic co-workers. Great though Faraday's output of work was, it seems to me that to match Rutherford's work in quantity as well as in quality, we must go back to Newton.

Voltaire once said that Newton was more fortunate than any other scientist could ever be, since it could fall to only one man to discover the laws which governed the universe. Had he lived in a later age, he might have said something similar of Rutherford and the realm of the infinitely small; for Rutherford was the Newton of atomic physics. In some respects he was more fortunate than Newton; there was nothing in Rutherford's life to compare with the years which Newton spent in a vain search for the philosopher's stone, or with Newton's output of misleading optical theories, or with his bitter quarrels with his contemporaries. Rutherford was ever the happy warrior—happy in his work, happy in its outcome, and happy in his human contacts.

It has been said by some men of high scientific attainments that Rutherford took small interest in science where definite measurements could not be made. This attitude arose less from disdain than from lack of appeal. He was living in an enthralling world of discovery, demanding much thought and close attention, and one which naturally evoked his enthusiasm. Other regions of science appeared to him uncertain and comparatively dull. It was probably for the same reason that he found even mathematical physics of secondary importance in his outlook. Metaphysics were to him a quaking morass or a closed door, while philosophy seemed an uncertain game where the outlook was necessarily obscure. Those statements and conjectures which could be tested by direct appeal to nature, and by repetition assured, were the foundations on which he built.

Prof. Alexander considered Rutherford "highly imaginative, but not speculative". "It is curious", he said, "that two of the greatest men of our day were both boys. Einstein was a merry boy, until sobered by recent tragedies, and Rutherford was a rowdy boy."

It is said of a great creative mathematician that, surveying his subject from a high pinnacle of abstract thought, he exclaimed "And we too are poets!" Indeed mathematics is akin to music and to poetry, since order, rhythm and melody are found in all three, and without imagination they cannot be created. The same remark may be applied with equal truth to astronomy and to physics. Two of the greatest men of our generation, Einstein and Rutherford, taking things as they are, have delved with poetic insight and imagination into the ultimate foundations of Nature.

It may seem at first almost ludicrous to compare Rutherford with Shelley, yet it is strange to find how many of the phrases of Francis Thompson, written about the great poet, are no less applicable to the great man of science:

He is still at play, save only that his play is such as manhood stops to watch, and his playthings are those which the gods give their children. The universe is his box of toys. He dabbles his fingers in the day-fall. He is gold-dusty with tumbling amidst the stars. He makes bright mischief with the moon. The meteors nuzzle their noses in his hand. He teases into growling the kennelled thunder, and laughs at the shaking of its fiery chain. He dances in and out of the gates of heaven; its floor is littered with his broken fancies. He runs wild over the fields of ether. He chases the rolling world. He gets between the feet of the horses of the sun. He stands in the lap of patient Nature, and twines her loosened tresses after a hundred wilful fashions, to see how she will look nicest in his song.

Finally, is it possible to summarise Rutherford's scientific work in simple language?

He discovered the magnetic detector of wireless waves, and began what Marconi completed. With Sir J. J. Thomson at Cambridge he determined the simple laws of the carriers (ions) of electricity in a gas, when radiated either by X-rays or ultra-violet light. He discovered and named the alpha and beta rays from uranium. At McGill he found thorium emanation—the first known radioactive gas. He discovered, with Soddy, that radio-elements break up when they eject a part of their atoms with considerable violence and that the residues form a new element. They obtained 'recovery and decay' curves. These changes are due to 'chance', that is, to unknown causes, which are internal, not external. In his Bakerian Lecture he described and gave the laws that he found concerning a whole chain of eight radioactive changes beginning with radium. He deflected alpha particles with electric and magnetic forces to a measured extent and calculated their speed, energy, mass and electric charge per unit mass. With Soddy he surmised that alpha particles were charged helium atoms, as the spectroscope confirmed. Later, with Royds at Manchester, he collected alpha particles and they *were* helium. It is not too much to say that at McGill he laid down the fundamental principles of radioactivity. He also measured with H. T. Barnes the heating effects of different rays and products and pointed out how such heat from radioactive matter in the earth might prolong the 'age of the earth'.

At Manchester, Rutherford and Geiger invented a simple detector which enabled them to count alpha particles, one by one, confirming the atomic theory of matter. He made accurate measurements of the properties of alpha particles. He used the known amount of helium in radioactive ores as a basis for the calculation of the age of such ores, running into hundreds of millions of years. He and his co-workers investigated the three great families, uranium-radium, thorium, and actinium with all their properties and radiations.

At Manchester, too, he noted that alpha particles might swerve to a large extent and even turn back when meeting matter. This remarkable effect led him to state and verify with Geiger that every atom has an exceedingly small nucleus within it, which contains nearly all the mass and a positive charge. Moseley showed that the number of unit charges in a nucleus is also the atomic number and gives the place among the ninety-two elements.

Bohr worked out a new mechanics of the electrons which move around the nucleus, so as to account for the optical and X-ray spectra in the simple cases. Rutherford, using alpha particles as projectiles, knocked protons (hydrogen nuclei) out of nitrogen atoms. What remained, after the alpha particle was swallowed and the proton released, was no longer nitrogen; it was oxygen. This was the first deliberate transmutation of matter. He also transmuted other light elements. He and his co-workers at Cambridge got powerful electrical fields so as to hurl protons, and later deuterons, at other elements.

The *annus mirabilis*, according to R. H. Fowler, was 1932, when Chadwick at the Cavendish discovered the neutron, Anderson in

California the positron, Cockcroft and Walton achieved artificial transmutation, while in the next year the Curie-Joliots in Paris established
artificial radioactivity. At this time the Cavendish was seething with
an activity energised by Rutherford.

All his life he played not with atoms, but mainly with the nuclei of
atoms: radioactivity was systematised, the nucleus discovered, and
transmutation achieved.

And what of the man? He was a man who loved children, and
children at once loved him. Young people stood in some little awe
of him, which thawed like hoar frost in the morning sun. He was
essentially sociable. Those who sat near him at dinner were
fortunate and they knew it. He might set the table in a roar, and
certainly he would knit the party together. He could get work out of
the laziest. He had high courage in a fight, but he saw more wisdom
in tact. He hated shams and an affected superiority. He was not
particular about minor points—whether of dress, speech, manners or
customs. Like the law, he was not mindful of trifles. He set great store
on earnest and thorough work. He grasped the essentials of a matter
and insisted on them. At meetings he naturally took the lead. He read
much and thought more. He would not let a point go past him in the
laboratory. He would repeat and vary an experiment until he was
certain. He made few mistakes; but we must feel thankful that he made
some. After making them, he cheerfully ignored them! He knew how
to follow the main scent and how not to hunt hares. His laboratory was
not an institution but a company of devotees with high ambitions and
resolve. He was proud of this company and their fame. There was very
little secrecy and still less grasping for priority. There was often wild
enthusiasm for some achievement almost beyond hope or belief.

"I remember", says Geiger, "when Rutherford came to the laboratory, obviously very happy, to tell me about the nucleus and I set to
work at once to verify it." "One of the great events of my life",
says Darwin, "was to have been actually present half-an-hour after
the nucleus was born."

Other people ask—what was the man like, how did he dress, did he really look like a farmer, what sort of accent had he, did he believe in immortality? None of these things really matters. Here was a king marching into the unknown. Who cares what his crown was made of, or what the polish on his boots?

In a moment he could switch off his science and take the keenest interest in all sorts of things. He was downright and critical when talking to a small group of friends. Then he threw caution to the winds; but always at the end—"he's not a bad sort of fellow when you come to know him". When he came to write, these things were smoothed out by his fair and deliberate judgment. High pressure needs a valve to blow off steam. He was at his best when the volcano was in eruption; the lava soon cooled.

With the death of Rutherford a great epoch in science came to an end. Galileo and Newton began and ended one such famous period— the discovery of the mechanism of the universe. Faraday and Maxwell achieved the foundations of electricity. Einstein and Planck have opened the doors of relativity and of quanta.

Rutherford was the Newton of the atom, dealing with the disintegration and the building of atoms. He was the king of the microcosm, leading an army to ever fresh conquests. The harvest of his work has not yet been fully revealed. Who can doubt that this will appear with time?

A man who knew him well for thirty years remarked, "Rutherford never made an enemy and never lost a friend." This is the more remarkable because he was an outspoken man; but what he said came from his whole kindly nature and not from his intellect alone.

He was a great lover of peace—but not at any price. He had a quick sympathy with the oppressed, and was a friend to the outcast.

Rutherford lived to be a centre of universal affection as well as esteem. His energy was volcanic and his enthusiasm intense; he had also an immense capacity for work.

He could not fail to be aware of this, but it left him quite unspoiled.

On the occasion of one of his discoveries, I said to him: "You are a lucky man, Rutherford, always on the crest of the wave!" To which he laughingly replied, "Well! I made the wave, didn't I?" and added soberly, "At least to some extent." Truly this wave went with him always, a fine wave, lit by the sunbeams, for he had the precious gifts of the poet—deep insight, powerful imagination, and a profound love of truth.

APPENDIX I

HONOURS

M.A. New Zealand, 1893;

B.Sc. and 1851 Science Scholarship, 1894;

B.A. Research Degree and Coutts Trotter Studentship, 1897;

D.Sc. New Zealand, 1901;

LL.D. Pennsylvania, Wisconsin, McGill, Birmingham, Edinburgh, Copenhagen, Glasgow;

Ph.D. Giessen, Yale;

D.Sc. Cambridge, Dublin, Durham, Paris, Oxford, Liverpool, Melbourne, Toronto, Bristol, Cape Town, London, Leeds;

D.Phys. Clark;

F.R.S., 1903;

Fellow of Trinity College, Cambridge;

Macdonald Professor of Physics, McGill University, Montreal, 1898–1907;

Langworthy Professor, University of Manchester, 1907–1919;

Cavendish Professor, University of Cambridge, 1919–1937;

Professor of Natural Philosophy, Royal Institution, London;

Rumford Medal, 1905;

Barnard Medal, 1910;

Copley Medal, 1922;

Franklin Medal, 1924;

Albert Medal, 1928;

Faraday Medal, 1930;

Bressa Prize, 1908;

Nobel Prize, 1908;

Order of Merit, 1925;

President of the Royal Society, 1925–1930;

President of the British Association, 1923;

Guthrie Lecturer, 1927;

President of the Institute of Physics, 1931–1933;

Knight, 1914;

Baron, 1931.

Honorary Member of:
Franklin Institute of Pennsylvania;
American Physical Society;
Washington Academy of Sciences;
Academy of Sciences of St Louis;
K. Vet. Soc. of Uppsala;
Deutsch. Chim. Ges. Inst. C.E.;
Royal Society of Edinburgh;
Manchester Lit. Phil. Soc.;
K. Acad. Vet. Amsterdam;
K. Danske Vid. Selskab;
American Academy of Arts and Sciences;
Royal Irish Academy;
Royal Society of Medicines.

Corresponding Member of:
Ges. Wiss. Göttingen;
Ges. Wiss. Bayer;
Akad. Wiss. München;
Phys. Med. Soc. Erlangen;
K. Svenska Vet. Akad.;
Soc. Ital. Sci.;
R. Accad. Sci. Torino;
Phil. Soc. Rotterdam;
Norske Vid. Akad.

Foreign Associate of:
Acad. Nat. Sci. Philadelphia;
R. Accad. Sci. Ist. Bologna;
R. Accad. Lincei.

Member of:
Inst. France;
Accad. Sci. Russia.

Hon. Member of:
Acad. Polonaise, Cracow;
Union Math. Phys. Czechoslov., Praha;
R. Phil. Soc. Glasgow.

PORTRAITS

Oswald Birley	Oil	Royal Society
,, ,,	Replica	Cavendish Laboratory
,, ,,	Replica	Wellington, N.Z.
P. A. László de Lombos	Oil	Trinity College, Cambridge
F. L. Emanuel	Oil	Nelson College, N.Z.
,, ,,	Sketch, pencil	A. S. Eve
Francis Dodd	Pencil	Fitzwilliam Museum
James Gunn	Oil Sketch on Canvas (1932)	National Portrait Gallery
Jannsens	Oil (1916)	Lady Rutherford
R. Schwabe	Pencil (1928)	Trinity College, Cambridge
,, ,,	,,	Athenæum
Sir Wm. Rothenstein	Pencil	Sir William Rothenstein

INDEX

Abraham, H., 256
Academic Assistance Council, 375–6, 381, 405
Adams, Walter, 376
Adelaide, 13, 307
Aden, 312
Adler, S., 414
Adrian, E. D., 399
Albert Hall, 374
Albert Medal, 325
Alexander, S., 240, 431
Alexander the Great, 366
Allen of Hurtwood, Lord, 405
alpha particles, 49, 90–1, 101, 129–31, 143–4, 146, 154, 174, 176–8, 180, 184, 188, 194–7, 200, 208, 274, 292–4, 320, 324, 345, 349, 368, 370–1, 392, 396, 432
American Association for the Advancement of Science, 99
American Physical Society, 80
Ames, J. S., 159, 256–7
Amsterdam, 185
Anderson, C. D., 138, 290, 367, 396, 433
Andrade, E. N. da C., 237, 244, 251, 295–6, 411
Antonoff, G. N., 228
Appleton, E. V., 315, 371
Arizona Canyon, 144
Armstrong, E. H., 84, 97, 140, 187, 315, 327, 353
artificial radioactivity, 378, 380, 393–4, 404
Asquith, H. H., Lord, 191, 213
Assouan, 312
Aston, F. W., 275, 291, 317, 327, 409, 419
Athenæum, the, 162, 424
Athenæum, The, 205
atmospheric radioactivity, 112
atom, the, 177, 210–42, 211, 429
atom, single, detected, 173–4
atomic disintegration, 97, 290–1, 299
atomic number, 211, 235, 289
atomic structure, 321, 355
Auckland, N.Z., 134, 310
Austen Leigh, A., 24
Austin, Lord, 407, 409
Australia, 305, 307–10
Ayrton, W. F., 49

Bacon, Francis, 316
Baeyer, J. F. W. A. von, 190, 213
Baldwin, Earl, 369, 370, 399, 407, 417, 427

Balfour, Andrew, 14
Balfour, Earl, 108, 157, 254, 321
Baker, N. D., 257
Bakerian Lecture, 100, 104, 108–9, 124, 360, 423
Ball, Sir Robert, 24, 27, 29–31, 54, 80
Banks, Sir Joseph, 322, 373
Barclay, James, 122
Barkla, C. G., 208
Barnard Medal, 193
Barnes, H. T., 69, 83, 106, 115, 231, 433
Barnes, W. S., 69
Barr Smith, 308
Baskerville, C., 119
Bateman, H., 238
Beattie, Sir Carruthers, 307, 329
Becquerel, H., 49, 60, 79, 80, 85, 141, 183, 193, 388, 428
Beilby, Sir George and Lady, 212, 215
Belief and Action, 377
Bell Telephone Co. of Canada, 373
Bellingham, 407
Berkeley, Cal., 147–50
Berlin, 183, 185, 326
beryllium, 406
Bessemer, Sir Henry, 422
beta rays, 49, 130, 207
Beveridge, Sir William, 376
Bickerton, A. W., 8, 9, 12, 191, 225
Birley, Oswald, 342
Birmingham, 173, 189, 209
Bjerknes, V. F. K., 43, 300
Bjerrum, N., 221
Black, Joseph, 403
Blackett, P. M. S., 95, 290, 293, 301, 306, 367, 418
Blondel, A. E., 355
Blondlot, P. R., 102
Board of Invention and Research, 249–51, 254, 345
Board of Invention, French, 256
Board of Trade, 382
Bohr, N. H. D. (Niels), 218–20, 223–4, 229, 236, 238, 243, 264–5, 289, 302–5, 314, 317, 319, 356–8, 361–3, 368, 378, 383, 419, 429, 430, 433
bolas, 6
Boltwood, R. B., 110–12, 125, 129, 132, 138–9, 142, 152–3, 155, 159, 164, 166–7, 173, 177, 180, 188, 190, 201, 214, 218, 223, 244–5, 249, 260, 321, 323
Boltzmann, L., 136

Tweedsmuir, Lord, 406
Tyndall, A. M., 117, 314, 323, 328
Tyrol, 223, 317
Tyrwhitt, Sir Reginald, 249

Uhlenbeck, G. E., 314
Umfalosi, River, 322–3
United Steel Companies, 387
Universities, Royal Commission, 318
Universities Congress, 213
University College, London, 92
Unwin, W. C., 354
Upper Chute, 397
uranium, atomic weight, 132; period of, 106
uranium X, 79, 201
uranium Y, 228
Urbain, G., 236–7
Urey, H. C., 356–7, 386

Vaughan, William, 67, 96
vibrations, 66
Vienna Academy of Sciences, 167–9, 172, 325
Vienna Radiological Congress, 225, 230
Vienna University, 114
Viljoen, J., 118
Villard, P., 74
Volta, A., 322
Volterra, V., 189
Vonwiller, O. U., 309

Walker, Sir Gilbert T., 421
Walker, J. W., 59, 68, 123
Wallich, G. C., 421
Walmsley, H. B., 244
Walton, E. T. S., 352, 360, 366
Walton, F. P., 59, 122, 371, 374, 434
Warburg, O. H., 263
Washington, 257, 259, 260
Watson and Sons, Messrs, 387

Watt, James, 403
wave-mechanics, 320, 335, 384, 390
Webster, A. G., 259
Wellington, Lord, 252
Wellington, N.Z., 244, 246, 310
Welsbach lights, 114
Wenckebach, K. F., 427
West, J., 251
Western Electric Co., 260
Westminster Abbey, 212, 427
Wheeler, H. L., 111, 125
Whetham, see Dampier, Sir W. C. D.
Whittaker, E. T., 401
Wien, W., 193, 215
Wilberforce, L. R., 31
Wills, H. H., 323
Wilson, C. T. R., 49, 68, 204–5, 246, 314–15, 324, 367, 396
Wilson, H. A., 49
Wilson, J. T., 44
Wilson, W., 214, 260
Wilson, Mt., 115
Windsor Castle, 190, 212–13
Winnipeg, 65, 187–8
wireless telegraphy, 67, 83, 339
wireless waves, 23
Withington, 163, 409
Wood, A. B., 238, 251
Wood, R. W., 103
Woolwich, 41–2
Wrinch, Prof. and Mrs, 148

X-rays, see Röntgen rays

Yale University, 110–11, 119, 120, 125, 129, 144, 167, 173, 214, 231, 261
York, Archbishop of, 410
Young, Thomas, 306

Zeleny, John, 48–9, 260
Zululand, 332, 335

Printed in the United States
By Bookmasters